ADVANCED TACTILE SENSING FOR ROBOTICS

WORLD SCIENTIFIC SERIES IN ROBOTICS AND AUTOMATED SYSTEMS

Editor-in-Charge: Prof T M Husband
(*Vice Chancellor, University of Salford*)

Vol. 1: Genetic Algorithms and Robotics — A Heuristic Strategy for Optimization (*Y Davidor*)

Vol. 2: Parallel Computation Systems for Robotics: Algorithms and Architectures (*Eds. A Bejczy and A Fijany*)

Vol. 4: Computer Vision, Models and Inspection (*A D Marshall and R R Martin*)

Vol. 5: Advanced Tactile Sensing for Robotics (*Ed. H R Nicholls*)

Vol. 8: Modelling and Simulation of Robot Manipulators: A Parallel Processing Approach (*A Y Zomaya*)

Forthcoming volumes:

Vol. 3: Intelligent Robotic Planning Systems (*P C-Y Sheu*)

Vol. 6: Intelligent Control: Aspects of Fuzzy Logic and Neural Nets (*C J Harris*)

Vol. 7: Visual Servoing (*Ed. K Hashimoto*)

World Scientific Series in Robotics and Automated Systems – Vol. 5

ADVANCED TACTILE SENSING FOR ROBOTICS

edited by

Howard R. Nicholls

University College of Wales
Aberystwyth, UK

World Scientific
Singapore • New Jersey • London • Hong Kong

Published by

World Scientific Publishing Co Pte Ltd
5 Toh Tuck Link, Singapore 596224
USA office: 27 Warren Street, Suite 401-402, Hackensack, NJ 07601
UK office: 57 Shelton Street, Covent Garden, London WC2H 9HE

British Library Cataloguing-in-Publication Data
A catalogue record for this book is available from the British Library.

ADVANCED TACTILE SENSING FOR ROBOTICS

ISBN-13 978-981-02-0870-7
ISBN-10 981-02-0870-7

Preface

This book covers a wide-ranging set of topics relevant to tactile sensing. Following an introductory chapter on tactile sensing, a survey of the important technologies used in tactile transduction is presented, in which a wide variety of tactile sensors are discussed. The findings are summarised in a reference table. The review of tactile sensing is extended to processing techniques in Chapter 3, which concentrates on basic tactile sensing issues such as determining parameters of contact, how tactile data can be used in stable grasping, and on object recognition using tactile data. A brief discussion of active tactile sensing is presented in this review, and more detailed coverage is provided in later chapters.

In Chapter 4, some of the mathematics of planar elasticity are presented, and the problems involved in extracting useful information from raw tactile data are discussed. Appendix A provides further detail, along with some derivations of the results used in the main part of this book.

Chapters 5 and 6 are concerned with tactile sensing in an industrial setting, and present the results of two European research projects which have involved touch sensing. Chapter 5 is about the integration of tactile and vision sensing for solving various localisation problems, whilst in Chapter 6, a low-cost sensing approach using touch probes in an automated assembly machine is described.

Chapter 7 begins a theme carried throughout the latter part of this volume, which is the study of human systems and their relevance to robotic tactile sensing. It must be stressed that the authors are not advocating an anthropomorphic approach, however, there are various lessons from the study of human touch which provide some useful indications as to how we may develop usable, versatile, dextrous manipulators with integrated tactile sensors.

The human tactile sensory system is reviewed by Moss-Salentijn in Chapter 7, providing the reader with the basic terminology used by Lederman and colleagues in Chapters 8 and 9. The results of new studies of human touch are presented in these chapters, and the relevance of these studies to robotics is emphasised throughout. Chapter 8 considers the use of data gathered from fingertip sensors in humans, and Chapter 9 discusses haptics: the combination of fingertip sensing with joint and muscle signals, an area of significance to roboticists concerned with dextrous manipulation.

One of the major points arising from the study of human tactile systems is that touch is an active sense: we extract information through contact by pushing, dragging, and poking our fingers over, around and into surfaces. In Chapters 10 and 11 by Allen and colleagues, the literature on active tactile sensing for object recognition is reviewed; they also present some results of their work using a robot-mounted dextrous hand, the Utah-MIT hand. They discuss the robotic use of exploratory procedures, which are a collection of methods of exploring surfaces through touch identified from studies of human systems. This provides a fitting end to the latter part of the volume, which takes the reader from the biology of human touch, through studies of how humans gather and use tactile information, to the implementation of exploratory touch in robotics. This volume ends by identifying some keys trends and outstanding issues in tactile sensing.

This book is more than a collection of papers from experts in the field, since it was designed at the outset to be an integrated volume covering both introductory material and new research results. Each chapter was written specifically for this volume. Whilst each chapter can be read independently of the rest, chapters are cross-referenced, and it is assumed in the latter part of the book, concerned with active sensing, that the reader is familiar with basic tactile terminology and concepts. It is also assumed that the reader has some basic knowledge of robotics and of sensing such as machine vision. Extensive reference lists are provided at the end of each chapter, enabling further detailed study of the literature.

Acknowledgements

Most importantly, I am indebted to the contributors for writing and submitting their chapters, and for being so responsive to my editorial comments. I am also indebted to the staff of the Department of Computer Science at the University of Wales, Aberystwyth, for their support. Special thanks go to Horst Holstein and Mark Lee for critical readings, to Nigel Hardy, Fred Long and Andy Ormsby for LaTeX and BIBTeX advice, and to Clive King for help with proof reading and for systems support.

This manuscript was produced using the LaTeX document preparation system written by Leslie Lamport.

Howard Nicholls
University of Wales
Aberystwyth, UK
October 1992

List of Contributors

Peter K. Allen, Paul Michelman, and Kenneth S. Roberts,
Department of Computer Science,
Columbia University, New York, NY 10027, U.S.A.

Ronald S. Fearing,
Department of Electrical Engineering & Computer Sciences,
University of California, Berkeley, CA 94720, U.S.A.

Roberta L. Klatzky,
Department of Psychology,
University of California, Santa Barbara, CA 93106, U.S.A.

Susan J. Lederman,
Department of Psychology,
Queen's University, Kingston, Ontario, Canada.

Letty Moss-Salentijn,
School of Dental & Oral Surgery,
Columbia University, New York, NY 10032, U.S.A.

Howard R. Nicholls and Nigel W. Hardy,
Department of Computer Science,
University of Wales, Aberystwyth, Dyfed, SY23 3BZ,
United Kingdom.

Dianne T. Pawluk,
Department of Mechanical Engineering,
Harvard University, Cambridge, MA 02138, U.S.A.

Zenon 'Zep' Rzepczynski,
MARI Applied Technologies Ltd.,
Old Town Hall, Gateshead, Tyne & Wear, NE8 1HE,
United Kingdom.

Summary of Table of Contents

1 **Introduction to Tactile Sensing** 1
Howard R. Nicholls

2 **Tactile Sensor Designs** 13
Howard R. Nicholls

3 **Processing and Using Tactile Sensor Data** 49
Howard R. Nicholls

4 **Planar Elasticity for Tactile Sensing** 75
Ronald S. Fearing

5 **Integrating Tactile Sensors — ESPRIT 278** 95
Zenon G. Rzepczynski

6 **Distributed Touch Sensing** 107
Howard R. Nicholls and Nigel W. Hardy

7 **The Human Tactile System** 123
Letty Moss-Salentijn

8 **Lessons from the Study of Biological Touch for Robotic Tactile Sensing** 151
Susan J. Lederman and Dianne T. Pawluk

9 **Lessons from the Study of Biological Touch for Robotic Haptic Sensing** 193
Susan J. Lederman, Roberta L. Klatzky and Dianne T. Pawluk

10 **Object Recognition Using Active Tactile Sensing** 221
Peter K. Allen

11 **Experiments in Active Haptic Perception with the Utah-MIT Dextrous Hand** 249
Peter K. Allen, Paul Michelman and Kenneth S. Roberts

12 **Future Trends in Tactile Sensing** 273
Howard R. Nicholls

Appendix. Basic Linear Elasticity 277
Ronald S. Fearing

Contents

Preface v

1 Introduction to Tactile Sensing **1**
1.1 Sensing and Robotics . . . 1
1.2 A Sensing Taxonomy . . . 2
1.3 What Is Tactile Sensing? . . . 4
1.4 Some Attributes of the Tactile Domain . . . 5
1.5 Haptic Sensing . . . 8
1.6 Tasks For Tactile Sensors . . . 9
References . . . 11

2 Tactile Sensor Designs **13**
2.1 Introduction . . . 13
2.2 Methods of Transduction . . . 15
2.2.1 Resistive . . . 15
2.2.2 Piezoelectric and Pyroelectric Effects . . . 20
2.2.3 Capacitive Techniques . . . 23
2.2.4 Magnetic Transduction Methods . . . 26
2.2.5 Mechanical Transduction Methods . . . 29
2.2.6 Optical Transduction Methods . . . 31
2.2.7 Other Techniques . . . 36
2.3 Summary . . . 37
2.3.1 Resistive and Conductive . . . 38
2.3.2 Piezoelectric and Pyroelectric Effects . . . 38
2.3.3 Capacitive Techniques . . . 38
2.3.4 Magnetic Transduction Methods . . . 39
2.3.5 Mechanical Transduction Methods . . . 39
2.3.6 Optical Transduction Methods . . . 40
References . . . 40

3 Processing and Using Tactile Sensor Data **49**
3.1 Introduction . . . 49
3.2 Basic Tactile Sensing Issues . . . 50

3.2.1 Parameters of Contact 50
3.2.2 Grasping and Tactile Sensing 51
3.2.3 Stable Grasping With Intrinsic Tactile Sensing 54
3.3 Object Recognition Using a Single Tactile Image 55
3.4 Active Tactile Sensing 60
3.4.1 Shape Description Without Specific Models 61
3.4.2 Model Matching 62
3.5 Haptic Sensing 64
3.6 Conclusion 67
References 67

4 Planar Elasticity for Tactile Sensing **75**
4.1 Introduction 75
4.2 Stresses and Strains in Plane Stress 77
4.2.1 Deflections for Plane Strain 80
4.2.2 Strain and Deflection Given Surface Traction 81
4.2.3 Determining Surface Tractions from Object Shape and Loading 83
4.3 Tactile Signal Processing 84
4.3.1 Deflection vs. Strain Measurements 85
4.3.2 Sensor Modelling 85
4.3.3 Filter Methods for Obtaining Surface Pressure and Deflection 85
4.3.4 Force Angle Estimation 86
4.3.5 Sampling Density 87
4.3.6 Finger Softness 90
4.4 Summary 91
References 91

5 Integrating Tactile Sensors – ESPRIT 278 **95**
5.1 Introduction 95
5.2 Project Structure 95
5.3 The Manufacturing Requirement 96
5.4 The Sensing Problem 97
5.5 An Integrated Approach 98
5.5.1 System Architecture 98
5.5.2 Tactile and Gripper System 99
5.5.3 Vision System 103
5.5.4 Communications Infrastructure 103
5.5.5 System Operation 104
5.6 Conclusion 105
References 106

6 Distributed Touch Sensing **107**
6.1 An Alternative To Regular Array Sensing 107
6.2 The InFACT Project: Flexible Assembly 108

6.3 The InFACT Machine Configuration 109
6.4 Sensing Philosophy and Sensor Integration 111
6.5 The Touch Probes 112
6.6 Assembly Sequence Specification 114
6.6.1 Measurement Tasks 115
6.6.2 Measurement Task Design 118
6.6.3 Uncertainties 118
6.7 Two Example Measure Tasks 120
6.7.1 The "Translate Until" Task 120
6.7.2 The "Find Hole Centre" Task 120
6.8 Conclusion 121
References 122

7 The Human Tactile System 123
7.1 Introduction 123
7.2 The Peripheral Nerve 124
7.2.1 Organisation of a Peripheral Nerve 124
7.2.2 Classification of Afferent Nerves 125
7.2.3 Afferent Fibre Distribution in a Peripheral Nerve 126
7.2.4 Nerve Conduction 126
7.2.5 Fast Adapting and Slowly Adapting Receptors 128
7.3 Nerve Terminals in the Skin 129
7.3.1 Structure of the Skin 129
7.3.2 Dermal-Epidermal Environment 129
7.3.3 Distinction Between Hairy Skin and Glabrous Skin 130
7.3.4 Glabrous Skin 131
7.4 Nerve Terminals in Glabrous Skin 131
7.4.1 Thermoreceptors and Nociceptors 133
7.4.2 Mechanoreceptors 134
7.5 Limb Proprioception 140
7.5.1 Afferents of the Joints 140
7.5.2 Afferents of the Muscles 142
7.6 Conclusion 143
References 144

8 Lessons From the Study of Biological Touch for Robotic Tactile Sensing 151
8.1 Introduction 151
8.1.1 What Is Touch Good For? 151
8.1.2 Definitions: Tactile, Kinesthetic, Haptic 152
8.1.3 Focus and Chapter Outline 152
8.1.4 What Can the Study of Biological Touch Contribute to Robotics 153

8.1.5 A Mini-tutorial on the Scientific Method in Biological Research 155
8.2 Skin Mechanics . . . 158
8.3 Cutaneous Sensitivity . . . 160
8.3.1 Amplitude Resolution . . . 160
8.3.2 Spatial Sensitivity . . . 162
8.3.3 Temporal Sensitivity . . . 165
8.3.4 Spatiotemporal Transfer Function . . . 168
8.4 Texture Perception . . . 169
8.4.1 Psychophysics . . . 169
8.4.2 Neural Basis . . . 171
8.4.3 Static Touch vs. Scanning . . . 173
8.5 Processing 2-dimensional Patterns . . . 174
8.5.1 Behavioural . . . 174
8.5.2 Neural Coding . . . 178
8.6 Thermal Sensing . . . 180
8.6.1 Psychophysics . . . 180
8.6.2 Neural Coding . . . 181
8.6.3 Robotic Applications . . . 181
8.7 Processing Spatially Distributed Inputs . . . 181
8.7.1 Fingers as Separate Processing Sites . . . 182
8.7.2 Are Two Hands Better Than One? . . . 182
8.7.3 Limited Cutaneous Field of View . . . 182
8.7.4 Selective Attention . . . 183
8.7.5 Preattentive vs. Attentive Tactile Processing . . . 183
8.7.6 Features for Computational Models of Early Tactile Processing 184
8.8 Conclusion . . . 185
References . . . 185

9 Lessons From The Study Of Biological Touch For Robotic Haptic Sensing 193
9.1 Introduction, Focus and Chapter Outline . . . 193
9.2 Haptic Perception . . . 194
9.2.1 Processing Raised 2-dimensional Spatial Patterns & Homogeneous 3-dimensional Objects . . . 195
9.2.2 Processing Common 3-dimensional Objects . . . 197
9.2.3 Patterns of Manual Exploration: "Exploratory Procedures" . . 198
9.2.4 Constraints on the Selection of Exploratory Procedures 200
9.2.5 Two-stage Sequence of Manual Exploration During Haptic Object Recognition . . . 205
9.2.6 Towards a Computational Model of Constraint-driven Exploration and Haptic Object Identification . . . 206
9.2.7 The Development of Manual Exploration and Haptic Object Perception . . . 207

9.2.8 Haptic Perception Through Wielding a Rod 207
9.3 Sensory-guided Motor Control 208
9.3.1 Cutaneous Information for Motor Control 208
9.3.2 Haptic Volumetric Attributes for Lifting an Unknown Object . 213
9.3.3 Contour Exploration Procedures (CEPs) for Grasping/Manipulation and the Role of the Second Hand 213
9.4 Conclusion 214
References 215

10 Object Recognition Using Active Tactile Sensing 221
10.1 Introduction 221
10.2 Early Work in Active Tactile Object Recognition 222
10.3 Interpretation Tree Methods 223
10.3.1 Eliminating Interpretations by Constraints 223
10.3.2 Strategies for Active Search 225
10.4 Intrinsic Tactile Sensing 228
10.4.1 Surface Recovery Using Intrinsic Tactile Sensing 229
10.5 Shape Interpretation Using Active Touch 232
10.5.1 Related Methods 233
10.6 Integrated Methods 234
10.6.1 Recognition Using Vision and Touch 234
10.6.2 Apprehension Using Vision and Touch 240
10.7 Summary 243
References 244

11 Experiments In Active Haptic Perception With The Utah-MIT Dextrous Hand 249
11.1 Introduction 249
11.2 A Haptic Perception Framework 250
11.3 The Columbia Hand-Arm System 253
11.3.1 Tactile Sensors 254
11.4 Exploratory Procedures for Object Recognition Tasks 256
11.4.1 Exploratory Procedure 1: Grasping by Containment 257
11.4.2 Exploratory Procedure 2: Planar Surface Explorer 261
11.4.3 Exploratory Procedure 3: Surface Contour Following 264
11.5 Summary 268
References 269

12 Future Trends in Tactile Sensing 273
12.1 Progress To Date 273
12.2 Some Key Trends in Tactile Sensing 274
12.3 Major Outstanding Issues 275
12.4 Conclusion 276
References 276

Appendix. Basic Linear Elasticity **277**
A.1 Definitions of Stress and Strain 277
A.2 Stress and Strain Relations 279
A.3 Coordinate Transformations for Stresses 281
A.4 Elastic Equilibrium: Navier's Equation 283
A.5 Solution for Line Load on Half-Space 287
A.5.1 Normal Strain for Line Load 289
A.5.2 Surface Displacement for Line Load 289
A.6 Solution for Load on Half Space 290
References 290

Index **291**

List of Figures

1.1 *Types of sensor.* 3
1.2 *Example tasks for tactile sensors.* 9

2.1 *Resistive tactile element: resistance measured through the rubber.* 16
2.2 *Resistive tactile element: resistance measured across the rubber.* . . . 16
2.3 *Piezoelectric/pyroelectric tactile element.* 20
2.4 *Capacitive tactile element.* 24
2.5 *Mechanical capacitive tactile element.* 25
2.6 *Magnetic tactile element.* 26
2.7 *Magnetoresistive tactile element.* 28
2.8 *Mechanical transducer: a linear potentiometer.* 30
2.9 *Optical tactile element: surface reflectance.* 32
2.10 *Optical tactile element: pressure to light transduction.* 33

4.1 *General tactile sensing problem.* 76
4.2 *Finger geometry for plane stress assumption.* 77
4.3 *Plane stress model and experimental data.* 79
4.4 *Comparison of impulse responses for normal load.* 80
4.5 *Comparison of impulse responses for inclined load.* 81
4.6 *Comparison of impulse response for half space and layer.* 82
4.7 *Surface deflection for uniform force distribution at 30°.* 84
4.8 *Frequency response $H(s)$ and aliased response $A(s)$.* 88
4.9 *Reconstruction error due to aliasing* 89

5.1 *Logical system configuration.* 99
5.2 *The gripper mounted tactile sensors and example images obtained from the valve-block shown.* 102

6.1 *The InFACT machine.* 108
6.2 *General view of the InFACT machine manipulators.* 110
6.3 *A sentinel touch probe from the InFACT machine.* 113
6.4 *A simplified railroad diagram of task classes.* 114
6.5 *Side and plan views of the knob assembly.* 116

7.1 *Organisation of a peripheral nerve.* 125

7.2 *Relative frequency of different classes of afferent fibres in a purely sensory nerve.* . . . 127
7.3 *Four different patterns of discharge in nerve fibres in response to peripheral stimulation (mechanical displacement).* . . . 128
7.4 *Diagram of a section of skin and hypodermis.* . . . 132
7.5 *Relative frequencies and sizes of the four types of mechanoreceptors in that part of the glabrous skin of the hand that is served by the median nerve.* . . . 135
7.6 *Mechanosensory terminals – Meissner's corpuscle and Pacinian corpuscle.* . . . 136
7.7 *Mechanosensory terminals – Ruffini corpuscle and muscle spindle.* . . . 137

8.1 *Examples of spatial resolution and hyperacuity tasks.* . . . 163

9.1 *Exploratory procedures and associated object properties.* . . . 199
9.2 *Schematic of the sequence of events during grasping and the associated tactile responses.* . . . 210

10.1 *Intrinsic tactile sensor developed by Brock and Chiu (1985).* . . . 228
10.2 *Locating contact point on a spherical sensor.* . . . 230
10.3 *Active vision/tactile system of Allen.* . . . 235
10.4 *Single finger tactile sensor.* . . . 235
10.5 *Organisation of tactile sensing.* . . . 237
10.6 *a) Image of pitcher; b) knot points and Coons' patch grid showing tactile traces; c) surface recovered from vision and tactile sensing.* . . . 239
10.7 *Experimental vision/tactile system.* . . . 240
10.8 *Flow of control for haptic exploration.* . . . 242

11.1 *A framework for robotic haptic perception.* . . . 251
11.2 *Utah-MIT hand.* . . . 254
11.3 *Columbia hand: system hardware.* . . . 255
11.4 *Object database.* . . . 259
11.5 *Recovered objects from Grasping by Containment EP.* . . . 260
11.6 *Tactile traces with recovered planar surfaces.* . . . 263
11.7 *Recovered solids of revolution from Surface Contour Explorer.* . . . 267

A.1 *Definition of Stress.* . . . 278
A.2 *Elastic cube (in projection) subject to pure normal stress.* . . . 279
A.3 *Elastic cube (in projection) subject to pure shear stress.* . . . 279
A.4 *Line load on elastic half space.* . . . 282
A.5 *Stress coordinate transformation.* . . . 283
A.6 *Equivalent stresses for surface traction.* . . . 284

List of Tables

5.1 *Command set for vision, tactile and gripper systems.* 104

7.1 *Classification of afferent nerve fibres serving sensory end organs.* . . . 126
7.2 *Morphologic classification of sensory terminals in glabrous skin.* . . . 132

8.1 *Summary of the properties of the four channels mediating the vibrotactile threshold characteristic.* 166

9.1 *EP-Property weightings, EP generality, and EP durations.* 201

Chapter 1

Introduction to Tactile Sensing

Howard R. Nicholls

1.1 Sensing and Robotics

Robots are currently found in a diversity of applications, such as welding, parts assembly, chemical laboratory experiments, mining, and manipulation of objects on spacecraft. Some of these robots are programmed to execute a sequence of motions and operations with little or no awareness of their environment. If the state of the robot's task is not as expected, for example, a component in an assembly operation is malformed, the robot may be unable to detect that something is wrong, which may lead to task failure or even damage to the robot or objects around it. Such robot systems are only suitable for a limited range of tasks, since they are unable to detect, diagnose and recover from errors in their environments or to respond to new situations.

One reason for this lack of versatility is that these systems have few, if any, facilities for sensing their environment. They are effectively blind, deaf, and without touch or smell sensation. If robots are to perform tasks reliably and adaptively, the incorporation of a variety of sensing modalities into the robot system is necessary. The control system must process and analyse sensor data concerning the state of the environment, and then initiate appropriate action based upon the outcome of the sensory processing. Ideally, the robot should determine its next action based upon sensory data indicating the result of the previous action.

Some robot systems are incorporating sensing to increase both their adaptability to state changes and their suitability for performing a variety of tasks. Typically, one, or just a few, types of sensor are found in these robot systems, and the programming methods used are usually fairly straightforward and make little use, for example, of techniques from the field of artificial intelligence. Such systems are gradually being implemented on the factory floor in commercial applications.

The goal of some robot systems, still mainly only under development in research laboratories, is that of intelligent control. This involves the ability to respond adaptively and correctly to a variety of situations. These may be previously unencountered and unpredicted situations, or ones in which there is perhaps only partial, incomplete or inconsistent information available. Such systems are often characterised by being knowledge-based, and it is common to find artificial intelligence techniques used in the programming of these robots.

A key aspect of recent developments in robotics is flexibility: the robot system must be able to adapt to changes in its environment or its task requirements without assistance from a human operator. If a robot is to be able to support such flexible task performance, there are three main categories of activity which it must perform (Lee, 1989):

1. Sensing, which involves the acquisition of data about the environment and the robot itself. The range and nature of sensors strongly affect the quality of the system's perceptions.

2. Thinking, i.e., symbolic computation directed at performing some activity. This can involve reasoning about both sensory data and actions.

3. Acting, which is not just the initiation of movements but also the planning of movement sequences and their control and coordination.

From this, it is evident that different types of sensor will be needed to supply reliable data about the state of the environment, and that sensing is a key issue. Before considering some sensing issues and then concentrating on tactile sensing, we need a definition of what a sensor is. In this volume, we define a sensor as a device or process which enables a computer system to determine properties of the physical world. This normally involves a conversion of real world parameters into a form which enables the computer system to process and analyse them.

1.2 A Sensing Taxonomy

There are many different parameters measured by robot sensors: internal parameters such as force, torque, velocity, acceleration, material behaviour, temperature and so on may be measured, as may external ones such as shape, dimension and location (Figure 1.1). The devices which measure such quantities take many different forms; for example, tactile sensors detect real world parameters through contact with an object, vision sensors through the use of a camera, and rangefinders through a triangulation system.

The system designer must decide which sensors should be used for a particular application; this process should be carried out in a top down manner, starting with task requirements, and going through several levels of analysis, eventually leading to the selection of a specific device. A taxonomy for sensing to aid this process has been developed by Lee (1989), who identified five levels of refinement leading to sensor selection:

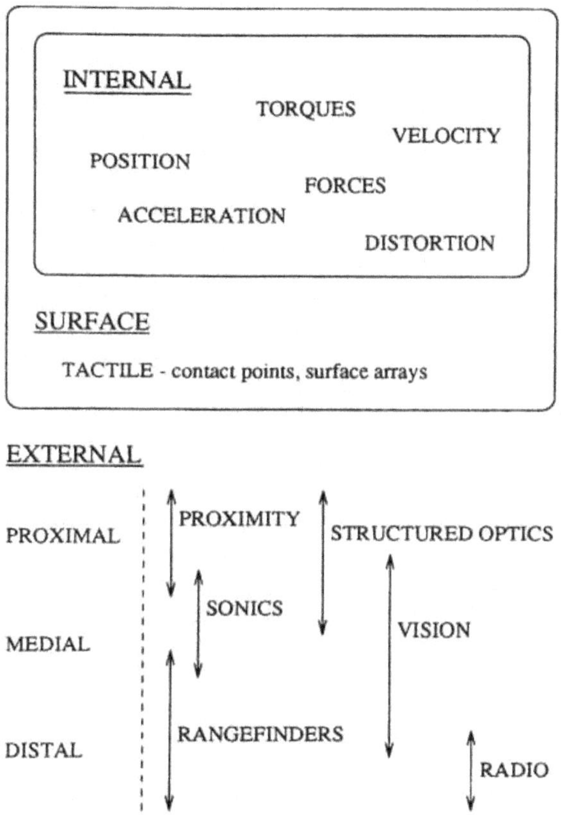

Figure 1.1: *Types of sensor (reproduced from Lee (1989)).*

1. Specification of task requirements; e.g., localisation, slippage detection, size confirmation, inspection, defect testing.

2. Choice of modality; e.g., vision, force, tactile.

3. Specification of sensor attributes; e.g., output complexity, discrete or continuous variable, imaging or non-imaging, local or global.

4. Specification of operational parameters; e.g., size, accuracy, cost.

5. Selection of mechanism; e.g., switching devices, inductive sensors, CCD vision imaging.

A single task requirement for sensing may be solved by using one of several different sensing modalities, but some of these modalities are better developed than others. For example there is a considerable body of experience in machine vision, whilst in other areas the technology and experience is immature.

This volume concentrates on a relatively new form of sensing, tactile sensing, which is concerned with interacting with the robot's environment through touch. Significant developments in this field have only occurred since the early 1980s, and there are lessons to be learnt about how and when tactile sensors should be used. To aid the coherent development of this emergent field, this volume presents detailed coverage of the technology of tactile sensing, some sample applications and also some new research results that indicate how we should proceed with research and use of tactile sensing.

1.3 What Is Tactile Sensing?

In this volume we define a tactile sensor to be a device which measures parameters of a contact interaction between the device and some physical stimulus. The interaction is normally confined to a touch sensitive region of the device's surface.

Tactile sensors provide data that is input to a computing system, and the acquisition, processing, and manipulation of this data constitutes tactile sensing. Some researchers limit their definition of tactile sensing to the measurement of forces at a set of discrete sites (e.g., Harmon (1984)), but this is a rather restrictive definition; there are various examples of sensing through touch that detect properties other than force (e.g., 3-dimensional shape (Sato *et al.*, 1986) and thermal conductivity (Russell, 1988)) and so a wider definition is used here.

Tactile sensors are used to sense a diversity of properties, concerning both attributes of a contacting stimulus and the relationship between the stimulus and sensor. Such a sensor may simply detect presence or absence of touch, whilst a more complex tactile sensor may provide data on the size, shape, position, thermal conductivity or distribution of forces of a contacting object.

Some of the properties which tactile sensors measure are:

Simple contact: the presence or absence of contact between sensor and stimulus.

Magnitude of force: a numerical value derived from each site represents the force (e.g., normal, torque, shear) applied by the stimulus at that site.

3-dimensional shape: the sensing process produces numerical values directly related to the 3-dimensional shape of an object touching the sensor.

Slip: the data indicates the movement of a contacting stimulus relative to the sensor surface. Typically, a slip sensor is mounted on a robot gripper.

Thermal properties: the thermal conductivity of a contacting stimulus, as well as its absolute temperature, can be measured through contact.

These are the main properties which are measured by tactile sensors and give rise to the raw data values output by the sensor prior to any kind of processing and analysis. One sensor may be capable of transducing more than one property. There are other properties which tactile sensors have the potential to transduce, such as texture, hardness, dryness, etc., but, as is shown in this volume, very few practical examples of these are documented.

A tactile sensor is usually made up of a set of individual touch sensitive sites. The dimensionality of such a sensor refers to the spatial arrangement of these sites, and falls into three basic classes:

Zero dimension: there is a single tactile sensitive site; for example, a limit switch or a gripper-jaw contact sensor.

One dimension: there are several sensing sites, all arranged co-linearly.

Two dimensions: there are sensing sites distributed over the area of the sensing surface, usually in a regular pattern.

Each discrete sensing site is known as a *tactel*, derived from *tactile element.* The term *taxel* is also sometimes used (not to be be confused with the texture analysis term *texel*), and the term *forcel* is sometimes applied to tactile elements which explicitly measure force. Tactile sensor designs vary notably in terms of their spatial resolution and sensitivity of elements.

Tactile sensing has received an increasing amount of attention since the early 1980s, with a number of new sensors and tactile data processing approaches publicised. It is still a sensing modality that is very much in its infancy when compared, for example, with developments in machine vision. One reason for the lack of sophisticated tactile sensing in robot applications is due to the scarcity of robust, reliable, accurate, and high resolution sensors. The technology of tactile sensors is now evolving into more suitable designs, but there is still much research to be done, both in developing robust sensors, and in processing the data and integrating it into a suitable control system.

1.4 Some Attributes of the Tactile Domain

The following attributes, originally identified by Overton (1984), and expanded upon here, characterise the main aspects of the low-level tactile domain. The tactile domain is also compared with some aspects of the visual domain, which has received much more research attention and is a fairly common sensing modality. The main tactile attributes are:

Explicit 3-dimensional information: the data from a tactile sensor can represent explicit 3-dimensional information about a contacting stimulus. The readings from a sensor can be interpreted in terms of force, but also in terms of a displacement normal to the sensor surface plane. For example, a linear potentiometer can serve as a force sensor if suitably calibrated, or alternatively can be used to measure linear displacement; an array of such sensors would enable a 3-dimensional surface map to be constructed.

No specularity: in machine vision, specularity is the occurrence of bright patches of light in the image because of direct reflection from the illuminating source. Tactile sensors do not have this problem, since they are not based on direct optical reflection from the object. The tactile analogue of visual specularity is perhaps electromagnetic interference from the surroundings, but this is only relevant to certain types of sensor hardware, and is not a fundamental tactile attribute.

No shadows: tactile sensors by their nature do not suffer from problems with shadows interfering with the tactile image, the intensity readings obtained being derived solely through contact between sensor and stimulus.

Direct measurements of stimulus: there is a direct mapping between stimulus shape and image, since non-background intensities are only present in the tactile image when there is actual contact between sensor and stimulus. This applies wherever the tactile sensor may be in space. The visual sensing case is again different, because of the mapping from three dimensions to two dimensions. Although size parameters can be calculated from a visual image, the sensor's distance from the viewing surface must be known to enable the calculation of the physical area one pixel occupies. If the camera is moved to a new location, the system must be re-calibrated. Tactile measurements are sensor position independent, visual measurements are not.

No perspective distortions: tactile measurements are direct, and do not introduce any distortion into the image. Vision sensing can have difficulties arising from distortion of the objects in the image due to perspective.

Ease of separation of stimuli from background: the segmentation of features in a tactile image from the image background is a straightforward task. If there is no contact between a sensor and any stimuli, then the background image intensity is at some uniform value. Any tactels having values different from this background level must therefore represent contact points. The task of segmenting the image into regions corresponding to areas of contact and regions representing no contact can be based upon a simple thresholding technique. The vision sensing case can be more complex, because of the specularity and shadow problems mentioned above, as well as illumination level variations, surface reflectance changes, and so on.

Tactile occlusion is different from visual occlusion: occlusion in tactile sensing is a situation in which two or more stimuli are in contact with the sensor, and one stimulus is preventing part of at least one other stimulus from touching the sensor surface. The resultant tactile image will be composed of regions corresponding to the surfaces of several stimuli, but the image of one of the stimuli will be overlaid on the region of another stimulus. For example, if a very thin object, say a ring washer, is placed on a tactile sensor and then a large rectangular block is laid on the washer, pushing the washer into the sensor, a composite tactile image will result. This image will contain a rectangular feature with a circular feature within it. The washer will occlude the rectangular block's image. Occlusion occurs in visual images also, but may be ambiguous. Visual occlusion occurs when one object blocks the line of sight between the camera and another object. This could be because one object is touching the other and so a region is obscured from sight. The same effect happens when they are quite separate from one another but are aligned with the camera. It is difficult to differentiate between these two situations.

Tactile sensors provide local views: the view of the environment provided by tactile sensors is very much localised, because such sensors only give responses to stimuli which are actually touching the sensor surface. Events happening close to, but not in contact with, a tactile sensor will not be registered at all by that sensor. The tactile sensor will often be smaller than the object being sensed, and will only detect one or a few faces of the object. Robot gripper-mounted sensors in particular are likely to be smaller than the object being grasped. The information provided by a single tactile image is therefore highly localised within the robot's environment, as opposed to vision sensing which can supply a global view of a workcell, for example. The local view can be supplemented by using multiple tactile images obtained through an active sensing strategy. This requires the integration of tactile images and manipulator trajectories. The local view can also be used to supplement global information supplied from vision sensors, providing a detailed image which is perhaps impossible to gain through global visual observation because of occlusion.

Tactile sensing may produce physical distortion: by its very nature, tactile sensing will involve physical interaction with a sensed object. This interaction can result in the physical deformation of the object under examination, or perhaps its movement in space. Experiments that sense hardness, for example, involve advancing the tactile sensor at small increments into the sampled object and recording tactile images at each stage, the changes in the images indicate the hardness of the sample. The act of poking the object of interest may deform it, in some cases irreversibly. Alternatively, the tactile sensing act could physically move the object under inspection. This interaction between tactile sensor and stimulus is in contrast to the passive nature of vision sensing.

Vision sensing has many applications and is used in practical applications. The above comparison shows that some of the problems present in vision are not present in tactile sensing, but tactile sensing has other limitations, such as only providing local views, and having the potential to distort a sensed object. The two sensing modalities complement one another rather than compete; for example, vision provides global views whilst tactile provides local views, sensing through touch can provide information such as force, hardness, and thermal conductivity whilst vision can provide data concerning colour, texture, and location.

It should also be noted that practical limitations in a particular tactile sensor may give rise to certain problems. Berger and Khosla (1991) make the point that some tactile sensors suffer from mechanical crosstalk between tactels and so require processing to compensate for errors in the raw data, and that the crosstalk may lead to background tactels having non-zero force readings. They also note that the low resolution of some tactile sensors makes some vision processing techniques, such as filters and neighbourhood operators, perform inappropriately, for example by leaving a low-frequency bias or by spreading edges in the image. When working in tactile sensing, therefore, it is important to have a good model of the behaviour of the sensor and to consider the appropriateness of likely processing methods.

1.5 Haptic Sensing

Early work on tactile sensing followed that of machine vision, by concentrating on the processing and analysis of single images, perhaps to achieve localisation or recognition of objects. More recent research has moved away from this scenario, towards treating tactile sensing as an active sense, whereby multiple tactile images are used, and data from the tactile sensor is integrated into an intelligent control system driving a hand or manipulator system. The results of processing a tactile image are used in selecting the next action to be taken, and the collection of data from many images are used to, say, construct a 3-dimensional representation of a touched object.

An examination of the human tactile sensory system reveals that information derived from touch is obtained in several ways (Loomis & Lederman, 1984). There are two constituents of human tactile perception, known as the cutaneous and kinesthetic components. The cutaneous information is derived from sensors on the fingertips, and so can be considered analogous to the data available from tactile sensor arrays. The passive collection and analysis of data from single robot tactile arrays is essentially cutaneous sensing. The second component of human perception, known as kinesthesia, arises out of muscle and joint signals. The whole hand and arm contribute information concerning the nature of the touched object. This kinesthetic information is important, since it is used in exploratory sensing. The kinesthetic component includes both afferent, or incoming, signals developed at the muscles and joints, and efferent, or outgoing, signals which are muscle action commands (Harmon, 1984). In a robot system, gripper finger and manipulator arm joint angles and forces provide kinesthetic data. The combination of cutaneous and kinesthetic

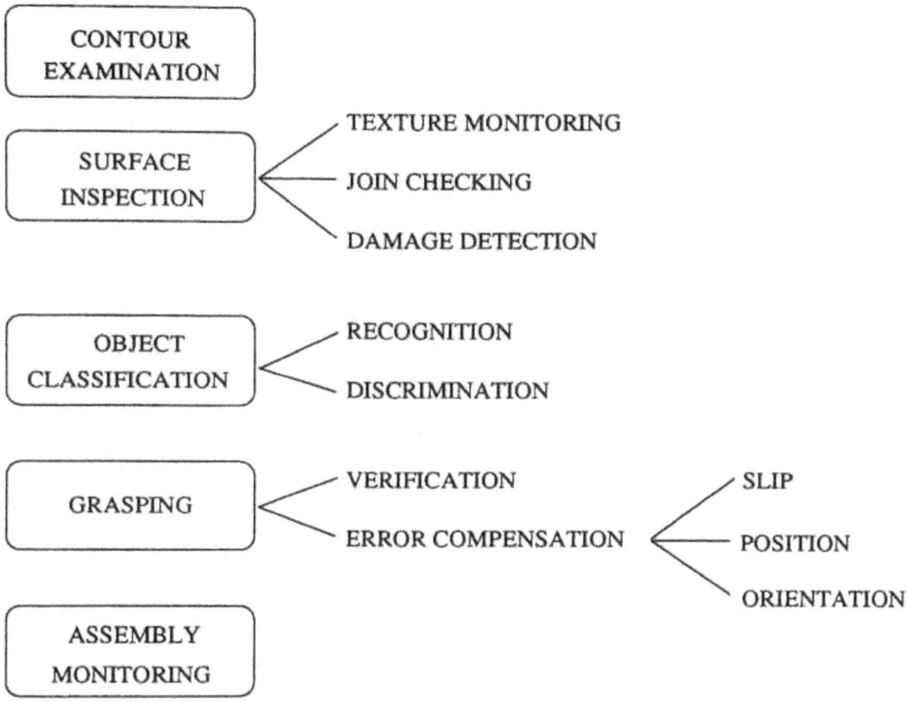

Figure 1.2: *Example tasks for tactile sensors.*

information is known as haptic sensing. The true haptic robot should be able to feel its way around an object, integrating tactile sensor image data, joint angles, other positional information and force-torque data in a dynamic manner.

The adoption of haptic sensing in robot systems clearly involves close integration of manipulator and sensor systems. Intelligent control is necessary to drive the feedback loop in active profile tracing experiments. The data from the cutaneous sensors must also be analysed bearing in mind the characteristics of the tactile domain outlined above. Particular emphasis is placed in this volume on the study of active touch and haptics, since they have considerable potential and are receiving increasing attention from robotics researchers.

1.6 Tasks For Tactile Sensors

There are various tasks to which tactile sensing may be applied in automation, which are summarised in Figure 1.2. An obvious task is the examination of contours to build up a model of an object, and a few researchers have experimented with such tasks using tactile sensors with some success (Bajcsy, 1985; Sato *et al.*, 1986; Allen, 1990). Recognition or discrimination of a grasped object is also a suitable application for gripper-mounted tactile sensors. Although this can be achieved, many industrial applications do not require recognition tasks since they operate in well constrained environments. However, in advanced robotics for unconstrained or unpredictable environments such as under water, in space, or in some hazardous industrial environments, such tasks are of major importance. The robot may encounter situations that require an ability to construct 3-dimensional representations of unknown objects, or to localise objects in unpredicted orientations.

Conversely, little work seems to have been done concerning the use of tactile sensors as inspection devices. Texture monitoring, join checking and damage detection, for example, could be performed using data from tactile sensor arrays. The lack of activity in this area is partly due to the comparatively poor spatial resolution of many sensor arrays. Fine details such as cracks and dents may not show up using some sensors. However, very high resolution tactile sensors, comparable to that of vision sensors, are being developed which have the potential to sense very detailed surface features.

Automated visual inspection is regularly used in industrial automated systems, but there are situations where the requisite data is ambiguous or even unobtainable through the use of vision. In such cases, adopting a tactile sensing method using a high resolution sensor could provide the necessary inspection data. Tactile sensing could be used as the sole method for inspecting parts, but another approach is to activate the tactile sensor to perform secondary inspection tasks, resolving details which cannot be determined by vision.

An advantage of using tactile sensors mounted on a robot gripper is that inspection and verification can be carried out during manipulation. A robot can grasp a part and, whilst moving, analyse the images from the gripper-mounted tactile sensors. Both grasp verification and surface inspection can thus be carried out "on the fly".

Tactile sensors on a robot gripper can provide manipulator-object relationship information as well as the more conventional object feature characteristics. Grasp monitoring concerns watching for changes in tactile data, to confirm object grasp or release, for example, but also involves location and orientation verification. The tactile image provides data on the position of the grasped object relative to the sensor, and in addition, the object's orientation. Typically, the object's centroid may be in the correct position, but the object has rotated from its desired orientation. Calculation of positional and rotational errors can be passed to the manipulator system, and appropriate trajectory adjustments made. This type of function is very important for autonomous systems, since it enables the automatic recovery from certain types of task errors. Sensitivity and accuracy requirements for tactile sensors used in such tasks are high, but some devices are capable of working successfully in this application area.

References

Allen, P.K. 1990. Issues in building intelligent grasping systems. *Proc. SPIE - Int. Soc. Opt. Eng. (Applications of artificial intelligence VIII, Orlando, FL, USA)*, **1293 pt.2**, 682–690.

Bajcsy, R. 1985. Shape from touch. *In:* G.N.Saridis (ed), *Advances in automation and robotics*. London: JAI Press Inc.

Berger, A.D., & Khosla, P.K. 1991. Using tactile data for real-time feedback. *Int. j. robotics research*, **10**(2), 88–102.

Harmon, L.D. 1984. Tactile sensing for robots. *Pages 109–158 of:* M. Brady, L. A. Gerhardt, H. F. Davidson (ed), *Robotics and artificial intelligence (NATO ASI series)*. New York: Springer-Verlag.

Lee, M.H. 1989. *Intelligent robotics*. London: Chapman Hall.

Loomis, J.M., & Lederman, S.J. 1984. Tactual pattern perception. *In:* Boff, K., & Thomas, J. (eds), *Handbook of human perception and performance*. New York: John Wiley.

Overton, K.J. 1984. *The acquisition, processing and use of tactile sensor data in robot control*. Tech. rept. COINS 84-08. Dept. Computer & Information Science, Univ. Mass. at Amherst.

Russell, R.A. 1988. Thermal sensor for object shape and material construction. *Robotica*, **6**(1), 31–34.

Sato, N., Heginbotham, W.B., & Pugh, A. 1986. A method for three-dimensional part identification by tactile transducer. *Pages 132–143 of:* Pugh, A. (ed), *Robot sensors vol.2: Tactile and non-vision*. Kempston, UK: IFS (Pubns).

Chapter 2

Tactile Sensor Designs

Howard R. Nicholls

2.1 Introduction

This chapter considers the main transduction methods used by tactile sensors, and examines various sensor designs. It aims to introduce the reader to the main technologies employed in tactile sensing, and to indicate the present state of development of tactile hardware. This text is an extension of a previously published comprehensive survey[1] (Nicholls & Lee, 1989), and it includes a considerable number of new references.

Sensors which provide reliable and accurate tactile data are critical to the development of tactile sensing. Consequently, much research in the tactile domain has been directed towards developing new designs for usable, robust tactile sensors. Most of these designs are for tactile sensor arrays which consist of a grid or matrix of sensing sites, each site called a *tactel.*

When developing any new sensor, a set of design criteria should first be established. An attempt at specifying some criteria for tactile sensors was carried out by the late Professor Leon Harmon of Case Western Reserve University, based upon the findings of a survey (Harmon, 1982). He sent a questionnaire to 55 persons in academic, industrial, government and private sectors asking about their views on tactile sensor specifications, systems, applications, problems and economics. As a result, he was able to devise a specification list for a practical tactile sensor, as well as identify the major application areas and problems within tactile sensing. His survey is a useful guide to tactile sensing, although it is becoming dated as his detailed literature search only covers up to November 1980. Nevertheless, the specification list is a useful guide for the designer; sensing requirements are also discussed in later chapters of this volume, with an emphasis on active sensing.

[1] All diagrams in this chapter are reproduced with permission from Nicholls & Lee (1989) and are copyright MIT Press.

The resulting set of tactile sensor requirements is summarised as follows:

1. The sensor surface should be both compliant and durable.
2. Spatial resolution should be 1 - 2 mm.
3. A range of 50 - 200 tactels is acceptable (e.g., 5×10, 10×20).
4. The sensor should be able to detect 5 g (0.049 N), ideally 1 g (0.0098 N).
5. A dynamic range of 1000:1 is satisfactory.
6. The sensor must be stable, repeatable and without hysteresis.
7. The response must be monotonic, though not necessarily linear.
8. The time resolution for the sensor should be at least 100 Hz.

In addition, it is desirable that a tactile sensor carries out local data processing to provide compact and high level information for any robot system to which it may be connected.

The above list is a mixture of absolute requirements and suggested likely specifications. Points 1, 6, and 7 must be present in any practical sensor, whilst the others are suggestions, particularly with respect to the number of elements and spatial resolution. They should certainly not be regarded as maxima, and the precise requirements are clearly affected by the intended application domain.

Harmon's list concentrates on sensor design parameters. However, an important point derived from the experience of designing the Utah/MIT dextrous hand (Jacobsen *et al.*, 1986), is that the tactile system as a whole must be considered: the desirable qualities of an experimental tactile system extend beyond the sensors themselves (Jacobsen *et al.*, 1987). Thus, the emphasis should shift from simply tactile *sensor* design, to tactile *system* design. These desirable qualities define system requirements for tactile sensing in machine manipulation research, and are identified as:

Multi-sensor support: a tactile system must be capable of managing not just a single array of a few dozen high resolution transducers, but large numbers of sensors distributed across a manipulator.

Speed: tactile data must be provided at rates which are well within the sensory feedback requirements for stable grasp and manipulation (may be > 500 Hz).

Reliability: a tactile system must be reliable in order to be useful. As well as durability, reliability involves minimising hardware failures, a problem which can partially be engineered out by reducing the number of wires running from the robot hand to the controller and minimising the electronic hardware that occurs *en route*.

Modularity: ideally, tactile sensing should exist as a subsystem within a larger manipulator sensing network, enabling substitution or addition of sensors.

Development of the tactile system for the Utah/MIT hand continues with these requirements in mind. There are several other examples of tactile system, rather than only sensor, design that reinforce this aspect of tactile sensing (Speeter, 1990b; Howe *et al.*, 1990; Berger & Khosla, 1991).

This survey examines the different transduction methods employed in tactile sensing and some significant sensors are described. Although the remainder of this chapter concentrates on individual designs, the reader should keep in mind the points made above concerning tactile system, as well as sensor, design.

Several other surveys of tactile hardware have been conducted, for example by Harmon (1984), Dario & De Rossi (1985), Pennywitt (1986), and Jayawant (1989). Another review concludes that tactile sensing should be combined with other senses for optimal adaptive robotics (Lauber *et al.*, 1988), whilst general-purpose manipulation is reviewed by Grupen *et al.* (1989). Also, two graduate-level books (Webster, 1988; Russell, 1990a) dedicated to tactile sensing provide detailed material concerning tactile sensor hardware design, as well as covering wider issues.

For ease of reference, some of the earlier papers (pre-1986) concerning passive and active force sensors, tactile array sensors, tactile transducers and ultrasonic sensors have been collected into a single volume by Pugh (1986b); a companion volume concerns vision sensors (Pugh, 1986a).

2.2 Methods of Transduction

Several techniques have been used to transform the interaction between a contacting stimulus and a tactile sensor into data suitable for computer analysis. The main classes and variants are described below.

2.2.1 Resistive

The transduction method which has received most attention in tactile sensor design is concerned with the change in resistance of a conductive material under applied pressure. This technique involves measuring the resistance either through or across the thickness of a conductive elastomer. The measured resistance changes with the amount of force applied to the material, resulting from the deformation of the elastomer altering the particle density within it. Most elastomers are made from carbon or silicon doped rubber, and the construction is such that the sensor is made up of a grid of discrete sites at which the resistance is measured (Figures 2.1 and 2.2). Designs developed as long ago as the late 1970s demonstrated that the principle could be incorporated into tactile arrays (Snyder & St.Clair, 1978; Briot, 1979).

Carbon fibres have been used in several tactile sensor designs (e.g., Larcombe (1981), Moore *et al.* (1987)). Carbon fibres are filaments of carbon created by baking unspun textile yarns, and are cylindrical in cross-section. When two fibres rest in contact with one another, the area of contact is very small, but when they are pushed together, the contact area increases, thereby increasing the electrical conductivity at the junction. The fibres are grouped into bundles, and these bundles

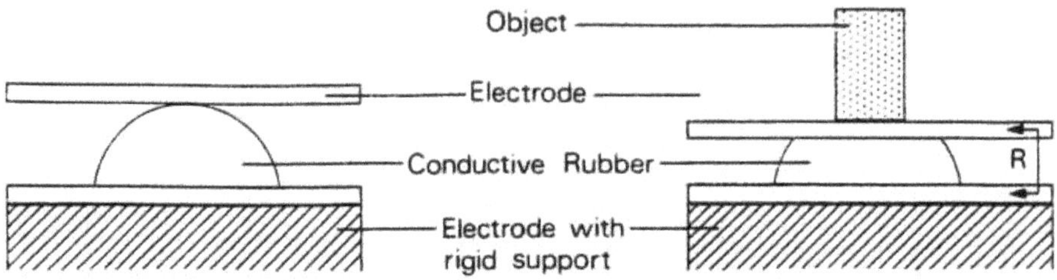

Figure 2.1: *Resistive tactile element: resistance measured through the rubber.*

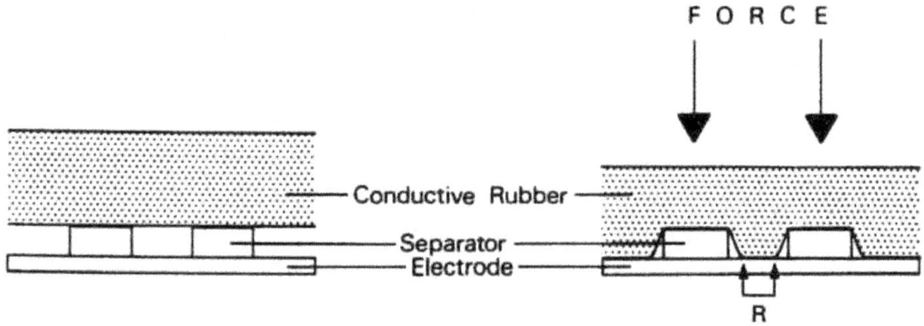

Figure 2.2: *Resistive tactile element: resistance measured across the rubber.*

can be arranged into an orthogonal grid to form a tactile sensor array. Typically, the unloaded resistance at bundle junctions is 2 kΩ, reducing to around 200 Ω for a 5 kg (49.05 N) load (Larcombe, 1981).

An example implementation developed in the UK uses a carbon fibre mat to separate the electrodes of a tactile sensor array (Robertson & Walkden, 1986). The 256 element array has an exponential output, demonstrates only a small level of hysteresis, and works over a practical load range of 0 – 600 kPa. Transducer life is in the order of two million cycles. This device is integrated into a sensing system that provides local image analysis, corrections for tactel non-uniformities and detection of overload.

A French-built tactile array uses carbon felt as the transducing medium (Pruski & Mutel, 1984). A significant aspect of this device is that it is used as a slip sensor. The sensor has a single layer of carbon felt with two electrodes attached to its lower face. As an object moves over the sensor surface, the long tufted carbon fibres are displaced and generate electrical noise. Appropriate circuitry detects the noise and, when the noise exceeds a predefined threshold, outputs a binary signal. A working sensor has been constructed and tested. Other slip sensors have been constructed using alternative technologies for example, mechanical (Dillman, 1982) and capacitance (Luo, 1984).

A design using carbon-loaded rubber originated by J. Purbrick at MIT formed the basis for several later designs (Purbrick, 1981). The sensor was constructed from a simple grid of silicon rubber conductors, each conductor, or "cord", being D-shaped in cross-section. Resistance at cord junctions was measured, and provided reliable output between 50 g (0.491 N) and 500 g (4.91 N) loads. However, some creep, the phenomenon of gradual increasing sensor output for constant load, was exhibited and pressure variations were difficult to discriminate.

Following on from Purbrick's work came a design using a flexible printed circuit board (PCB) etched in one direction with a sheet of anisotropically conductive rubber (ACS) placed above it (Hillis, 1982). The ACS material is conductive along one axis, which is placed orthogonal to the PCB etchings to form a grid pattern. A separator is used to keep the ACS away from the PCB in the unloaded state. An important point to note about this design is that the elastomer is squeezed onto the electrode, hence the contact resistance is measured, not the internal resistance through the elastomer. The separator, therefore, not the elastomer, determines the sensitivity and dynamic range of the sensor.

One of the few commercial tactile sensors developed was influenced by the work of Purbrick and Hillis. The Barry Wright Corporation Sensoflex tactile sensor is based around a matrix of elastomeric conductive rods, the two layers being separated by non-conductive elastomers. A 16×16 array with 2.5 mm centre to centre spacing over an active area of 39.6×39.6 mm has been developed, as well as an 8×16 array with 1.3 mm spacing over 9.6×19.8 mm (Peterson, 1984). The sensor is repeatable, has low hysteresis and has minimal set (non-recovered deformation after removal of load).

M. Raibert has produced tactile sensor designs incorporating novel features. His first design is both a tactile processing computer and a tactile array sensor (Raibert & Tanner, 1982). In this device, a conductive rubber sheet is placed over a printed circuit board and the resistivity of the sheet measured at discrete sites. The significant feature of this design is that the printed circuit board incorporates VLSI circuitry so that each tactel not only transduces its data but processes it as well. Each site performs transduction and processing operations at the same time as all the others. A prototype 6×3 element sensor was built. The main problems were that the Dynacon rubber exhibited some hysteresis, was non-rugged, and produced a blob-shaped receptive field; however, the VLSI approach was demonstrated as being viable.

In another design by Raibert (1984), Hillis's approach is taken a stage further. In the Hillis (1982) sensor, the separator between the layers of conductors determines the sensitivity and spatial resolution of the device. In his 1984 design, Raibert adopts a similar approach, which allows the elastic material to be optimised for mechanical properties independently of its electrical properties. This is because it is the contact resistance, not the internal resistance, of the elastomer which is significant here. A printed circuit board at the base of the sensor has an array of notch-like holes, with electrodes at the bottom of the holes. The elastomer rests on the top, and applied pressure forces the rubber into the holes and into contact with the electrodes. Different shaped, but basically triangular, holes produce different responses, typically either logarithmic, linear or exponential. VLSI circuitry is used to multiplex sensing elements, consequently only five wires into the sensor are needed. An experimental design of 48 elements, each one 0.3×0.6 mm and capable of supplying four bits of data, was built.

The integration of sensing elements with CMOS processing circuits has been considered by some Japanese researchers (Sugiyama *et al.*, 1990), who have used polysilicon piezoresistors for transduction on a silicon substrate. Inter-cell spacing is 250 μm for their 32×32 element array, and CMOS processing circuits are formed around the array, on the same chip, making it very compact. These designs illustrate that LSI and VLSI techniques can be applied to tactile sensing.

A tactile sensor that has been used extensively in various experiments at the University of Pennsylvania is built in the form of a probe (Bajcsy & Goldberg, 1984). The sensor was donated by G. Giralt of LAAS, Toulouse. The finger has 133 electrodes covered by a variable-resistance rubber layer. As pressure is applied, the voltage drop at each electrode is measured and turned into a binary value. The sensor acts as a probe rather than as a planar device, since it has twelve faces, connected to twelve facets that meet at the sensor tip. Each face has ten electrodes arranged in a straight line, each facet has a single electrode, and there is an electrode on the tip as well. The probe was mounted on an (x, y, z) positioning device (Wolfeld, 1981) and was used in profile tracing experiments.

Various other sensors constructed using silicon rubber are described in the literature (Christ & Sanderson, 1982; Chalupa *et al.*, 1982; Overton, 1984; van Brussel & Belien, 1986; Russell, 1987; Kah-Bin & Yoon-Song, 1988; Ghani, 1988).

The piezoresistive sensor described by Ghani (1988) is a 32×32 array; the columns are copper tracks on a PCB, and each row is formed from a piezoresistive rubber tube with a flexible conducting wire down the inside. Experiments using this sensor, and integrating it with a vision system, are discussed in Chapter 5 of this book.

A design based on the piezoresistivity of a conductive polymer film called a force sensing resistor (FSR) was developed by B. Tise at Sandia National Laboratories, New Mexico (Tise, 1988). He identified FSR as being particularly suitable for tactile sensing since it is most sensitive in the low force ranges, but still gives a response when very large loads are applied. Another advantage of using this polymer film approach is that the sensor can be bent and moulded to non-planar surfaces without damage, and is suited to robot fingertip mounting.

A single sensing element is constructed by putting a layer of FSR on top of a metal conductor. This is done on two separate surfaces, and the two are brought into face to face contact. Electrical resistance between the metal conductors, through the FSR polymer, is measured. A thin Mylar film provides the top and bottom layers of the sensor. Tise fabricated an array of such sensors, and developed an analogue multiplexer arrangement such that an $n \times n$ array only requires a $2n$ input multiplexer, providing an efficient array scanning mechanism.

As mentioned at the beginning of this chapter, it is important that the tactile system as a whole must be considered, and that it should incorporate multi-sensor support. T.H. Speeter, at AT&T Bell Labs, notes that many designers have concentrated on sensors but not on sensing systems; consequently, he developed a system capable of handling data from up to 16 separate tactile arrays, each typically of 16×16 tactels (Speeter, 1990a). Previously, he developed an FSR-based tactile array (Speeter, 1989); this was integrated into his tactile sensing system, although the system will work with sensors using other transduction methods.

Some problems with abrasion and noise between the two FSR layers of the tactile sensor were encountered, but were solved with the imposition of a thin, low resistance sheet of anisotropic conductor, called Conductive Polymer Interconnect (CPI), between the FSR layers. These sensors were mounted on the Utah/MIT dextrous hand (Jacobsen *et al.*, 1986).

Speeter's tactile sensing system (TSS) enables multiple arrays to be accessed in parallel with achieved sampling rates in excess of 60 Hz. The resolution of the sensors can be dynamically modified by a simple switching technique. Thus, high resolution sensing at lower processing rates, or low resolution at faster rates can be selected dynamically according to robotic task requirements. He suggests that it may be more important to optimise sensor sampling frequency rather than sensor element density, since low resolution, high frequency data is needed for robotic control. However, some applications may demand feature recognition from high density tactile images. The TSS allows the tuning of sensor sampling to give optimal performance.

One other notable resistive design uses a conductive fluid for transduction (Curtin, 1987). Each sensing element consists of a reservoir of conductive fluid leading into which is a capillary tube containing a bare resistive wire. The reservoir is covered with a compliant skin. When force is applied to the sensor surface, the fluid is displaced within the capillary tube, thereby changing the resistance of the enclosed wire. A 4×4 array was constructed and experiments performed; minimal hysteresis was experienced, output was stable, and a dynamic range of 500:1 was achieved, with a lower threshold of 2 g (0.0196 N). Performance characteristics are controlled by choice of resistive wire and capillary tube, enabling a product range of such devices to be constructed.

Approaches based on measuring resistance have probably received the most amount of attention in tactile sensing, and several designs have been quite successful. A wide dynamic range can be attained with considerable durability. Hysteresis and creep are present in most designs, probably because of the reliance on elastomers. Scanning techniques can reduce the number of wires required by an $n \times m$ sensor to be $n+m$ (e.g., Overton (1984), Tise (1988)). The resistive transduction method is also compatible with printed circuit technology, as demonstrated notably by Raibert and Tanner (1982).

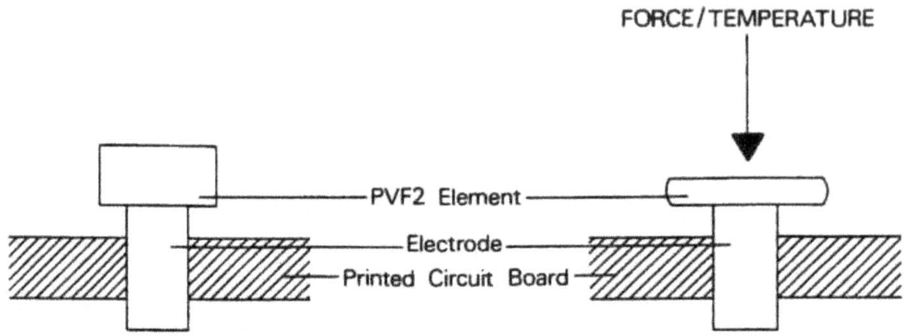

Figure 2.3: *Piezoelectric/pyroelectric tactile element.*

2.2.2 Piezoelectric and Pyroelectric Effects

Sensors based around the piezoelectric effect are rather different from the resistive devices described above. The piezoelectric effect is the generation of a voltage across a sensing element when pressure is applied to it (Figure 2.3). The voltage generated is proportionally related to the applied pressure. No external voltage is required, and a continuous analogue output is available from such a sensor. Correspondingly, the pyroelectric effect is the generation of a voltage when the sensing element is heated or cooled. An important feature of such sensors is that they are inherently dynamic. If a load is maintained, or the temperature is constant, then the sensor output decays to zero. Such sensors are most suited for sensing pressure changes or thermal variations.

Polymeric materials with piezoelectric and pyroelectric properties are appropriate for use with tactile sensors. Quartz and some ceramics have piezoelectric properties but are fragile; however, a polymer known as polyvinylidene fluoride (PVF_2 or PVDF) has been used successfully in several sensors, and proposed for novel applications (e.g., robot medical assistance (Dario *et al.*, 1987) and aircraft refuelling (Pirolo & Kolesar, 1989)). PVF_2 has good mechanical properties and is a durable material, thereby making a good tactile sensing surface. It is not normally piezoelectric in its raw state, but can easily be made so by suitably orienting the PVF_2 sample and applying a high electric field (Dario *et al.*, 1984). It then becomes highly piezoelectric and pyroelectric with a wide dynamic range. It can be made as thin as 5 μm and up to 2 mm thick, and may be moulded into various forms; hence PVF_2 has a number of attractions when considering tactile sensor materials. Although the mechanisms by which PVF_2 thin films convert temperature change into surface charge are not well understood (Smith, 1987), it is certainly usable for thermal sensing.

A problem with piezoelectric sensors is scanning an array of sensing elements. A row-column scanning system for accessing data from each tactel, which minimises the number of wires required, is difficult to implement on a piezoelectric sensor. Discrete charge amplifiers, one per tactel, may be required, and screening is necessary to reduce electrical interference. Another problem with PVF_2 is separating the piezoelectric effects from the pyroelectric effects since both may be present, thus protection from thermal variations may be necessary if pressure variations are important.

One of the primary exponents of piezoelectric and pyroelectric tactile sensors is the group at the Centro "E. Piaggio", University of Pisa. They developed a PVF_2-based sensor to detect both pressure and thermal changes, with multiple sensory layers analogous to the dermal and epidermal layers in human skin.

The sensor consists of eight layers, the outermost layer being a protective film (Dario *et al.*, 1984; Dario & De Rossi, 1985). The base of the sensor is a PCB on which metal electrodes are fixed. A layer of PVF_2 is bonded to the PCB, and is metal plated on its upper surface. This forms the deep *dermal* sensor. Deformation of this layer causes the piezoelectric effect to generate a voltage which is sensed via the electrodes. On top of the dermal sensor is an elastomer layer above which lie a resistive coating and another PVF_2 layer, the *epidermal* sensor. The epidermal PVF_2 layer has electrodes above and below it connected to conductors.

The epidermal layer is protected from a contacting stimulus by a thin Mylar film. The layer is highly sensitive but it is also susceptible to thermal fluctuations as well as pressure variations. Simple contact data is provided by the epidermal layer, and in addition, when rubbed against an object, the output signal variations give a measure of the roughness of the object's surface.

The resistive coating on the underside of the epidermal sensor is used to form a thermal sensor. The sensor is connected to a power supply causing the device to heat up. When an object touches the sensor surface, heat flows from the resistive coating through the epidermal PVF_2 and into the object. The heat flow is related to the thermal conductivity of the object.

Although the pyroelectric component of the epidermal sensor output is a problem, the elastomer layer helps to alleviate the problem for the dermal sensor. The elastomer has low thermal conductivity and so there is a delay of around one second before temperature variations affect the dermal sensor output.

The problem of scanning an array of piezoelectric sites was solved by using multiplexing circuits which connect each tactel output to a single charge amplifier. A 64 element matrix is scanned by four 16-channel analogue multiplexers which are switched by another multiplexer. The output from each tactel is fed into a single, high quality, FET-input operational amplifier.

The multilayer sensor of 8×16 elements behaved encouragingly in laboratory tests. A dynamic range of 4000:1, over 1.02 - 407.76 g (0.01 - 40 N) loads was established. Slips of a few hundred micrometres lasting a few milliseconds were detected. The output from both layers is linear, with negligible hysteresis.

A variation on this multilayered sensor is incorporated into an anthropomorphic robotic testbed. The sensor is mounted on a tendon-actuated finger and is intended for exploratory tactile sensing of complex surfaces. One possible use for this set-up is in a robot workstation for assisting a disabled person (Dario *et al.*, 1987). In this design, two additional sensing elements are part of the multilayer fingertip. The first is equivalent to a fingernail, consisting of an elastic structure incorporating a PVF_2 layer acting as a strain gauge. This is intended for sensing textured surfaces by rubbing the fingernail along the pitted surface and detecting the resultant forces. The second additional sensor is a curved array of ultrasonic transducers made out of PVF_2 film, enabling proximity as well as tactile sensing.

Another PVF_2-based design, incorporating six piezoelectric polymer elements, is aimed at resolving shear stress components in a tactile sensor (Domenici *et al.*, 1989). Experimental results demonstrated significant qualitative agreement between measured sensor output and predicted theoretical results for the shear stress component resulting from loading the sensor with a rigid indenter. Stresses and strains in tactile sensors have also been considered by Fearing (1985), who compared a plane-stress model with the response from an actual sensor, achieving close agreement between experimental and predicted results (Fearing, 1990). Fearing presents an introduction to planar elasticity in Chapter 4 and Appendix A of this book.

A device based on PVF_2 has been constructed and tested which has an overall size of an adult fingertip, greater than the human tactile spatial resolution, and a high sensitivity (Reston & Kolestar, 1990). These features were achieved by coupling a PVF_2 layer with the gate-electrode contacts of MOSFET amplifiers, to boost sensor output whilst maintaining high spatial resolution. No coupling between the sensor elements was detected, indicating that even higher spatial resolution could be achieved. The 5×5 sensor array has 600 μm inter-element spacing, each element occupying 600×600 μm. A linear response over 0.8 - 76.0 g (0.0078 - 0.75 N) was achieved.

Researchers at Stanford University have been involved in novel forms of tactile sensing, and are particularly interested in slip and texture perception (Howe & Cutkosky, 1989). They have developed two unusual devices, known as the stress rate sensor and the skin acceleration sensor. The stress rate sensor is designed to measure the rate of change of the stresses induced near the surface of the sensor as it slides over irregularities. This is achieved by embedding small strips of PVF_2 just below the sensor surface, and amplifying the electrical current produced by the PVF_2 resulting from changes in stress.

The skin acceleration sensor consists of a hard plastic cylinder at its core, a layer of soft polyurethane foam and an outer skin of relatively stiff silicon rubber, which is textured. The outer skin partially compresses the foam layer and a small accelerometer is mounted on the inner surface of the skin. The sensor was used in experiments to detect slip and to sense textures.

The justification for an accelerometer-based device came from a study of humans which indicates that people can detect displacements of fingertip skin as small as one or two micrometres through sensing accelerations. Howe and Cutkosky found that sensing skin acceleration is an effective way to detect incipient slip. They also found that sensor surface texture is important, since a very smooth skin produces a very large coefficient of friction, which causes a stick-slip motion as the sensor slides over smooth surfaces. They found that an effective texture consisted of parallel ridges a few hundred micrometres wide and up to a few hundred micrometres high. However, some ridges would catch on surface irregularities and snap back, producing high local accelerations. Smoother textured skins slide better but produce lower accelerations. The stiffness of the sensor skin is also significant: using a hard outer skin with a soft inner enhances sensitivity, durability and compliance.

Howe and Cutkosky also incorporated small PVF_2 strips in the rubber skin, so that both stress-rate and skin acceleration could be detected by the same sensor. They are actively involved in further experiments with these devices (Howe *et al.*, 1990).

Various other sensors employing the piezoelectric effect have been published (Abramowitz *et al.*, 1984; Pirolo & Kolesar, 1989; Brandolini & Gandelli, 1990), all of which provide useful tactile data.

2.2.3 Capacitive Techniques

Tactile sensors within this category are concerned with measuring capacitance, which is made to vary under applied load. The capacitance of a parallel plate capacitor depends upon the separation of the plates and their area, so that a sensor using an elastomeric separator between the plates provides compliance such that the capacitance will vary according to applied load (Figure 2.4). Although Harmon dismisses such sensors as being too susceptible to external fields and dependent upon specific materials (Harmon, 1982), some researchers have produced practical designs.

Various factors affect the performance of capacitive sensors. Sensitivity is determined by permittivity (which should be as high as possible), the Young's modulus of the elastomer, and the distance between electrodes (i.e., the thickness of the dielectric material) (Seekircher & Hoffman, 1988). Dependence on elastomers means there will be some hysteresis and creep, however. Sensitivity can also be enhanced by embedding conductive particles in the dielectric layer. These particles decrease the effective distance between the electrodes, enabling a higher capacitance and hence a higher absolute sensitivity for the sensor.

An example of a capacitive sensor was built by R. Boie at Bell Labs; his 8×8 tactel sensor uses an elastic outer layer for the contacting surface (Boie, 1984). This is on top of an elastic dielectric layer, which has orthogonal conductive strips above and below it. The lower strips are etched on a printed circuit board. The familiar matrix pattern of sensing sites is thus established, each site acting as a capacitor. Force readings are obtained by measuring capacitance through the elastic dielectric layer. The capacitive impedance is linearly proportional to the mechanical displacement of the elastic layer. Each element occupies 2.5 mm^2 and can measure pressures up to 5 kPa. It is possible to construct a 32×32 array with current

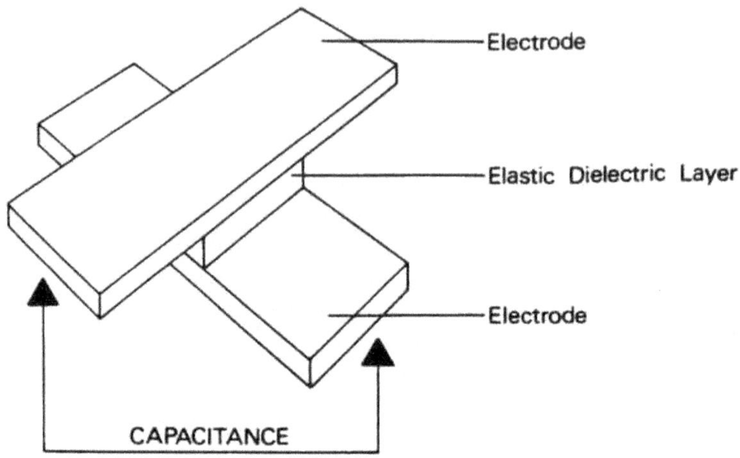

Figure 2.4: *Capacitive tactile element.*

technology, the sensor producing a composite video output. Boie contends that much of the interference attributed to capacitance sensors is actually man made, and can be screened out, although there is some low intrinsic noise. The materials used have both suitable mechanical properties and dielectric properties, and give a wide dynamic range. The widths and separation of the conductive strips determine the spatial resolution, limited by the degree of spread around applied force points.

An extension of Boie's design is an 8×8 element sensor constructed with an electronically insulating dielectric of silicon rubber (Siegel *et al.*, 1987). A detailed analysis is presented in the paper; the sensor gives a linear response and there is no notable hysteresis. This sensor is good for detecting changes but not for absolute values due to repeatability problems. This team also successfully designed and constructed a combined force and thermal sensor (Siegel *et al.*, 1986).

A design for a 9×9 capacitive tactile sensor is described by Wolffenbuttel and Regtien (1989). An interesting feature of this design is that edge detection using the Laplace operator is readily achievable in hardware. By applying an appropriate voltage to the neighbours of a selected tactile element, the output of the associated a.c. operated charge amplifier is proportional to the convolution of that element with the Laplace operator. A degree of local processing by the sensor is therefore possible.

R.S. Fearing, when at the Standard AI Laboratory, developed a 64 element capacitive sensor in the form of a cylinder (Fearing, 1987). The sensor was designed to be mounted on the Stanford/JPL tendon-actuated hand (Salisbury, 1982). The cylinder has 7×8 tactels, and a further eight tactels are placed on the hemispherical fingertip. The lower plates of the capacitors are eight stripes of copper at 45°

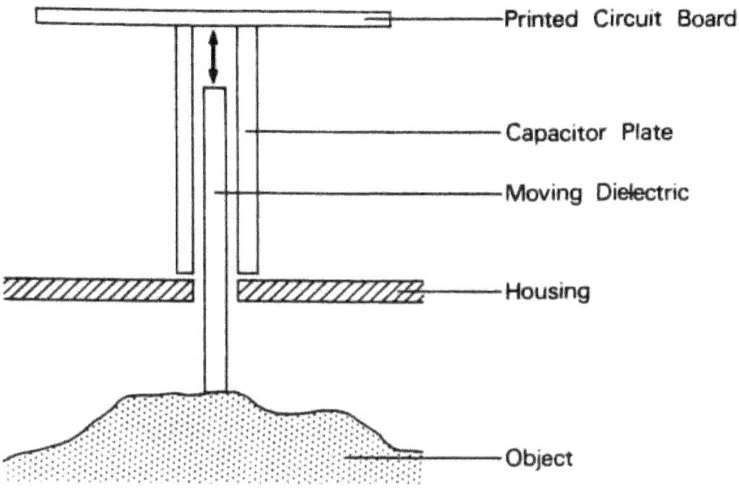

Figure 2.5: *Mechanical capacitive tactile element.*

intervals around the cylinder axis. The upper plates are seven rings of copper spaced along the length of the cylinder. This sensor was used in various experiments, including twirling a small bar held in the three fingers of the robot hand.

A novel sensor uses the change in capacitance caused by relative contact movement between sensor and object to detect slip (Luo, 1984). The contacting sensor surface comprises a set of parallel rollers, where each roller is a half cylinder of conductive material and a half cylinder of non-conductive material. The rollers are mounted in a non-conductive material such that the casing and rollers act as a variable capacitor. A slipping object will rotate the rollers, causing the capacitance to change which is then measured, thereby facilitating slip sensing. The sensor measures the change of phase angle, with the amount of phase shift providing a measure of the scale of slip. A highly linear relationship between detected phase shift angle and sensor output was established.

A tactile sensor array based on a moving dielectric element has been constructed at the University of Sussex, UK (Jayawant *et al.*, 1986). Each sensing element has two coaxial capacitor cylinders, acting as plates, fixed to a printed circuit board. A dielectric element is in the space between the cylinders. The dielectric is spring mounted and displaced by contact with an external stimulus, hence it moves up and down between the capacitor plates as contact loads vary. A force-displacement relationship is thereby established. A simple 4×4 matrix was built, using 2-dimensional scanning techniques to reduce the number of wiring connections necessary. One drawback is that the sensor surface is composed of cylindrical pins which are connected to the moving dielectric. The sensor surface is not smooth and small objects could fall between the pins and not be detected. Figure 2.5 illustrates

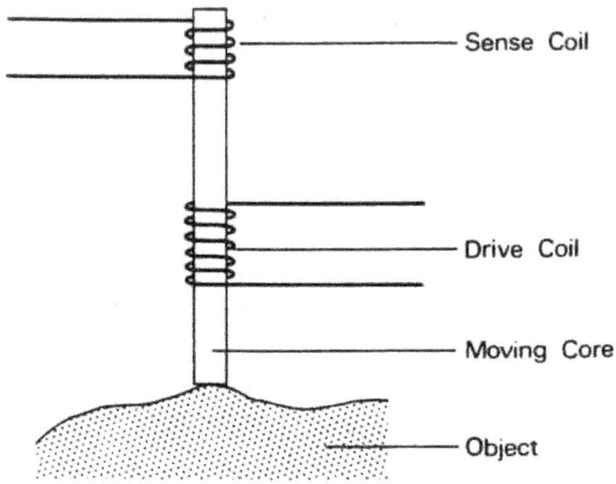

Figure 2.6: *Magnetic tactile element.*

a mechanical capacitive tactile element from this sensor.

Other capacitive designs include a 32×32 element array with six bit resolution (Suzuki *et al.*, 1990), and a novel electrorheological fluid-based device (Kenaley & Cutkosky, 1989). In the latter design, the fluid lies between a grounded elastomer skin and a positively charged electrode, providing simple tactile sensing based on capacitance. This device has the added advantage that the fluid acts as a conformal pad when the voltage is turned off, but when energised it turns into a plastic, thus interlocking the gripper fingertip-mounted sensor with a grasped object.

Capacitance sensors show some promise. The noise problem seems to be partly soluble, and the sensors give wide dynamic range and are robust. However, capacitance decreases with reduced physical tactel size and so stray signals will influence the sensor output for very small sites. Also, some dielectric materials are temperature sensitive, although the papers described above do not mention this problem.

2.2.4 Magnetic Transduction Methods

Sensors using magnetic transduction are divided into two basic categories. The first category groups together sensors which use mechanical movement to produce a change in magnetic flux (Figure 2.6). The second category concerns magnetoelastic materials which show a change in magnetic field when subjected to mechanical stress.

One of the earliest robot tactile sensors is based on the first category of magnetic sensors. First published in 1977, the paper is reprinted in Sato *et al.* (1986). The sensor consists of an 8×8 matrix of probes, each probe being energised by a drive coil. The probes are displaced perpendicularly when in contact with an object, thus forming the contours of the object. The displacement of a probe causes an induced voltage which is measured by a sense coil, and the sensor output is then calibrated in terms of probe deflection in millimetres. The sensor is rather cumbersome, and suffers from probes jamming when on steep slopes due to side loads and friction. However, it was used successfully to differentiate objects based upon their contours, and was one of few tactile sensor arrays in use in the 1970s.

Another sensor employing magnetism comprises a line of twenty Hall effect integrated circuits, each opposite a magnet (Kinoshita *et al.*, 1983). Contact with a stimulus displaces the integrated circuits, producing a change in magnetic flux, indicated by the output of the sensing element. This data is calibrated and used to form a profile of the stimulus. Despite being bulky (each element is 19×14 mm), the sensor has been mounted on a robot gripper and used in an active profile-tracing experiment.

Magnetoelastic materials undergo changes in their magnetic field when subjected to stress, and therefore suggest themselves as possible transducers in tactile sensors. A material is magnetoelastic if there is a relationship between changes in its internal magnetic moment, changes in the physical forces applied to, or by, it and changes in its physical length (Vranish *et al.*, 1982).

Researchers at the Gould Research Centre have carried out investigations using a nickel-based core wrapped with wire in a manner similar to that of a transformer, having primary and secondary windings (Checinski & Agrawal, 1986). The direction of the vector of magnetic induction affects the secondary magnetic flux and hence the secondary voltage generated. The amount of secondary voltage is directly related to the amount of applied force, thus providing the functionality of a tactile sensor. The experimental single element sensor has wide dynamic range, a linear response and hysteresis of about 1%.

A design extending Checinski's, having 256 elements with 2.5 mm centre to centre spacing, has been successfully built (Luo & Henderson, 1986). The core material is Vitrovac 4040, based on nickel and iron, and the sensing system incorporates a microprocessor, bandpass filter, analogue-to-digital conversion and video display. Tactile image processing can be performed by the microprocessor. This sensor is highly linear over 0 - 50 g (0 - 0.49 N) loads, with a slight tailing off up to 70 g (0.687 N). Hysteresis is low, and the sensor is temperature stable. No figures are given for durability cycle times, although the inherent robustness of a design requiring no moving parts or dependency upon elastomer properties suggest it is rugged. Another important feature is that the temperature error is low.

Another magnetoelastic tactile sensor, functioning like a strain gauge, has been constructed using a material called Metglas (Mitchell & Vranish, 1985). This is made as a thin ribbon. In the sensor, the Metglas is bonded to a specimen backing, and a coil is wound around the composite. The coil generates a field which encloses the Metglas. When the composite is stressed, the Metglas changes it susceptibility, which is measured as a change in inductance of the coil. Experiments showed that the sensor is two orders of magnitude more sensitive than a conventional strain gauge, and has good temperature stability. However, a problem inherent

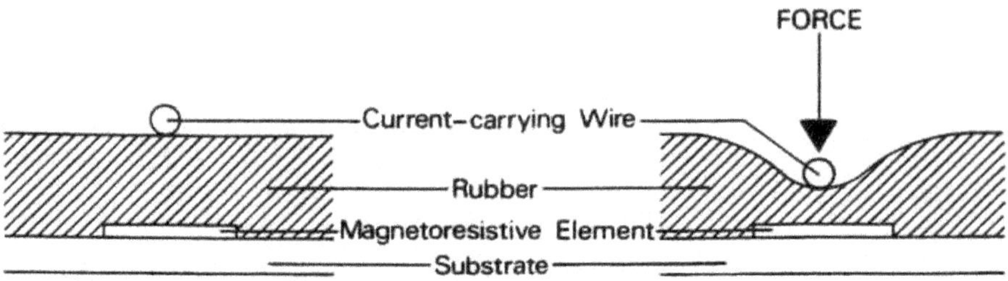

Figure 2.7: *Magnetoresistive tactile element.*

in the design is the bonding of the Metglas strip to the backing specimen. A perfect bonding between the two would restrict the rotation of the Metglas magnetic domain and the sensor would not work. Despite its other advantages, the inevitable compromise on bonding will be a problem.

J.M. Vranish, who has conducted considerable investigations into magnetoelastic materials, has designed a magnetoresistive tactile sensor (Vranish, 1986b). A material which is magnetoresistive undergoes a change in electrical resistance when subjected to a magnetic field (Figure 2.7). This design has 8×8 elements of the magnetoresistive material permalloy (81-19 Ni-Fe) etched onto a suitable substrate with connectors to form an array. Above this is a thin rubber layer on top of which is a Mylar film with copper conducting wires on it. These wires provide the magnetic field which will affect the permalloy resistance underneath. Covering the whole structure is a sheet of rubber which forms the outer surface of the sensor array. Contacting objects cause the copper and Mylar to be displaced, thereby altering the magnetic field strength around the permalloy elements. The field strength at the permalloy is inversely proportional to the distance between permalloy element and flat wire above it, and a linear response is achieved. The elements are 2.5 mm apart over a 25 mm^2 surface.

The sensor has been built and theoretical predictions suggest that it will be able to detect depressions as little as 0.025 mm, with a dynamic range over 30 - 300 Pa. Durability also seems reasonable, and magnetic shielding is not required — a point the other researchers do not mention. Vranish states that the sensor can be made spatially denser, and that the dynamic range can also be improved.

One other design proposed by Vranish is for a magnetoinductive tactile array with 0.5 mm spacing and 8 bits per tactel resolution. The paper design for this sensor and its theoretical performance is given in detail (Vranish, 1986b).

A major digression from the approach of most researchers in tactile sensing was taken by a team at AT&T Bell Labs, New Jersey. This group asserts that normal force sensing is not the most important parameter to be measured if tactile sensors are to be fully utilised. They believe that local shear and torque are of more use (Hackwood *et al.*, 1986), and state that tactile sensors are best used for manipulative tasks where the dynamics of gripper-stimulus interaction are of prime importance. This is opposed to some current approaches which use tactile sensors for shape recognition. To illustrate this, they designed and built a tactile array sensor to measure shear and torque as well as normal force.

Each element in the array has a magnetic dipole embedded in an elastic medium which rests on a substrate containing magnetoresistive detectors. As applied forces deform the elastic material, the dipoles are displaced producing a change in the surrounding area's magnetic field. Magnetoresistors detect this change and produce an electrical signal which can then be measured; shear and torque components are then extracted from this data. One example of this design was built with a 7×7 matrix (Hackwood *et al.*, 1986), this was for illustration only as the size is determined by fabrication criteria; element sizes as small as 0.5 mm are possible. No hysteresis was observed, and sensitivity was in the range 1 - 100 g (0.0098 - 0.98 N) in this example. The force sensitivity required for a particular application can be achieved by appropriate selection of the elastic medium.

Magnetic sensors have had only limited exploitation in robotics. Magnetoelastic and magnetoresistive sensors have a wide dynamic range, little hysteresis, basically linear responses and are robust. Shear and torque as well as normal forces can be sensed. Disadvantages are susceptibility to stray fields and noise, although filtering circuits can be incorporated. The choice of suitable materials is narrow. Generally, these sensors deserve more attention, but, surprisingly, very little has been published in this area since 1986.

2.2.5 Mechanical Transduction Methods

Some sensors rely on a mechanical displacement caused by an applied force. The simplest example is a spring loaded switch giving on-off contact readings. A linear potentiometer provides a graded scale of deflection, and the output can be considered in terms of either force or linear displacement. An example of such a sensor is used to detect travel in the z-axis (central approach axis) of a robot gripper (Warnecke & Haaf, 1981). Other devices measure angular displacement, such as optical shaft encoders, angular potentiometers, etc.

A common example of mechanical displacement in simple touch sensors is the movement of a linear probe. For example, Presern *et al.* designed a three degrees of freedom probe for arc welding applications. It is intended for seam tracking, the robot moving along an ideal trajectory and the probe giving (x, y, z) data on the actual seam position (Presern *et al.*, 1981).

R.A. Russell, formerly at the University of Wollongong, Australia and then at Monash University, has experimented with probe-based sensors. An early experiment involved a whisker assembly mounted on the end of a robot manipulator which was used in an object recognition system (Russell, 1985a). The probe proved very susceptible to damage, and so Russell developed a miniature three degrees of freedom tendon actuated manipulator, with a tactile sensor mounted on the

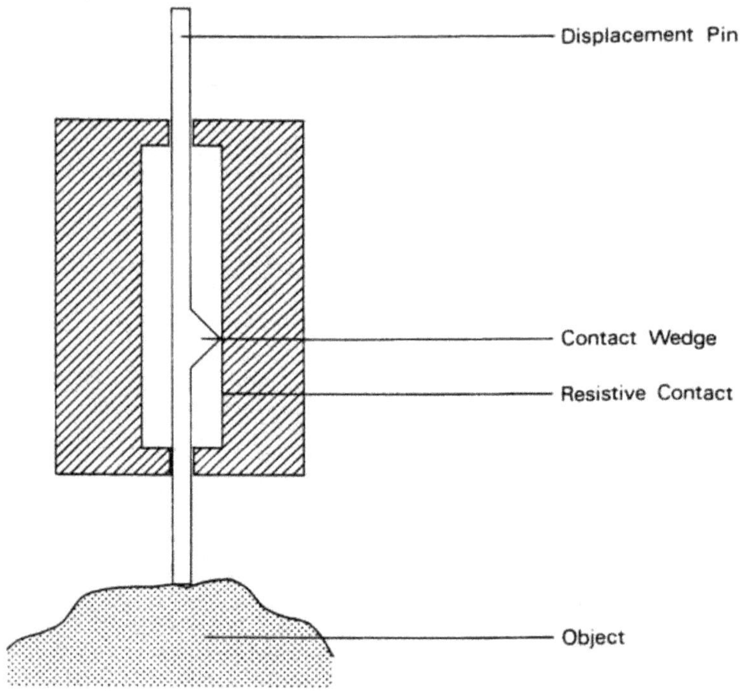

Figure 2.8: *Mechanical transducer: a linear potentiometer.*

end. He has experimented with simple contact, and has also mounted thermal and contact sensors on the probe (Russell, 1988a; Russell, 1990b). The probe and sensor arrangement is intended for sensing 3-dimensional surface features.

Some other novel mechanical devices detect the sound generated by movement against the sensor surface. Such a sensor, developed at the University of Karlsruhe, Germany, uses a microphone containing a piezocrystal. Microphones are mounted on opposing jaws of a sensory gripper, so that the signals generated by the sensor are the result of actions within the gripping area. Noise from the gripper motor is filtered out by a rubber material covering the crystal. Slip can also be detected by this sensor (Dillman, 1982).

Figure 2.8 illustrates a mechanically-based sensor. Many of the devices detailed in other sections could technically be described as mechanical sensors, for example (Sato *et al.*, 1986), but they have been included elsewhere to illustrate an alternative design within a transduction class.

2.2.6 Optical Transduction Methods

The development of optical fibre technology and solid state cameras has led to some interesting new tactile sensor designs. The capability for high spatial resolution images, freedom from electrical interference and ease of separation of sensor from processing electronics are some of the attractions of incorporating optical transduction methods into tactile sensors.

The Lord Corporation marketed tactile sensors using electro-optical transduction with mechanical displacement (Rebman & Morris, 1986), in two versions, the LTS-100 and the LT-200. In these two designs, the sensor surface is made of a compliant material which has on its underside a grid of elongated portions forming pins. When the surface receives an applied force, the pins on the underside undergo a mechanical deflection normal to the surface. Each pin is moved into the light path of a photo emitter-receiver pair. The amount of movement determines the level of light reaching the photoreceiver, which is digitised and represents the amount of force applied at that site.

The LTS-100 sensor was an 8×8 array with 7.62 mm centre to centre spacing, whilst the LTS-200 had 10×16 tactels 1.8 mm apart. Each deflection increment represents 3.18 g (0.031 N), facilitating a dynamic range of 0 - 681 g (0 - 6.682 N). The LTS-100 sensor was connected to a Motorola 68000 processor which analysed the tactile image and provided position and orientation data to a computer system. This was used successfully in the assembly of a flexible diaphragm and plastic cap.

A simple but effective sensor using optical fibres works by detecting the amount of reflected light from an illuminated surface (Crosnier, 1986). Light from a light-emitting diode (LED) is passed down a fibre-optic cable to illuminate any proximal surfaces. A second cable picks up any reflected light from illuminated surfaces within a detection zone and directs it onto a photodiode (Figure 2.9). This simple technique was built into a finger which can sense contacts perpendicular to the finger axis, and also radially. Radial contact is detected using a mobile ring around the finger and six evenly spaced fibre-optic transmitter/receiver pairs. Movement of the ring cuts the beam of light coming from one or several of the fibre-optic transmitters, which is detected by the corresponding receivers. Axial contact detection is provided by a transmitter/receiver pair in the end section of the finger, which can be displaced upwards slightly. A contact force of 10 g (0.098 N) is the threshold for detection. Three of these fingers were built and incorporated into a gripping system employing contact data from the sensors.

Several planar designs also use this technique of detecting reflected light from a surface. These designs have a compliant elastomer surface layer, which is deformed by contacting objects. Various kinds of photosensors are used to measure reflected light on the underside of the elastomer, the amount of reflected light being used to determine contact parameters. For example, a design by Schneiter and Sheridan illuminates and detects light on the underside of a silicon rubber layer using a matrix of fibre-optic transmitter/receiver pairs (Schneiter & Sheridan, 1984). Ristic *et al.*

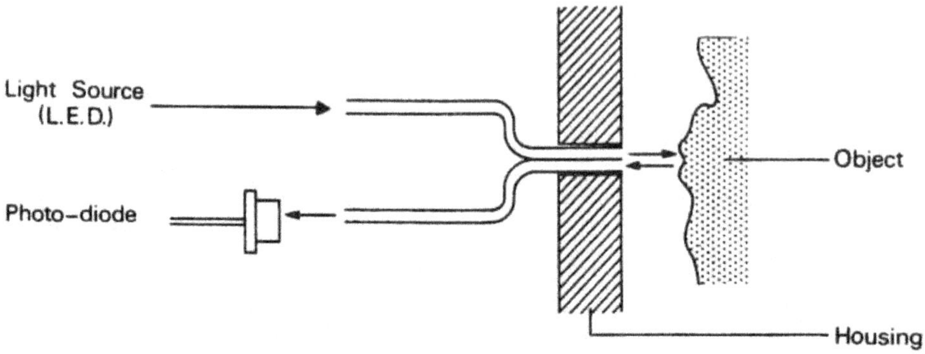

Figure 2.9: *Optical tactile element: surface reflectance.*

(1989) also demonstrate the viability of fibre-optic pairs in their experiments, whilst another sensor (Collins & Hoover, 1987) uses an optical RAM to capture the tactile image and achieve high resolution. Potentially high resolution and linear responses are features of these designs.

Another fibre optic tactile sensor uses the phenomenon of light being radiated normal to the axis of an optic fibre at places where there are inhomogeneities in the fibre (Schoenwald *et al.*, 1987). The sensor consists of four parallel, evenly spaced optic fibres mounted on a substrate. A second array is mounted on top of these, but oriented orthogonally to form a grid pattern. The fibres are surface abraded at the grid junctions. The rows can be separated from the columns by a clear elastomer, or by an elastomer with holes at the junctions between fibres. The rows are sequentially illuminated by LEDs and the columns are connected to photodiodes, which are sequentially scanned during the time that a single row fibre is illuminated. An efficient scanning mechanism is thus achieved. When a row fibre is excited by a LED, light is radiated from the places where it was abraded. By reciprocity, radiated light from the row fibre incident on the corresponding column fibre will be coupled into that column fibre along its axis. The light from this coupling is detected by the photodiode. Force applied to the sensor surface compresses the elastomer and increases the coupling between fibres by decreasing their separation.

A design significantly different from Schoenwald's, but which is superficially similar, has been developed at Purdue University (Jenstrom & Chen, 1989; Emge & Chen, 1991). In this design, a tactile array is formed by arranging optical fibres into four overlapping layers, two active layers sandwiched between two corrugation layers. When force is applied to a fibre intersection, small distortions, or microbends, appear in the stressed fibres, decreasing the transmitted light in these fibres. This sensor, although constructed from a grid of optical fibres, achieves tactile sensing through light attenuation, whereas Schoenwald's sensor depends on the coupling of

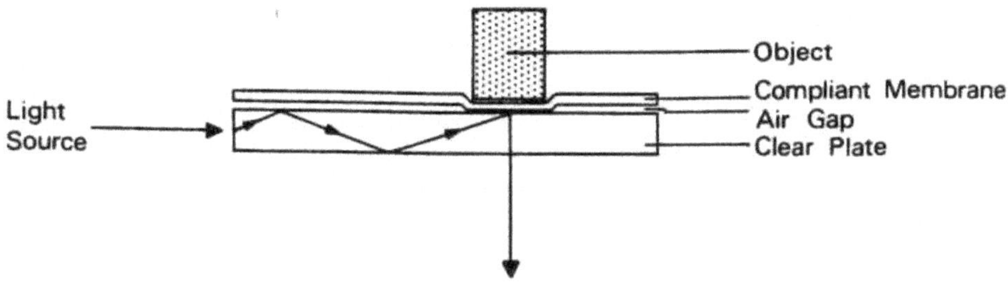

Figure 2.10: *Optical tactile element: pressure to light transduction.*

light between fibres. Jenstrom and Chen's 5×6 tactel sensor has a dynamic range of 5 - 400 g (0.049 - 3.9 N) and a force resolution of 0.2 g (0.0019 N), but exhibits 20% drift of response in 60 seconds. It gives a largely linear response to applied force, with 2.4 g (0.0023 N) hysteresis.

Several designs using another optical transduction technique show promise. Such sensors use the properties of reflection between media of different refractive index (Figure 2.10). The sensor comprises a clear plate illuminated on its edge by a light source, and an opaque elastic membrane above the face of the plate. The plate has a higher refractive index than the air surrounding it, and the membrane has a higher refractive index than both air and the clear plate. The light directed into the plate remains within it due to total internal reflection. This occurs because the plate is surrounded by air having a lower refractive index than the plate. When an object is placed on the rubber membrane, the rubber is brought into contact with the plate. At the places where this happens, light is diffusely reflected out of the lower face of the plate, because the rubber has a higher refractive index than the plate. When viewed from below, bright areas show up that correspond to the shape and position of a contacting object. The intensity of the bright areas is proportional to applied force.

S. Begej, formerly of the University of Massachusetts at Amherst, has developed various sensors using this transduction technique. After experimenting with a planar tactile table, having a tactile element density of 1500 tactels/cm^2 and using a CID camera for image capture (Begej, 1984), he developed a 32×32 tactel planar sensor. This device has a 0 - 40.8 g (0 - 0.4 N) dynamic range, and uses a coherent fibre-optic cable to direct the tactile image to a purpose built optical-to-digital converter incorporating a CCD array, analogue to digital converter, and microprocessor (Begej, 1988b). Simple control and feature extraction (e.g., image transfer, calculation of average intensity and first moments) is supported by the microprocessor software.

An advantage of this transduction method is that various sensor geometries are feasible, and Begej has demonstrated this by constructing a fingertip shaped sensor about the size of an adult human thumb (Begej, 1988a; Begej, 1988b). A first design utilises a 25 mm diameter Pyrex test tube (designated as the waveguide element) covered with a white polyethylene sheet, which is held in place by a black neoprene covering. The test tube is illuminated through the edge by an array of optical fibres, and the resultant image collected via a fish-eye lens. This fingertip sensor was made more compact by replacing the fish-eye lens with optical fibres epoxied to the inside of the waveguide. The optical fibres used to capture the tactile image are distributed in two densities to approximate the variation in density observed in human fingertips. 38 fibres illuminate the waveguide, and a further 256 fibres transport the tactile imprint to a remote display.

White and King, working at the University of California, Berkeley, also investigated suitable skin materials for an optical tactile sensor (White & King, 1985). They found that neoprene rubber exhibits significantly less hysteresis than both latex and silicon rubber. They experimented with hemispheric, pyramidal and smooth surface textures. All gave a response that had a linear section, and pyramidal elements had the widest range of response.

They also examined tactile elements for measuring shear, tensile stress and torque. The most versatile element was based on detecting the compression of a rubber ring resting on the clear plate surface. The element is pre-stressed to create a response without an applied force, so that tensile forces will produce a change in the sensor output. Shear force is measured by determining the difference in response between one side of the ring and the other. Torque can be detected by marking the ring with a pattern and tracking the rotation of that pattern. A prototype element has been built and gives promising results.

A sensor developed at the University of Wales, Aberystwyth uses solid state camera technology (Nicholls, 1990). A CCD imaging array is incorporated into the body of a robot gripper finger, and the gripping face of the finger contains a membrane and illuminated plate arrangement. Via a mirror and lens, the image from the plate is directed onto the CCD array. A cable running from the finger connects the CCD array to an electronics module, which generates drive signals and also produces a video output. This sensor has an active area of 2.3 cm^2, and a dynamic range (using 0.43 mm thick latex rubber) of 3 - 15 g (0.029 - 0.147 N). Thinner latex rubbers (e.g., 0.18 mm) are better at picking out textures, but have a narrower dynamic range.

The sensor has been used in grasp error compensation experiments, and for measurement and inspection of bicycle axle threads. An early prototype was able to perform image processing in a local electronics module (Mott *et al.*, 1986), but with only 64×64 tactel images.

A Japanese team have also experimented with this total internal reflection technique, and direct the tactile image onto an array of discrete phototransistors in a table-top mounted sensor (Tanie *et al.*, 1986; Park *et al.*, 1987), whilst researchers at Temple University, Philadelphia, have used a CCD array for image capture in their table-top mounted sensor (Makki *et al.*, 1988). Another planar device uses LEDs for edge illumination, but photodetectors are also mounted into the edge of the plate, each detector sited directly opposite a LED (Kawashima, 1987). This device thus measures optical attenuation along the path between light source and detector. A strongly textured rubber is used to achieve a monotonic, but non-linear response.

Another sensor using this transduction technique has been developed at the Electrical Engineering Department of the University of Newcastle-upon-Tyne, UK (Harris *et al.*, 1987; Ghani, 1988). This sensor directs the optical tactile image onto a CCD array, producing a video output. One version of the sensor has been integrated into a robot gripper finger which is able to rotate the sensing pad on the fingertip. This enables rotational grasp errors to be corrected without moving the whole manipulator or re-grasping. Work using this sensor for grasp position and orientation tasks is described in Chapter 5 of this book.

The principle of photoelasticity has been used in several tactile sensors which use a layer of photoelastic material as the primary transduction element. The force pattern at the surface of the photoelastic layer results in a characteristic stress pattern within the material; the pattern can reveal forces, shears, and torques. A sensor developed at the University of Oxford typifies such designs (Cameron *et al.*, 1988; Loh *et al.*, 1989). Circularly-polarised monochromatic light is shone through a layer of photoelastic plastic onto a compliant, reflective surface, which provides the interface to the sensed object. Reflected light passes back through the photoelastic material and a circular polariser to a camera, where an image can be observed. When stressed, the photoelastic material becomes birefringent[2], which can be observed with polarised light. The resultant light pattern can be interpreted in terms of stress. An interesting side effect is that edges are enhanced in the tactile image, since large changes in force profiles produce large stresses in the photoelastic material. This produces an effect similar to that of a high-pass filter, emphasising edges in the image.

The Oxford researchers built such a sensor into a robot finger, and also experimented with non-planar sensors, since the photoelastic plastic can be moulded. A detailed mathematical model for describing the tactile sensor was developed (Cameron *et al.*, 1988), showing how forces and stresses may be extracted from the data. Cameron *et al.* also observe that using motion in tactile sensing, similar to the way humans use active touch, enhances the capabilities of this device.

Other examples include a photoelasticity-based sensor employing a CCD array for image capture which could be used as a slip sensor (Eghtedari & Morgan, 1989), and a double-polarisation interferometer used in a strain gauge (Fürstenau, 1990).

The advantages of these optical techniques are simplicity, very high resolution, freedom from electrical interference, low interconnection and cabling requirements, and ease of remote processing. The video output from these designs enables straightforward connection to existing image processing systems. A disadvantage of this approach is the dependency on elastomers, which may affect the robustness of

[2]Birefringence is the property whereby a material has a distinct optical axis, and light travels through the medium at a different speed when polarised in the direction of the axis than when polarised orthogonal to it, hence producing optical interference (Loh *et al.*, 1989).

the sensor, and give hysteresis and creep problems. The designs based on photoelasticity have the advantage of enabling forces, stresses and torques to be extracted.

2.2.7 Other Techniques

The main categories of tactile sensor have been examined in the previous sections. There are several other techniques which deserve mention but which do not clearly fit into any of the previous categories.

For example, a force sensor using pulse-echo ranging consists of a layer of silicon rubber below which is a layer of the piezoelectric material polyvinylidene fluoride (PVF_2) (Grahn & Astle, 1984). The PVF_2 has electrodes on its upper and lower surfaces, forming the tactels, and is mounted on a substrate. Objects touching the sensor deform the silicon rubber layer. The tactels are subjected to 5 V, 200 ns-wide excitation pulses which, by the reverse piezoelectric effect, deforms them. This deformation produces ultrasonic pulses which reflect off the upper surface of the silicon rubber back to the transmitting tactels, which then act as receivers. The time taken to receive the echo is linearly proportional to the amount of deformation of the rubber. A 12 element array was fabricated, and the sensor exhibited a 2000:1 dynamic range. A significant point about this sensor is that the response time is equal to the ultrasonic pulse transit time, around 5 μs, enabling high bandwidth. Spatial resolution of forces 0.5 mm apart is also possible. Ultrasonic pulse-echo ranging using PVF_2 is also incorporated in Dario's sensing system at Pisa (Dario *et al.*, 1987).

Researchers at Pisa also considered polyelectrolyte gel as a transduction mechanism for tactile sensing (De Rossi *et al.*, 1989). In their analysis, a sensor is represented as a semi-infinite block of gel whose surface is covered by a thin, impermeable membrane. Application of a point load to the surface causes a solid polymer within the gel to be perturbed from its uniform state. This results in uneven distribution of the charge density in the gel, accompanied by spatial changes in concentrations of various types of mobile ions. These concentrations can be measured by ion-selective electrodes, and it is shown that the measured values are a direct function of the applied point stress. A linear response is achievable over a 10.194 - 1019.4 g (0.1 - 10 N) force range. This technique has promise for skin-like sensors, which can be moulded to non-planar configurations.

Silicon strain gauges were experimented with by Transensory Devices Inc., who were able to incorporate all the relevant circuitry and transduction mechanism on a single piece of silicon (Kowalski, 1985). The output from the sensor is an analogue voltage that is proportional to applied force; the response is linear with a dynamic range of 100:1 over a 10 - 1000 g (0.098 - 9.8 N) load.

A unique design developed at the University of Florida (Patterson & Nevill, 1986) is known as the induced vibration touch sensor. The sensor consists of two layers of PVF_2 piezoelectric film covered with a ridged silicon rubber skin. The PVF_2 layers are sensitive in orthogonal axes. The unusual aspect of this sensor is that it detects relative motion between the sensor and a contacting object. The motion between sensor and touching object produces vibrations that induce a voltage in each sensing layer. The output from each layer is related to the speed, orientation and shape of the object in contact. This design is dynamic as opposed to most other designs which are inherently static. Various experiments were conducted, and successful recognition of cylinders and spheres differing by 1.6 mm in diameter was achieved.

Finally, thermal properties can be sensed through contact with an object. For example, R.A. Russell has developed several thermal sensor arrays (Russell, 1985b; Russell, 1988b). An early design has 2×10 elements spaced 2.5 mm apart, consisting of a heat source (a power transistor) soldered to a copper block (Russell, 1985b). The copper has a known rate of thermal conductivity and spreads the heat from the transistor evenly across its bulk. A thermistor measures the temperature of the copper and a feedback circuit is used to maintain a constant temperature. A 3 mm outer layer of silicon rubber containing another thermistor acts as the contacting surface. The temperature drop caused by the flow of heat into an object is measured by the thermistor in the outer layer. In experiments, the sensor was able to distinguish between aluminium, wax, cork and plastic foam. A later, similar, design has 10×10 elements, spaced 3 mm apart on a brass block (Russell, 1988b). The typical response time for these sensors is about 3 seconds. Images representing the material constitution of touching objects provide a novel form of tactile data, and may enable correspondingly novel applications.

2.3 Summary

This chapter has dealt with the major categories of tactile sensors, classified in terms of the transduction method used. It is clear that there are opportunities for innovative designs, since tactile sensing has only seen significant research activity since the early 1980s and there are very few commercially available sensors. In addition to the requirements proposed by Harmon (1982), it is apparent from examining sensor applications that tactile sensors which can be moulded to non-planar shapes will be of particular use. For example, robot hands with tactile arrays moulded to the fingers will enable more dextrous manipulative tasks to be undertaken. Also, a robot and parts of its environment could be covered with sections of tactile sensitive material to detect and localise contact.

More emphasis needs to be placed on tactile system design, since it apparent from this survey that most researchers have concentrated on developing individual sensors. The following chapter considers the processing and use of tactile sensors, indicating some of the problems in integrating tactile sensors.

No single technology is appropriate for tactile sensing. Below are summarised the main advantages and disadvantages of each transduction method and some typical designs are noted.

2.3.1 Resistive and Conductive

Advantages:
Wide dynamic range.
Durability.
Good overload tolerance.
Compatibility with integrated circuitry, particularly VLSI.

Disadvantages:
Hysteresis in some designs.
Elastomer needs to be optimised for both mechanical and electrical properties.
Limited spatial resolution compared with vision sensors.
Large numbers of wires may have to be brought away from the sensor.
Monotonic response but often not linear.

Typical Designs:
Raibert (1984), an extension of Hillis's 1982 design.
Tise (1988).

2.3.2 Piezoelectric and Pyroelectric Effects

Advantages:
Wide dynamic range.
Durability.
Good mechanical properties of piezo/pyroelectric materials.
Temperature as well as force sensing capability.

Disadvantages:
Difficulty of separating piezoelectric from pyroelectric effects.
Inherently dynamic: output decays to zero for constant load.
Difficulty of scanning elements.
Good solutions are complex.

Typical Designs:
Dario *et al.* (1984).
Domenici *et al.* (1989).

2.3.3 Capacitive Techniques

Advantages:
Wide dynamic range.
Linear response.
Robust.

Disadvantages:
Susceptible to noise.
Some dielectrics are temperature sensitive.
Capacitance decreases with physical size ultimately limiting spatial resolution.

Typical Designs:

Siegel *et al.* (1987).
Fearing (1987).

2.3.4 Magnetic Transduction Methods

a) Mechanical Displacement

Advantages:

Wide dynamic range.
Large displacements possible.
Simple.

Disadvantages:

Poor spatial resolution.
Mechanical problems when sensing on slopes.

Typical Designs:

Sato *et al.* (1986).

b) Magnetoelastic and Magnetoresistive

Advantages:

Wide dynamic range.
Low hysteresis.
Linear response.
Robust.
Normal force, shear and torque capability.

Disadvantages:

Susceptibility to stray fields and noise.
A.c. circuitry required.

Typical Designs:

Hackwood *et al.* (1986).
Vranish (1986a).

2.3.5 Mechanical Transduction Methods

Advantages:

Well known technology.
Good for probe applications.

Disadvantages:

Complex for array constructions.
Limited spatial resolution.

Typical Designs:

Russell (1990b).

2.3.6 Optical Transduction Methods

Advantages:

Very high resolution.
Compatible with vision sensing technology.
No electrical interference problems.
Processing electronics can be remote from sensor.
Low cabling requirements.

Disadvantages:

Dependence on elastomer in some designs - affects robustness.
Some hysteresis.

Typical Designs:

Begej (1988b).
Loh *et al.* (1989).
Nicholls (1990).

References

Abramowitz, J.D., Goodnow, J.W., Paul, B., & Whitney, A. 1984. *Pennsylvania articulated hand - an end effector to determine shape by touch.* Tech. rept. MS-CIS-84-56, GRASPLAB 28. Dept. Computer & Information Science, Moore School, Univ. Pennsylvania, Philadelphia, PA.

Bajcsy, R., & Goldberg, K.Y. 1984 (May). *Active touch and robot perception.* Tech. rept. MS-CIS-84-42, GRASPLAB 14. Dept. Computer & Information Science, Moore School, Univ. Pennsylvania, Philadelphia, PA.

Begej, S. 1984 (Nov). *An optical tactile-array sensor.* Tech. rept. COINS 84-26. Dept. Computer & Information Science, Univ. Mass. at Amherst.

Begej, S. 1988a. Fingertip-shaped optical tactile sensor for robotic applications. *Pages 1752–1757 of: Proc. IEEE int. conf. on robotics and automation (Philadelphia, PA).* Washington, DC: IEEE Computer Soc. Press.

Begej, S. 1988b. Planar and finger-shaped optical tactile sensors for robotic applications. *IEEE j. robotics and automation*, **4**(5), 472–484.

Berger, A.D., & Khosla, P.K. 1991. Using tactile data for real-time feedback. *Int. j. robotics research*, **10**(2), 88–102.

Boie, R.A. 1984. Capacitive impedance readout tactile image sensor. *Pages 370–378 of: Proc. IEEE int. conf. on robotics (Atlanta, GA).* Washington, DC: IEEE Computer Soc. Press.

Brandolini, A., & Gandelli, A. 1990. Signal evaluation and measurement techniques in a single cell hybrid tactile sensor for large object manipulation. *Pages 126–131 of: Proc. IEEE instrumentation and measurement technology conf. (San Jose, CA).* Washington, DC: IEEE Computer Soc. Press.

Briot, M. 1979. The utilisation of an artificial skin sensor for the identification of solid objects. *Pages 529–547 of: Proc. 9th int. symposium on industrial robotics.* Kempston, UK: IFS(Pubns).

Cameron, A., Daniel, R., & Durrant-Whyte, H. 1988. Touch and motion. *Pages 1062–1067 of: Proc. IEEE int. conf. on robotics and automation (Philadelphia, PA).* Washington, DC: IEEE Computer Soc. Press.

Chalupa, V., Marik, V., & Volf, J. 1982 (May). Tactile matrix for shape recognition. *Pages 339–348 of: Technological & methodological advances in measurement – proc. 9th IMEKO congress of the Int. Measurement Confed. (Berlin).*

Checinski, S.S., & Agrawal, A.K. 1986. Magnetoelastic tactile sensor. *Pages 229–235 of:* Pugh, A. (ed), *Robot sensors vol.2: Tactile and non-vision.* Kempston, UK: IFS (Pubns).

Christ, J.P., & Sanderson, A. 1982 (Sept). *A prototype tactile sensor array.* Tech. rept. CMU-RI-TR-82-14. Robotics Institute, Carnegie-Mellon Univ., Pittsburgh, PA.

Collins, A.S., & Hoover, W.A. 1987. A prototype for an image-based tactile sensor. *Pages 1760–1765 of: Proc.IEEE int. conf. on robotics and automation (Raleigh, NC).* Washington, DC: IEEE Computer Soc. Press.

Crosnier, J.J. 1986. Grasping systems with tactile sense using optical fibres. *Pages 209–217 of:* Pugh, A. (ed), *Robot sensors vol.2: Tactile and non-vision.* Kempston, UK: IFS (Pubns).

Curtin, M.J. 1987. New force/displacement transducer doubles as a precision tactile sensor. *Pages 177–189 of: Proc. sensors expo (Detroit).* Peterborough, NH: Helmers Publishing.

Dario, P., & De Rossi, D. 1985. Tactile sensors and the gripping challenge. *IEEE spectrum*, Aug, 46–52.

Dario, P., De Rossi, D., Domenici, C., & Francesconi, R. 1984. Ferroelectric polymer tactile sensors with anthropomorphic features. *Pages 332–340 of: Proc. IEEE int. conf. on robotics (Atlanta, GA).* Washington, DC: IEEE Computer Soc. Press.

Dario, P., Bergamasco, M., Femi, D., Fiorillo, A., & Vaccarelli, A. 1987. Tactile perception in unstructured environments: A case study for rehabilitative robotics applications. *Pages 2047–2054 of: Proc. IEEE int. conf. on robotics and automation (Raleigh, NC)*. Washington, DC: IEEE Computer Soc. Press.

De Rossi, D., Lazzeri, L., Domenici, C., Nannini, A., & Basser, P. 1989. Tactile sensing by an electromechanochemical skin analog. *Sensors & actuators (Switzerland)*, **17**(1-2), 107–114.

Dillman, R. 1982. A sensor controlled gripper with tactile and non-tactile sensor environment. *Pages 159–170 of: Proc. 2nd int. conf. on robot vision and sensory controls (Stuttgart)*. Kempston, UK: IFS (Pubns).

Domenici, C., De Rossi, D., Bacci, A., & Bennati, S. 1989. Shear stress detection in an elastic layer by a piezoelectric polymer tactile sensor. *IEEE trans. electr. insul*, **EI-24**(6), 1077–1081.

Eghtedari, F., & Morgan, C. 1989. A novel tactile sensor for robot applications. *Robotica*, **7**(4), 289–295.

Emge, S.R., & Chen, Chin-Lin. 1991. Two-dimensional countour imaging with a fiber optic microbend tactile sensor array. *Sensors & actuators B. chem. (Switzerland)*, **B3**(1), 31–42.

Fearing, R.S. 1987. Some experiments with tactile sensing during grasping. *Pages 1637–1643 of: Proc. IEEE int. conf. on robotics and automation (Raleigh, NC)*. Washington, DC: IEEE Computer Soc. Press.

Fearing, R.S. 1990. Tactile sensing mechanisms. *Int. j. robotics research*, **9**(3), 3–23.

Fearing, R.S., & Hollerbach, J.M. 1985. Basic solid mechanics for tactile sensing. *Int. j. robotics research*, **4**(3), 40–54.

Fürstenau, N. 1990. Investigation in a tactile sensor based on fiber-optic interferometric strain gauges. *Proc. SPIE - Int. Soc. Opt. Eng. (Fiber optic and laser sensors VII, Boston, MA)*, **1169**, 531–539.

Ghani, N. 1988. A collaborative sensors project. *Pages 59–70 of: Proc. 4th. CIM Europe conf. (Madrid)*. Kempston, UK: IFS (Pubns).

Grahn, A.R., & Astle, S. 1984 (June). Robotic ultrasonic force sensor arrays. *Pages 21.1 - 21.18 of: Proc. Robots 8 (Detroit, MI)*.

Grupen, R.A., Henderson, T.C., & McCammon, I.D. 1989. A survey of general-purpose manipulation. *Int. j. robotics research*, **8**(1), 38–62.

Hackwood, S., Beni, G., & Nelson, T.J. 1986. Torque-sensitive tactile array for robotics. *Pages 123–131 of:* Pugh, A. (ed), *Robot sensors vol.2: Tactile and non-vision*. Kempston, UK: IFS (Pubns).

Harmon, L.D. 1982. Automated tactile sensing. *Int. j. robotics research*, **1**(2), 3–31.

Harmon, L.D. 1984. Tactile sensing for robots. *Pages 109–158 of:* M. Brady, L. A. Gerhardt, H. F. Davidson (ed), *Robotics and artificial intelligence (NATO ASI series)*. New York: Springer-Verlag.

Harris, A.J., Allen, C.R., Glass, W., Shepherd, A., Cooper, L., & Moss, S. 1987. An integrated smart tactile sensor for 2-axis robot grippers. *Proc. SPIE – Int. Soc. Opt. Eng. (Intelligent robots and computer vision, Cambridge, MA)*, **726**, 164–170.

Hillis, W.D. 1982. A high resolution imaging touch sensor. *Int. j. robotics research*, **1**(2), 33–44.

Howe, R.D., & Cutkosky, M.R. 1989. Sensing skin acceleration for slip and texture perception. *In: Proc. IEEE int. conf. on robotics and automation (Scottsdale, AZ)*. Washington, DC: IEEE Computer Soc. Press.

Howe, R.D., Popp, N., Akella, P., Kao, I., & Cutkosky, M.R. 1990. Grasping, manipulation and control with tactile sensing. *Pages 1258–1265 of: Proc. IEEE int. conf. on robotics and automation (Cincinnati, OH)*. Los Alamitos, CA: IEEE Computer Soc. Press.

Jacobsen, S.C., Iversen, E.K., Knutti, D.F., Johnson, R.T., & Biggers, K.B. 1986. Design of the Utah/MIT dextrous hand. *Pages 1520–1532 of: IEEE int. conf. on robotics and automation (San Francisco, CA)*. Washington, DC: IEEE Computer Soc. Press.

Jacobsen, S.C., McCammon, I.D., Biggers, K.B., & Phillips, R.P. 1987. Tactile sensing system design issues for machine manipulation. *Pages 2087–2096 of: Proc. IEEE int. conf. on robotics and automation (Raleigh, NC)*. Washington, DC: IEEE Computer Soc. Press.

Jayawant, B.V. 1989. Tactile sensing in robotics. *J. physics E: Sci. instrum.*, **22**, 684–692.

Jayawant, B.V., Onori, M.A., & McK.Watson, J.D. 1986. Robot tactile sensing: A new array sensor. *Pages 199–205 of:* A.Pugh (ed), *Robot sensors vol.2: Tactile and non-vision*. Kempston, UK: IFS(Pubns).

Jenstrom, D.T., & Chen, Chin-Lin. 1989. A fiber optic microbend tactile sensor array. *Sensors & actuators (Switzerland)*, **20**(3), 239–248.

Kah-Bin, L., & Yoon-Song, C. 1988. Low cost tactile gripper using silicone rubber sensor array. *Robotica*, **6**(1), 23–30.

Kawashima, T. 1987. An optical tactile sensor using the CT reconstruction method. *Electronics and communications in Japan, part 2*, **70**(10), 35–43.

Kenaley, G.L., & Cutkosky, M.R. 1989. Electrorheological fluid-based robotic fingers with tactile sensing. *Pages 132–136 of: Proc. IEEE int. conf. on robotics and automation (Scottsdale, AZ)*. Washington, DC: IEEE Computer Soc. Press.

Kinoshita, G., Hajika, T., & Hattori, K. 1983 (Sept). Multifunctional tactile sensors with multielements for fingers. *Pages 195–202 of: Proc. int. conf. on advanced robotics.*

Kowalski, C. 1985. Silicon succeeds in touch sensing. *Robotics today*, April, 29–36.

Larcombe, M.H.E. 1981. Why carbon fibres can give robots a strong sense of grip. *Sensor review*, **1**(1), 58–59.

Lauber, A., Sandell, B., Holmbom, P., & Pederson, O. 1988. Tactile sensors for industrial robots. *Sensor review*, **8**(2), 84–88.

Loh, K.W., Durrant-Whyte, H., & Daniel, R. 1989. *A photoelastic tactile sensor.* Tech. rept. OUEL 1792/89. Dept. Engineering Science, Univ. Oxford, Parks Road, Oxford, UK.

Luo, R. C., & Henderson, T.C. 1986. A servo-controlled robot gripper with multiple sensors and its logical specification. *J. robotic systems*, **3**(4), 409–420.

Luo, R.C. 1984 (Dec). Design and implementation of hand-based tactile sensors for industrial robots. *Pages 423–432 of: Proc. int. conf. on robotics and factories of the future (Charlotte, N.Carolina).*

Makki, A.Z., Collins, A.S., & Alexander, C.K. 1988. An alternative geometrical setup for the image-based tactile sensor. *Pages 440–447 of: Proc. 20th South Eastern symposium on system theory*. Washington, DC: IEEE Computer Soc. Press.

Mitchell, E.E., & Vranish, J. 1985. Magnetoelastic force feedback sensors for robots and machine tools - an update. *Pages 187–194 of: Proc. 5th int. conf. on robot vision and sensory controls (Amsterdam)*. Kempston, UK: IFS (Pubns).

Moore, T.N., Jeswiet, J., Nshama, W., & Grotenhuis, R.L. Ten. 1987. Development of advanced sensors for industrial robots. *Pages 79–84 of: Proc. 2nd int. conf. on computer-aided production engineering (Edinburgh)*. Bury St. Edmonds, UK: Mech. Eng. (Pubns).

Mott, D.H., Lee, M.H., & Nicholls, H.R. 1986. An experimental very high resolution tactile sensor array. *Pages 179–189 of:* Pugh, A. (ed), *Robot sensors vol. 2: Tactile and non-vision.* Kempston, UK: IFS (Pubns).

Nicholls, H.R. 1990. Tactile sensing using an optical transduction method. *Pages 83–95 of:* Henderson, T.C. (ed), *Traditional and non-traditional robot sensors (NATO ASI series).* New York: Springer-Verlag.

Nicholls, H.R., & Lee, M.H. 1989. A survey of robot tactile sensing technology. *Int. j. robotics research*, **8**(3), 3–30.

Overton, K.J. 1984. *The acquisition, processing and use of tactile sensor data in robot control.* Tech. rept. COINS 84-08. Dept. Computer & Information Science, Univ. Mass. at Amherst.

Park, M., Tanie, K., & Lee, S.B. 1987. An array tactile sensor system for robot hand. *Pages 925–929, vol. 3 of: Proc. Tencon '87: IEEE Region 10 conf. on computers and communications technology toward 2000 (Seoul, S. Korea).* New York: IEEE.

Patterson, R.W., & Nevill, G.E. 1986. Performance of an induced vibration touch sensor. *Pages 220–228 of:* Pugh, A. (ed), *Robot sensors vol.2: Tactile and non-vision.* Kempston, UK: IFS (Pubns).

Pennywitt, K.E. 1986. Robotic tactile sensing. *Byte magazine*, Jan, 177–200.

Peterson, R.R. 1984 (Aug). A compliant tactile sensor utilising conductive elastomers. *In: Proc. first world conf. on robotics research (LeHigh Univ., Bethlehem).*

Pirolo, D.G., & Kolesar, E.S. 1989. Piezoelectric polymer tactile sensor arrays for robotics. *Pages 1130–1135 of: Proc. IEEE national aerospace and electronics conf. (Dayton, OH).* New York: IEEE.

Presern, S., Dacar, F., & Spegel, M. 1981. Design of three active degrees of freedom tactile sensor for industrial arc welding robots. *Pages 207–213 of: Proc. 4th British Robot Assoc. conf. (Brighton, UK).* Kempston, UK: IFS (Pubns).

Pruski, A., & Mutel, B. 1984 (Dec). Direct contact sensors based on carbon fibre. *Pages 409–415 of: Proc. int. conf. on robotics and factories of the future (Charlotte, NC).*

Pugh, A. (ed). 1986a. *Robot sensors vol. 1: Vision.* Kempston, UK: IFS (Pubns).

Pugh, A. (ed). 1986b. *Robot sensors vol.2: Tactile and non-vision.* Kempston, UK: IFS (Pubns).

Purbrick, J.A. 1981. A force transducer employing conductive silicone rubber. *Pages 73–80 of: Proc. 1st int. conf. on robot vision and sensory controls (Stratford, UK).* Kempston, UK: IFS (Pubns).

Raibert, M.H. 1984. An all digital VLSI tactile array sensor. *Pages 314–319 of: Proc. IEEE int. conf. on robotics (Atlanta, GA).* Washington, DC: IEEE Computer Soc. Press.

Raibert, M.H., & Tanner, J.E. 1982. Design and implementation of a VLSI tactile sensing computer. *Int. j. robotics research*, **1**(3), 3–18.

Rebman, J., & Morris, K.A. 1986. A tactile sensor with electrooptical transduction. *Pages 145–155 of:* Pugh, A. (ed), *Robot sensors vol.2: Tactile and non-vision.* Kempston, UK: IFS (Pubns).

Reston, R.R., & Kolestar, E.S. 1990. Robotic tactile sensor array fabricated from a piezoelectric polyvinylidene fluoride film. *Pages 1139–1144 of: Proc. IEEE 1990 national aerospace and electonics conf. (Dayton, OH).* New York: IEEE.

Ristic, R., Benhabib, B., & Goldenberg, A.A. 1989. Analysis and design of a modular electro-optical tactile sensor. *IEEE trans. on robotics and automation*, **RA-5**(3), 362–368.

Robertson, B.E., & Walkden, A.J. 1986. Tactile sensor system for robotics. *Pages 89–97 of:* Pugh, A. (ed), *Robot sensors vol.2: Tactile and non-vision.* Kempston, UK: IFS (Pubns).

Russell, R.A. 1985a. Object recognition using articulated whisker probes. *Pages 605–612 of: Proc. 15th int. symposium on industrial robots (Tokyo).* Kempston, UK: IFS(Pubns).

Russell, R.A. 1985b. A thermal sensor array to provide tactile feedback for robots. *Int. j. robotics research*, **4**(3), 35–39.

Russell, R.A. 1987. Compliant-skin tactile sensor. *Pages 1645–1648 of: Proc. IEEE int. conf. on robotics and automation (Raleigh, NC).* Washington, DC: IEEE Computer Soc. Press.

Russell, R.A. 1988a. A 3 degrees of freedom tactile probe. *Pages 898–907 of: Proc. 19th int. symposium on industrial robots (Sydney, Aus.).* Kempston, UK: IFS (Pubns).

Russell, R.A. 1988b. Thermal sensor for object shape and material construction. *Robotica*, **6**(1), 31–34.

Russell, R.A. 1990a. *Robot tactile sensing.* Englewood Cliffs, NJ: Prentice-Hall.

Russell, R.A. 1990b. Tactile sensing of 3-dimensional surface features. *Robotica*, **8**(2), 111–115.

Salisbury, J.K. 1982 (July). *Kinematic and force analysis of articulated hands.* Tech. rept. STAN-CS-82-921. Dept. Computer Science, Stanford Univ., CA.

Sato, N., Heginbotham, W.B., & Pugh, A. 1986. A method for three-dimensional part identification by tactile transducer. *Pages 132–143 of:* Pugh, A. (ed), *Robot sensors vol.2: Tactile and non-vision.* Kempston, UK: IFS (Pubns).

Schneiter, J.L., & Sheridan, T.B. 1984. An optical tactile sensor for manipulators. *Robotics and computer integrated manufacturing*, **1**(1), 65–71.

Schoenwald, J.S., Thiele, A. W., & Gjellum, D.E. 1987. A novel fiber optic tactile array sensor. *Pages 1792–1797 of: Proc. IEEE int. conf. on robotics and automation (Raleigh, NC).* Washington, DC: IEEE Computer Soc. Press.

Seekircher, J., & Hoffman, B. 1988. Improved tactile sensors. *Pages 20.1–20.6 of: Proc. IFAC symposium on robot control (Karlsruhe, West Germany).* Oxford, UK: Pergamon Press.

Siegel, D.M., Garabieta, I., & Hollerbach, J.M. 1986. An integrated tactile and thermal sensor. *Pages 1286–1291 of: Proc. IEEE int. conf. on robotics and automation (San Francisco).* Washington, DC: IEEE Computer Soc. Press.

Siegel, D.M., Drucker, S.M., & Garabieta, I. 1987. Performance analysis of a tactile sensor. *Pages 1493–1499 of: Proc. IEEE int. conf. on robotics and automation (Raleigh, NC).* Washington, DC: IEEE Computer Soc. Press.

Smith, P.R. 1987. Use of polyvinylidene fluoride (PVDF) thin film as pyroelectric sensors. *In: Proc. sensors expo (Detroit).* Peterborough, NH: Helmers Publishing.

Snyder, W.E., & St.Clair, J. 1978. Conductive elastomers as a sensor for industrial parts handling equipment. *IEEE trans. instrumentation and measurement*, **IM-27**(1), 94–99.

Speeter, T.H. 1989. Flexible piezoresistive touch sensing array. *Proc. SPIE - Int. Soc. Opt. Eng. (Optics, illumination and image sensing for machine vision III, Cambridge, MA)*, **1005**, 31–43.

Speeter, T.H. 1990a. Control of the Utah/MIT dextrous hand: Hardware and software hierarchy. *J. robotic systems*, **7**(5), 759–790.

Speeter, T.H. 1990b. A tactile sensing system for robotic manipulation. *Int. j. robotics research*, **9**(6), 25–36.

Sugiyama, S., Kawahata, K., Yoneda, M., & Igarashi, I. 1990. Tactile image detection using a 1k-element silicon pressure sensor array. *Sensors and actuators*, **A**(22), 397–400.

Suzuki, K., Najafi, K., & Wise, K.D. 1990. A 1024-element high-performance silicon tactile imager. *IEEE trans. on electronic devices*, **ED-37**(8).

Tanie, K., Komoriya, K., Kaneko, M., Tachi, S., & Fujikawa, A. 1986. A high resolution tactile sensor. *Pages 190–198 of:* Pugh, A. (ed), *Robot sensors vol.2: Tactile and non-vision*. Kempston, UK: IFS (Pubns).

Tise, B. 1988. A compact high resolution piezoresistive digital tactile sensor. *Pages 760–764 of: Proc. IEEE int. conf. on robotics and automation (Philadelphia, PA)*. Washington, DC: IEEE Computer Soc. Press.

van Brussel, H., & Belien, H. 1986. A high resolution tactile sensor for part recognition. *Pages 49–60 of: Proc. 6th int. conf. on robot vision and sensory controls (Paris)*. Kempston, UK: IFS (Pubns).

Vranish, J.M. 1986a. Magnetoinductive skin for robots. *Pages 599–631 of: Proc. 16th int. symposium on industrial robots (Brussels)*. Kempston, UK: IFS(Pubns).

Vranish, J.M. 1986b. Magnetoresistive skin for robotics. *Pages 99–111 of:* Pugh, A. (ed), *Robot sensors vol.2: Tactile and non-vision*. Kempston, UK: IFS (Pubns).

Vranish, J.M., Mitchell, E., & DeMoyer, R. 1982. Outstanding potential shown by magnetoelastic force feedback sensors for robots. *Sensor review*, **2**(4), 200–205.

Warnecke, H.J., & Haaf, D. 1981. Components for programmable assembly – grippers, sensors, conveying systems for an industrial robot engaged together with fixed automation. *Pages 225–233 of: Proc. 2nd int. conf. on assembly automation (Brighton)*. Kempston, UK: IFS (Pubns).

Webster, J.G. 1988. *Tactile sensors for robotics and medicine*. New York: John Wiley.

White, R.M., & King, A.A. 1985 (June). Tactile array for robotics employing a rubbery skin and a solid-state optical sensor. *Pages 18–21 of: Proc. 3rd int. conf. on solid state sensors and actuators (Philadelphia, PA)*.

Wolfeld, J.A. 1981 (Aug). *Real time control of a robot tactile sensor*. Tech. rept. MS-CIS-81-04. Dept. Computer & Information Science, Moore School, Univ. Pennsylvania, Philadelphia, PA.

Wolffenbuttel, R.F., & Regtien, P.P.L. 1989. Integrated tactile imager with intrinsic contour detection option. *Sensors and actuators*, **16**, 141–153.

Chapter 3

Processing and Using Tactile Sensor Data

Howard R. Nicholls

3.1 Introduction

This chapter examines the various approaches taken in the processing and use of tactile data. Early phases of tactile sensing research concentrated on developing new sensors. Many of the processing techniques were derived from computer vision and image processing, and little emphasis was placed on the use of sensors in practical tasks. Currently, tactile sensors are used for tasks such as simple feature extraction, grasping and manipulation, as well as for recovery of 3-dimensional shape and for object recognition. More emphasis is now being placed on integrating tactile sensing into complete systems, the tactile sensor being only one component of a complex system typically involving multiple sensors, a manipulator, and a controller (see, for example, Jacobsen *et al.* (1987), Berger and Khosla (1991)). Also, a survey conducted in 1988 (Nicholls & Lee, 1989) concluded that there was a trend away from static, single tactile image analysis towards treating tactile sensing as an active modality, suited to exploratory situations in unstructured environments. When conducting the literature search for this text, it was found that the trend towards active tactile sensing is continuing.

The following sections examine some basic issues in tactile processing, including determination of various contact parameters, grasping, and basic feature extraction. Progress made towards active tactile sensing is also briefly reviewed here, and is considered in more detail in this volume by Lederman *et al.* (Chapters 8 and 9) and by Allen *et al.* (Chapters 10 and 11), whilst the human tactile system is considered by Moss-Salentijn in Chapter 7.

3.2 Basic Tactile Sensing Issues

3.2.1 Parameters of Contact

Various basic parameters can be sensed concerned with contact between a tactile sensor and an object in the environment. The most elementary information is simply the presence or absence of contact, as typified by the use of binary touch probes (Williams & Lill, 1987; Seeley & Williams, 1990). Although very basic information is provided from such sensors, when combined with information from other sensors on the robot, quite complex tasks can be accomplished and uncertainties resolved. Chapter 6 by Nicholls and Hardy examines in detail the use of a set of binary probes distributed around a flexible assembly machine; the binary probes enable feeding confirmation, and resolution of positional uncertainties.

In another example of sensing with touch probes, simple presence/absence data from a binary probe combined with information from joint angle sensors was used to achieve localisation of objects in a robot workspace (Zhou & El-Zorkany, 1987). Plane equations, edges, and vertices could be recovered using contact sensing by integrating information from these sensors.

A distinction has been made between tactile sensing and force sensing, combining the two modes under the general heading of contact sensing (Dario, 1989). Dario defines force sensing as the measurement of global mechanical effects of contact, and tactile sensing as the detection of a wide range of local parameters affected by contact. The global versus local distinction is important, and Dario extends this distinction, taking a systems view that force sensing is essentially related to low-level compliant motion control, whilst true tactile sensing implies higher level (perceptual) processing.

Many scenarios in contact sensing will involve a rigid object indenting the compliant covering of a tactile sensor array. The indentation of the compliant skin can be analysed in two conceptually different ways (Dario, 1989), depending upon the task being executed. If the sensor were mounted on the fingertip of a robotic hand, it might be appropriate to measure the contact stresses inside the compliant skin to enable the control of fine manipulation tasks. Alternatively, the deflection profile of the indented skin could be measured and analysed, enabling recovery of local features and possibly object recognition. The same physical effect, indentation of a compliant skin, can thus give rise to quite different processing and control approaches.

The problems of recovering the local shape of a contacting body (or the applied forces) given a set of sub-surface strain (or displacement) measurements in an elastic body are considered by Fearing in Chapter 4 of this volume. Essentially, if an object is pressed into a compliant finger, contact stresses (σ_x, σ_y, σ_z, τ_{xy}, τ_{xz}, τ_{yz}) will be distributed over the contact area. The tactile sensor will measure either the displacement (u, v, w) of the elastic material or the strains (ϵ_x, ϵ_y, ϵ_z, γ_{xy}, γ_{xz}, γ_{yz}) within it. The indentation and sub-surface deformation are functions of the contact and finger shape, applied loading and material properties of both sensor and object. This relationship is complex, but the theory of elasticity can be used to model sensor-object interactions, enabling recovery of usable contact-derived data. Problems can be encountered in object shape reconstruction due to noise in the data derived from the transduction process. However, methods such as Bayesian *a posteriori* estimation (Nowlin, 1991) can be used to recover more reliable data.

Further consideration of recovery of data and of planar elasticity for tactile sensing is deferred to Chapter 4 by Fearing, who also presents some background material concerning elasticity in Appendix A.

3.2.2 Grasping and Tactile Sensing

Advanced robots often have end effectors with multiple jointed fingers, and attempt grasping and dextrous manipulation, such as rolling and twirling of objects, in unstructured environments. This gives rise to many problems, such as determining a stable grasp, modelling the kinematics, and incorporating closed-loop feedback into the control of the end effector.

Grasping an object can be considered as the exertion of force and moment on an object to keep it in stable equilibrium or to move it. This is known as force-closure (Nguyen, 1988). Grasping can also involve the constraint of motion of an object by a set of contacts, irrespective of the magnitude of the contact forces. This is known as form-closure.

Force-closure grasping was investigated by Nguyen, who developed simple algorithms for directly constructing such grasps based on the shape of a grasped object (Nguyen, 1988). The theorems which he presents help to explain why certain types of grasps carried out by humans are successful. For example, he shows that a soft finger contacting an acute vertex can be approximated as a point contact with a wide friction cone, and that the larger the friction cone, the greater is the likelihood that the grasp is force-closure. This partially explains why people grasp objects at edges and corners, and why the contacting surface of human fingers is soft.

This leads to the problem of selection of appropriate skin materials for robot grippers and fingers. Cutkosky *et al.* (1987) examined experimentally skin materials for optimising grasping. They found that treads, grooves, and surface textures substantially improve the consistency of surface friction, and hence grasp security, under conditions of moisture. Fearing and Hollerbach (1985) also examined the effect of having ridges on a sensor surface, and reported that ridges can increase the strain amplitude by a factor of four. The sensitivity of a tactile sensor can therefore be increased by placing sensing elements beneath and between ridges to detect the maximum-amplitude strain peaks, whilst also improving grasp security.

Some studies of grasping have assumed that contact friction forces will be sufficient to prevent an object from sliding from the gripper fingers, but Trinkle *et al.* (1988) considered the mechanical analysis of robot hands where the friction is not large enough to prevent sliding. They developed the mathematics of frictionless grasping, and implemented a planner and simulator for the grasping of convex polygons assuming negligible friction and dynamic effects. In their simulation, they were able to execute grasps under position control without the need for force control or for tactile sensing, although they assumed that accurate descriptions of the important physical bodies in their system were available.

Rebman and Kallhammer (1986) also examined the conditions for slippage by grasping objects in a parallel jaw gripper, applying an external force to the object, and measuring object displacement using a tactile sensor array. They were concerned with identifying the precursors to slip, and found that there is a non-linear increase in displacement for a linear increase in applied (external displacing) force. The motion of the displaced object increases slowly at first and accelerates until the object breaks away from the gripper.

Sensor output during a robot task was also considered by Vaidyanathan and Wood (1990), who were concerned with identifying significant events in a robot task. They examined the output of eight tactile elements during grasp-release cycles and developed a system which could indicate the robot task status by filtering and thresholding a tactile image.

Some approaches to stable grasping do not utilise tactile sensors, but instead use compliance in the finger joints to achieve stable grasps. For example, Kaneko and Tanie (1990) consider the problem of grasping objects, where exact object position and shape are not known. Instead of using tactile sensors distributed over the robot hand and fingers, their system uses joint compliance to detect contact between the gripper and object. Firstly, there is an approach phase, whereby the fingers of the gripper are opened widely until part of a finger link touches an object. Through compliance control, the system is able to determine which finger link is in contact with the object, and this posture noted. Then, a small posture change is introduced whilst maintaining contact with the object; an approximate contact point can be computed as the intersection of these two postures. They call this method self-posture changeability (SPC); one advantage of it is that no additional sensing hardware need be added to the multi-jointed fingers of the gripper, the only modification is in the software of the control system.

The Cost Sensitive Learning system (CSL) developed by M. Tan is also able to sense objects and to grasp them (Tan, 1990). CSL operates in two stages: a training stage, where sensing procedures are run on sample objects and object descriptions built up, and an execution phase, where descriptions are compared with sensed parameters and appropriate predefined grasp procedures are selected. A vision system is used to locate both a mobile robot and an object to be grasped, and a wrist-mounted sonar sensor is used to extract object features. The system is designed such that the execution phase is cost-sensitive and aims to maximise use of gathered sensor data in the selection of grasp procedures.

A system constructed by Sharon Stansfield at Sandia National Laboratories creates a representation of a sensed object using visual processing. This representation is then used in generating and executing grasp procedures (Stansfield,

1991b). Rather than taking an algorithmic approach to grasping, Stansfield adopts a heuristic method derived from the human model of grasping. This model divides the grasp into two stages: the transport stage where object properties such as spatial location are used to guide the hand to the object, and the manipulation stage, where properties such as object size and shape are used to pre-shape the hand ready for a grasp. Stansfield proposes that the final stage, when contact has been established, involves using haptic information relating to touch in a feedback manner to adjust the grasp for properties that have not been extracted visually (e.g., material construction and mass). The heuristics relate features to specific hand pre-shapes, such as grip, wrap, and pinch, and are generalised. An important aspect of this method is that objects do not have to be recognised in order to be grasped, making it particularly suited to situations where there are sensor errors, and for grasping in unstructured environments.

The human tactile system was also considered by robotics researchers at Stanford University (Howe *et al.*, 1990). They conducted various robotic manipulation tasks and found several common properties between forces, tactile data and motions in a robot gripper and in the human hand. To study basic issues in control and grasping, they developed a simple manipulator consisting of two fingers, each equipped with force sensors and tactile sensors and having two degrees of freedom. Two kinds of tactile sensors were employed: the stress rate sensor (Cutkosky & Howe, 1988) for detecting local contact stresses, and the skin acceleration sensor (Howe & Cutkosky, 1989). The latter sensor uses an accelerometer to measure vibrations at the finger-object contact, and is particularly suited for detecting the making or breaking of contact, incipient slip, and collisions between the grasped object and other objects in the environment.

The skin acceleration sensor proved useful for detecting vibrations in grasped objects and for signalling transition between phases in a manipulative task, such as from pre-contact to loading, and from loading to manipulation. This sensor was able to detect incipient slip, enabling grasp forces to be adjusted. Similarity between the role of certain human receptors and the skin acceleration sensor was noted. It was also found that the force control strategy used in the robot gripper was similar to that employed by humans. Humans tend to increase or decrease grasp forces and load (lifting) forces in parallel to minimise the chance of slip, maintaining a constant ratio of grasp force to load force despite variations in the load force (Howe *et al.*, 1990). This strategy was used in the control system for their manipulator system. They concluded that for manipulative control, fingertip tactile sensors should be used only to report changes in contact status and the motion of the object, rather than absolute state.

The above experiments demonstrate that there is value in considering the human tactile system when developing systems for robotic dextrous manipulation, without necessarily adopting an anthropomorphic approach. Certain aspects of the human system have relevance to robotics, and this is pursued as a theme in the latter part of this volume (Chapters 7 - 11). Various manipulation issues, including properties of objects, manipulators and tasks are also discussed in an earlier survey by Grupen *et al.* (1989).

3.2.3 Stable Grasping With Intrinsic Tactile Sensing

This section[1] is concerned with intrinsic tactile sensing, which is the measurement of forces and torques at the fingertips of a robot hand. In this method, sensors are used to read the force and torque resultant vectors generated by the distributed contact pressure acting between two touching bodies (fingertip and object surface). This has particular use for maintaining stable grasps.

The essence of a stable grasp is to maintain a grasped object in equilibrium, thus being able to withstand small disturbance forces acting on the object. In general, for a given grasp geometry, there are an infinite number of forces and torques that can be applied that satisfy this equilibrium requirement. Hence, an optimisation approach is used to choose from among these many possible contact force systems, one which satisfies certain optimality constraints. These constraints, outlined by Bicchi *et al.* (1989) are:

1. The contact forces and torques generated by friction must comply with limitations given by Coulomb's law. The further the contact conditions are from slippage danger, the more robust the grasp.

2. The stresses due to contact forces must be less than the maximum allowable in order not to damage or deform the object to be grasped.

3. Contact forces must be maintained at a minimum level in order to guarantee continuity of contact over time and overcome sensor noise.

4. The finger actuator torques necessary to produce the required contact forces must be within their operating envelope.

Bicchi (1990) has described a method to find the contact position of the sensor using certain assumptions in order to constrain a solution. If a hard finger contact is assumed, then the location of the centroid that can be determined may be different from the actual centroid depending upon frictional forces on the surface. The hard finger model should therefore not be used in high friction environments. If high friction is indicated, a soft finger model can be employed that will allow a more accurate determination of the contact centroid and normals. (See also Chapter 4 by Fearing for a discussion of finger models.)

The main idea here is to set down constraint equations involving forces and normals and solve for the contact centroid. The surface of the fingertip can be

[1]Contributed by Peter Allen of the University of Columbia, New York.

modelled as an implicit equation:

$$S(\mathbf{r}) = 0 \tag{3.1}$$

where $\mathbf{r}$ is a point on the surface. A useful model is a quadratic form, which encompasses planar, spherical and ellipsoidal finger shapes. A unit surface normal at the point of contact can be defined as:

$$\mathbf{n} = \frac{\nabla S(\mathbf{r})}{\|\nabla S(\mathbf{r})\|}. \tag{3.2}$$

A contact centroid C can be defined as a point such that the set of contact forces is equivalent to a resultant force P applied at C plus a moment Q normal to the surface at C. Using the force ($\mathbf{f}$) and torque ($\mathbf{m}$) measurements from the intrinsic tactile sensor, the equilibrium force and moment balance equations are:

$$\mathbf{f} = \mathbf{p} \tag{3.3}$$
$$\mathbf{m} = \mathbf{q} + \mathbf{c} \times \mathbf{p} \tag{3.4}$$

where $\mathbf{p}$ is the contact force, $\mathbf{c}$ is the centroid of contact and $\mathbf{q}$ is the spin torque of the contact. The contact centroid is assumed to be on the surface yielding this constraint equation:

$$S(\mathbf{c}) = 0. \tag{3.5}$$

In addition, if the spin torque vector $\mathbf{q}$ is assumed proportional to the unit normal on the contact surface, it can be approximated by:

$$\mathbf{n} \propto \mathbf{q} = \frac{K}{2}\nabla S(\mathbf{c}). \tag{3.6}$$

This yields a set of non-linear equations that can be solved by a variety of methods. Certain assumptions about the kinds of contacts of the fingers on the object are needed to constrain the solution set. Two methods of solution are presented by Bicchi for hard and for soft finger contacts (Bicchi, 1990).

3.3 Object Recognition Using a Single Tactile Image

Having briefly examined issues in grasping and tactile sensing, techniques for processing single tactile images will now be considered.

Object discrimination or recognition using single images is often the goal of simple tactile sensing systems. The first stage is to process the image, extracting significant parameters such as shape quantifiers, and then to apply a pattern recognition paradigm which attempts to match sensed features with properties of known objects. The image processing required may be extensive, necessitating filtering to remove unwanted noise in the image, segmentation of individual features from the image background, and classification of features into known categories.

Various processes have been used to match sensed features with those of known objects, ranging from stochastic techniques to knowledge-based heuristics. Typically, they are based on the notion of a feature vector which records all the measured features concerning the object in contact. The feature vector is then compared with previously computed vectors from a set of known objects; the matching criterion is usually the Euclidean distance between the vectors (Schneiter & Sheridan, 1990).

One of the earliest attempts to classify the nature of surface features via tactile sensing used a pattern recogniser for distinguishing features such as pits, spurs, and cracks (Snyder & St.Clair, 1978). The processing was straightforward, the first stage being the operation of a region grower to group tactels into regions of high and of low pressure. The second stage was the application of simple predicates to classify regions as one of pit, point, ridge, crack, spur or edge by using rules concerning the size, location and pressure type (high or low) of a region. In addition, presence of touch and also location and magnitude of centre of pressure were calculated.

In another example, M. Briot developed a system capable of recognising the face, one of eight, of a known object which was touching the sensor (Briot, 1979). A stochastic classification method was used, based on assessing the probability of certain structural features of the sensed object. The program also calculated the object's position and location using the centroid and second moment (see below) about the major axis of the region present in the tactile image. In another recognition experiment, Chalupa *et al.* (1982) used linguistic pattern recognition parameterised by features such as corner positions, edge lengths, and angle between edges to classify sensed shapes including rectangles, squares, triangles, and ellipses. The parameters to the string grammar were all based on boundary features, derived from the Freeman chain code (Freeman, 1974).

Another boundary-based method is described by Ozaki *et al.* (1982). The plane closed curve of a contour can be represented as a chain code, which can be considered as a set of unit-vectors with standard angles $k.\alpha$ (k=1,2,..,n) relative to some baseline. The angle α is the chosen angular increment. The curve can then be stored as a unit-vector count histogram with n entries, one per angle. Such a histogram is called the unit-vector distribution (UVD). Ozaki *et al.* use UVDs in a tactile pattern recognition system. Firstly, they record an original pattern of each sample object, recording the contour as a UVD; a second pattern is then made from a deliberately blurred version of the original pattern. Similarity between each blurred pattern and a sampled, unclassified pattern is determined by a difference operator; the blurred pattern is used to compensate for digitisation errors in the sampled pattern. The blurred pattern with the shortest distance from the sampled, unclassified pattern is taken as the result, enabling discrimination between cylinders and prisms. This method works well for convex curves using a small vector length and a small incremental angle, but the larger these two parameters, the less reliable the method. It is not suitable for handling concave curves. Such curves would have to be broken into a sequence of curves, the divisions being at the boundaries of the concave and convex parts.

Quantitative measures of image features are often used in recognition tasks. One class of measure, called moments, can describe properties of a feature regardless of that feature's position and orientation in a digital image, and so is a method particularly useful for recognising randomly oriented objects. The general technique is to use filtering, thresholding and perhaps other processing to leave a single region representing an object in an image. Moments can then be calculated to quantify the region of interest.

An introduction to moments is provided by Gonzalez (1987), summarised below, whilst the original development of the concept is presented by Hu (1962).

For an image f, the moment of order $(p+q)$ is given by:

$$m_{pq} = \sum_x \sum_y x^p y^q f(x,y) \tag{3.7}$$

for p, $q = 0, 1, 2, \ldots$ and where $f(x,y)$ is the value of a pixel (or tactel) at position (x,y). The centroid $(\overline{x},\overline{y})$, i.e., the centre of gravity, is given by $\overline{x} = \frac{m_{10}}{m_{00}}, \overline{y} = \frac{m_{01}}{m_{00}}$, and is used in the definition of the *central moments*, which are invariant under translation:

$$\mu_{pq} = \sum_x \sum_y (x-\overline{x})^p (y-\overline{y})^q f(x,y). \tag{3.8}$$

The *normalised central moments* denoted by η_{pq} are defined as:

$$\eta_{pq} = \frac{\mu_{pq}}{\mu_{00}^{\gamma}} \tag{3.9}$$

where

$$\gamma = \frac{p+q}{2} + 1 \tag{3.10}$$

for $p+q = 2, 3, \ldots$.

Whilst centralised moments are invariant to translation changes, normalised moments allow for scaling and rotation. Further, a set of seven invariant moments can be derived from the second and third order moments. This set of moments has been shown to be invariant to translation, rotation, and scale change (Hu, 1962).

A feature set made up of area and normalised first and second moments was used in a simple recognition system in which objects were placed on the planar sensor and a tactile image recorded (Christ & Sanderson, 1982). The objects taught and recognised included tapes, tubes, batteries and cans, all smaller than the sensor; 48 out of 50 classifications were correct. The objects were uncomplicated and diverse, but the experiments demonstrated that simple feature extraction and a straight-forward decision rule (based on distance of each measured feature from each taught feature) are of utility. Similarly, measures based on radii, square root of the area, and invariant moments of features were successfully used in a recognition task conducted at the King Fahd University of Petroleum and Minerals, Saudi Arabia (Al-Mohammad & Al-Mouhamed, 1987). Also, an extension to the moment-based approach, using complex-moments, is discussed by Luo and Loh (1987).

Some other image processing techniques were used in experiments with the Lord Corporation LTS-200 10×16 element tactile sensor array (Muthukrishnan *et al.*, 1987). Non-linear median filtering was used for noise suppression, then the image was thresholded using a bimodal histogram to achieve a binary image. For objects less than the sensor size, local gradient operators (the Sobel and Roberts operators) were used to extract edges, and template matching for recognition carried out. Contour tracing using a variation on a raster tracking method was possible in the images tested. Segmentation and region growing were also experimented with to aid extraction of edges associated with objects greater in size than the sensor.

A Lord LTS-200 sensor was used for object recognition tasks at Case Western Reserve University (Jurczyk & Loparo, 1989). The approach adopted was to transform the raw tactile image from the spatial domain into a new domain, and then attempt object classification by correlation with similarly transformed data from a set of reference objects. The reference objects were selected to represent local geometric features, including horizontal and vertical edges, vertices, spherical surfaces, etc. Three transforms were used: Walsh-Hadamard (WHT), discrete cosine (DCT), and fast Fourier (FFT) (full definition of these is given by Jurczyk and Loparo (1989)). An advantage of using such transformations is that the transformed data is unaffected by the position of features in the raw data: a contacting object's image features will have the same transform data independent of its orientation and location on the tactile sensor array. All the transforms possessed good correlation characteristics; the FFT was usually the superior, although the slowest to compute. The R (rapid) transform and the rationalised Haar transform were also tried, but were not consistent enough to warrant further investigation.

The large-field Lord LTS-300 80×80 element sensor array has been used in object recognition work both in the UK (Grant *et al.*, 1988), and in the US, for example by Roach *et al.* (1989). Single images were used to achieve recognition, despite column streaking inherent in the tactile image, a problem reported by both groups of researchers. This necessitated development of suitable filtering algorithms to remove the streaks and other noise. Whilst Grant *et al.* used shape attributes and rule-trees for object model matching, Roach *et al.* calculated geometric features such as perimeter, diameter, radius, convex perimeter, convex area, and elongation. A spherical dual representation is used for Roach's object model, a technique also used in computer vision. This representation, a combination of Gaussian sphere encoding and duality has several advantages: it is invariant under rotation, translation is easily achieved, and objects can be reconstructed in less than quadratic time on the number of faces of a polyhedral model. Non-convex polyhedra can also be represented. A full description of the dual spherical representation is provided by Roach in an earlier paper (Roach *et al.*, 1987).

Other examples of tactile processing include the work of Basu *et al.* (1988) and of Guiheux *et al.* (1988). The latter group of workers used attributes including compactness, contour of a slice, volume, 3-dimensional moment invariants, and surface normals in model matching of data from two gripper-mounted tactile sensor arrays. In this system, measured features are plotted in a multi-dimensional space and a classification made. This traditional pattern recognition approach based on a feature space is augmented by storing the features in a blackboard architecture (Engelmore & Morgan, 1988) and using a cross-checking module to attempt to resolve ambiguities in classifications. In a simpler system, Basu *et al.* used a 5×5 tactel sensor based on strain gauges which provided images clearly distinguishing edges, flat surfaces and vertices. They intend using this sensor in an active sensing strategy.

Filtering, thresholding, segmentation, etc. are all techniques commonly found in image processing, particularly vision sensing; several thorough introductions are available for the reader requiring more information on this subject (e.g., Ballard and Brown (1982), Gonzalez and Wintz (1987)).

The above examples illustrate that the application of image processing techniques can yield useful results. However, as mentioned in Chapter 1, there are differences between the tactile domain and the visual domain and careful consideration should be given to the particular scenario in which image processing is used. For example, the spatial resolution of the image, mechanical crosstalk between tactels, and background noise all need to be considered. Berger and Khosla (1991) in their real time feedback system driven by data derived from tactile images go to considerable lengths to establish reliable data. This involves adapting the threshold setting for each tactel, rather than applying a fixed threshold setting for the whole image. The fixed threshold operation may fail to filter out tactels whose non-zero readings are derived solely from crosstalk, but which should have a zero value. Reliable thresholding using an adaptive technique then enables accurate edge detection to be accomplished in Berger and Khosla's system using a simple algorithm based on the values of each tactel's eight neighbours.

Berger and Khosla also note that vision images have a substantially higher resolution than most tactile images, and that the discrete approximations used in creating local neighbourhood operators in machine vision do not hold very well for tactile images. This is because a 3×3 window on a tactile image often represents a large physical area and could correspond to a substantial change in the image (Berger & Khosla, 1991). Applying such operators can cause significant spreading of edges in a tactile image. This is also noted by De Groot (1988) who discusses the significance of sensor (tactel) size in tactile sensing, and how this relates to errors between the measured areas and the actual areas of sensed objects. He concludes that a tactel area of about 50% of the square of the resolution yields an optimal result[2]. Thus, the choice of image processing techniques will thus be dictated by both the characteristics of the sensor and of the objects to be sensed.

---

[2]De Groot defines resolution as the reciprocal of the distance between adjacent tactels.

3.4 Active Tactile Sensing

Recognition of objects through the analysis of a single tactile image is essentially a static mode, whereby the sampled object is placed on a sensor or grasped in a tactile gripper and the resultant image examined. It is analogous to some types of 2-dimensional vision processing, and uses techniques developed in the visual domain. The static mode of tactile analysis is known as cutaneous sensing, analogous to human fingertip sensing.

The tactile domain differs from the visual domain in a number of ways, but one major constraint on the utility of a tactile image is that it provides only a local view of the contacting stimulus. Particularly in gripper-mounted tactile sensors, the active area of the sensor is often small compared with the size of the objects being handled. Very likely, only a small number of faces out of the total on an object will contact a tactile sensor.

A way to improve this situation is to use an active sensing strategy. Instead of relying on a single static image, a number of images are recorded from different parts of the object under examination. The set of such images coupled with position data from the hand and manipulator system can be used to construct a 3-dimensional representation of a sampled object. Other parameters concerning the object, such as texture and hardness, may be sensed in this way. This active approach is termed kinesthesia, and the combination of cutaneous and kinesthetic sensing is called haptic perception.

Active sensing is used to achieve various goals, for example, to localise an object in the robot's world (i.e., to establish its position and orientation relative to a frame of reference), to recover a description of a (possibly unknown) object's shape, or to recognise a touched object. Usually two main issues are involved here: strategy and control. A strategy is needed for the exploration of an object by moving a robot finger and/or hand around a touched object and gathering data. Such gathered data might include shape, surface texture, or material properties. The data obtained from contact between sensor and object will also typically be used to determine the next move to make such that further information will be gathered about the object through touch.

The acquired data may be integrated into an object representation, and perhaps model matching attempted if recognition is the goal. In order to be able to support such an active search strategy, a suitable control algorithm is needed. Advanced control systems incorporating complex data in the control loop are difficult to build; this is one of Brady's stated problems of robotics, and is "a tough nut to crack" (Brady, 1989). Such active sensing has also been called *dynamic exploration* by Berger and Khosla (1991), who note that the processing of the raw data from the tactile sensor into an object representation can also be a major issue.

Tactile sensors need not have high resolution, or be complex to achieve some of these tasks. For example, a probe-like sensor mounted on a robot was used to compute the location of an object relative to the robot (Zhou & El-Zorkany, 1987).

The probe was advanced normal to the surface of an object at several points, and the robot's location recorded on contact. These locations were used to determine the equation of the feature being touched. By touching a plane at three locations, three points are obtained which are sufficient to determine the equation of that plane. Edges and vertices can also be recovered with this technique.

It is also possible to use a tactile sensor array both to detect contact between a grasped object and some surface in the robot's world, and to classify the nature of that contact (Houshangi & Koivo, 1989). So far, object grasping and the analysis of the resultant tactile images has been examined, assuming that the object is constrained by the gripper and that the object is not in contact with any other surface. However, if the grasped object touches a surface then the tactile image will change, enabling the nature of that contact to be determined from the new image.

For example, a rectangular parallelepiped object grasped in a tactile gripper may be in point, line, or plane contact with a planar surface. The tactile image of the grasped object in each of these contact states is different, providing data by which the object/external surface contact can be classified. To perform such a classification, the image of the grasped object when not in contact with the planar surface is subtracted from the image obtained when contact is achieved, leaving an image representing the resultant forces. Thus for a line contact, the angle between the object and the planar surface can be calculated from the resultant image; the angle can then be incorporated in a manipulator servo loop. Houshangi and Koivo (1989) cite an example of bringing a grasped object into point contact with a plane, then servoing the manipulator to achieve a line contact and finally rotating the object about the line, bringing the object into planar contact with the surface so that it can be released from the gripper.

3.4.1 Shape Description Without Specific Models

In the literature, it was found that two main methods of determining shape and properties of an object from sensory data have been used (Berkemeier & Fearing, 1990): shape description without specific models, and model matching.

First, shape description will be considered, whereby sensory measurements and geometric constraints are used to derive a representation of an object. Although specific models are not used, objects may be assumed to be of a certain class (Berkemeier & Fearing, 1990), such as polygonal, generalised cylinder, or surface of revolution.

For example, Allen (1986) describes objects in terms of surface patches and builds up a representation of an object by surface exploration aided by both tactile and vision sensors; this work has been extended by using superquadrics for object representation (Allen & Roberts, 1989) (see also Chapters 10 and 11). Simple polygonal object representations have been used for localisation (e.g., by Cole and Yap (1987) and by Zhou and El-Zorkany (1987)), such systems typically use tactile probes to determine contact location.

Some researchers have experimented with locating points on an object's surface using a tactile sensor and construct a surface shape representation of any general 3-dimensional object. Tsujimura and Yabuta (1989) use only a single probe connected to a force/torque sensor from which contact position can be recovered; this method is similar to the intrinsic tactile sensing proposed by Bicchi (1990). Although measured points on the surface of the touched object do not have to belong to a particular type of object classification, Tsujimura and Yabuta's work focuses on methods for gathering the localisation data and minimising shape measurement errors rather than on storing the data in some uniform object model.

A method for recovering the curvature form of an unknown object was also investigated by Montana (1988); this system performs rotational probes and measures the motion of the point of contact across the surface. The curvature form is estimated as that which fits these measurements in a least squares way. In this system, the contact data is also used to close the loop around the kinematics of contact, enabling tracing of the touched object's surface.

Other approaches assume objects are of some particular shape class; for example, Printz (1987) assumes that the objects will be cones or cylinders of circular or elliptical cross-section, whilst Berkemeier and Fearing (1990) use shape descriptors based on surface of revolution. Their system is able to recover the axis of the surface of revolution using only sparse sensing — it requires contact location, surface normal and principal curvature directions at only three points of contact. Although these systems are designed to sense shape parameters within a restricted class of shape (e.g., polygonal, generalised cylinders, etc.), these representations are of value because many man-made objects fit these types. It may thus be possible to execute shape recovery sensing in certain man-made environments using only relatively simple object representations.

3.4.2 Model Matching

Model matching is typically used to localise and identify objects by comparing relations between sensed features and features represented in specific object models. For a collection of measured features, there is a tree of possible interpretations of this sampled data. Various constraints are used to prune this interpretation tree, hopefully leading to a match with a specific object's representation. The efficacy of the pruning affects the active sensing strategy — if a match cannot be made then further data must be gathered. This technique was extensively developed by Lozano-Pérez and colleagues and documented in a series of papers (Gaston & Lozano-Pérez, 1984; Grimson & Lozano-Pérez, 1984; Grimson, 1986; Grimson & Lozano-Pérez, 1987); it is further explained by Allen in Chapter 10 of this volume, who examines in depth active tactile sensing for object recognition.

Other workers have extended this interpretation tree method (e.g., Ellis (1987), Schneiter & Sheridan (1990)) in an active sensing strategy, whilst in contrast, Cameron (1989) uses a statistical decision theoretic approach to guide the search for points on an object's surface. This latter approach has several benefits, notably that at each stage of the matching process information is provided regarding the current probabilities of each interpretation, thus effectively assigning a probability to each branch of an (implicit) interpretation tree.

Some of the first researchers to recognise the importance of active sensing used a sensor with wide dynamic range to gather data suitable for input to surface curvature calculations (Gurfinkel *et al.*, 1974). The classifications used were: vertex, spherical surface, plane, cylindrical surface, edge, and curved edge, derived from the readings from a single tactile image. Experiments for recognising cubes, spheres, cones, etc. were conducted using two sensor arrays, one on each finger of a parallel jaw gripper. This was then extended to active sensing experiments. Models of the above objects were stored including surface curvatures and details of presence or absence of parallel planes and edges. The recognition strategy consisted of moving the sensor into contact with an object and calculating surface curvature. The curvature data was then compared with a set of object models and a hypothesis made about the class of object; this choice determined the next move to make. The object's surface was traced using this hypothesis-verification loop until recognition was achieved.

This piece of work was significant for its time. Gurfinkel *et al.* studied human methods of tactile perception and realised the importance of active sensing as well as static analysis. The hypothesis-verification loop, or generate and test, is a common technique in artificial intelligence and it is unusual to find it in use in robotics in the early 1970s.

The top-down approach of the hypothesis-verification loop was used by Hillis (1982) when at MIT. He mounted his sensor on a tendon actuated finger having torque and position feedback. Sample objects were placed on a compliant surface beneath the finger, and were touched and pushed to extract data. The sample objects were small industrial fasteners which were smaller than the sensor's active area. These were screws, cotter pins, dowel pins and washers, all no more than 12 mm in size. The distinguishing parameters used were: aspect ratio to represent global feature shape, local pressure anomalies to represent bumps and depressions, and finally stability. This latter was determined by lowering the finger onto the sampled object and then monitoring the force required to push it along in various directions.

The hypothesis-verification loop used in active sensing is important because data from a single image is often insufficient for recognition. By making a hypothesis and gathering data, then modifying the hypothesis on the basis of new readings, reliable conclusions can be made by an iterative process. Hillis designed his system around this mechanism; if the image obtained was insufficient to extract relevant data, or additional data were required, then further images were gathered until a firm classification could be made. Although successful, the sample objects were all smaller than the sensor and only a simple parameter set was necessary for recognition. However, he demonstrated the feasibility of this approach.

Researchers at the University of Pennsylvania have investigated various aspects of tactile sensing, and advocate an active approach. Most of their work has been conducted using either a Lord Corporation tactile sensor (Rebman & Morris, 1986) or a French-built finger-like probe known as the French finger (Bajcsy & Goldberg, 1984).

Using the French finger on an (x, y, z) positioning frame (detailed in Brown (1980)), recognition by tracing was attempted (Bajcsy & Goldberg, 1984). The examined objects were rigid and static and their surfaces could be linear, curved, convex or concave. All the objects had vertical non-concave sides. The recognition experiments involved moving the probe around the workspace until contact with an object was made, and then recording the coordinates of the contact point. The probe was traced over the object's profile and the trajectory recorded. All the contact points were then connected together and the centroid, volume and second moments were calculated. These parameters were compared with object descriptions stored in a database and matching attempted. The sample objects included jars, coffee cups, and cardboard boxes. Recognition was successful for the dozen objects in the database, and volume calculations were accurate to within 6%.

3.5 Haptic Sensing

Some recent tactile sensing research has focused on issues in exploratory sensing, and several researchers have looked at results in psychology concerning the use of the human hand as part of this study. The human haptic system combines information such as pressure, vibration, and thermal properties obtained from cutaneous (finger-tip) sensors, with information from the kinesthetic system that measures position and movement from muscles, tendons, joints, etc. The parallels between human haptic systems and robotics were remarked on in a review paper by Harmon (1984), and have been investigated notably by Klatzky *et al.* (1987b), Allen and Roberts (1989), and Stansfield (1991a).

One of the key aspects of haptic perception is that it is active and purposive, thus the motor subsystem and the sensory subsystem must be closely integrated (Lederman & Klatzky, 1987; Stansfield, 1991a). It is not only the processing of the sensory data which is important, but also the way in which the motor system is used to extract the data that is of significance. A common theme is the development of a set of "Exploratory Procedures" (EPs), which are directly linked to the extraction of specific haptic object properties such as weight, local surface shape, hardness, elasticity, and plasticity. Klatzky *et al.* have concentrated on examining the psychology and psychophysics of human touch and consider its relevance to robotics (Klatzky *et al.*, 1987a; Lederman & Klatzky, 1987; Chapters 8 and 9, this volume) whilst Allen and colleagues at the University of Columbia, New York, have used EPs for shape recovery and object recognition in a robotic system (Allen & Michelman, 1990; Chapter 11, this volume). Stansfield has also carried out significant work in this field and has developed robotics implementations of EPs, notably for an articulated sensate robot hand (Stansfield, 1991a).

In Stansfield's system, the EPs are organised according to the specific haptic property which they are designed to extract, although several EPs were implemented for extracting data about a single property. For example, hardness is sensed either by squeezing an object in a robot hand, or by poking it with a gripper finger, and EPs were developed for each of these strategies. Another property, the elasticity or plasticity of a surface, is measured by pushing with a tactile sensor against a surface then slowly removing the sensor and monitoring the data obtained. If contact is maintained on the sensor as it is withdrawn then the surface is elastic, but if cutaneous (fingertip) contact is lost before the sensor has retracted to its position of initial contact then the surface has retained its deformation and is classified as plastic.

A related piece of work is that of Buttazzo *et al.* (1991), who decompose complex tactile operations into elementary sensory-motor acts known as tactile subroutines. They have developed subroutines for sensing hardness, texture, and friction, and use a probe-like force/torque sensor and an artificial fingernail-like sensor mounted on a PUMA 560 robot arm for experimentation. Hardness is measured by advancing the probe into an object and recording the manipulator position when an initial force F_1 threshold is crossed, then continuing advancing the probe until a second force threshold F_2 is crossed, at which moment the position is again recorded. From the two recorded positions, the displacement D of the object's surface is calculated and hardness H computed as:

$$H = \frac{F_2 - F_1}{D} \; . \tag{3.11}$$

The subroutine for measuring texture was executed by rubbing the fingernail sensor over an object surface at a predetermined force and sampling the sensor at 3.2 kHz. Experiments were conducted to sense the texture of a planar surface having parallel wires laid across it. Spacing between the wires and thickness of the wires were used as parameters for controlling surface roughness. After each exploration, the measured signal was processed and two parameters computed: the distance d between spikes on the signal and the amplitude A of the spikes. The spacing between spikes can be computed as:

$$d = \frac{nv}{f} \tag{3.12}$$

where n is the number of samples acquired between two spikes, v is the velocity of the sensor, and f is the sampling frequency. The system was able to discriminate between wires 0.5 mm apart, and between wires 0.2, 0.6, 1.0 and 1.5 mm thick.

The friction subroutine assesses static and dynamic friction coefficients. Its method is derived from the observation that most humans assess friction by touching an object with a finger, pressing moderately hard and then exerting a force on the finger which tends to slide it over the object's surface. This force is increased until the fingertip actually slips, after which the operation is complete (Buttazzo *et al.*, 1991). Their experiments were carried out using intrinsic tactile sensing (see sections 3.2.3 and 10.4) enabling recovery of location of the contact point on the finger surface, the magnitude and direction of the contact, and the values of the normal and tangential (friction) components of the contact force.

In these friction experiments, a robot-mounted fingertip with an intrinsic tactile sensor was brought into initial contact with an object, then the robot arm attempted to move the fingertip in the tangential direction, increasing force linearly with time until the fingertip slipped. The values of the normal and tangential components of the contact were sampled at 10 Hz and recorded in a buffer during this period and also after slippage occurred. This data enabled assessment of the static friction (μ_s) and dynamic friction (μ_d) coefficients: the friction ratio R_f (the ratio between the tangential and normal components of contact force) was computed for each sample, and then plotted on a graph. R_f was observed to increase almost linearly to the static friction limit μ_s, after which motion (slip) starts, and R_f drops to μ_d. Some problems due to perturbations of the mechanical system and of the sensor measurements were encountered, making exact quantitative estimates of μ_s and μ_d difficult, but the data gathered from different surfaces was sufficiently distinct to enable qualitative surface classification according to friction characteristics.

Some earlier work, conducted at the University of Pennsylvania, also investigated object properties through active touch, using the French finger tactile probe (Bajcsy, 1985). Experiments in material hardness evaluation, texture analysis, and surface normal calculation were conducted. Hardness was sensed by advancing the sensor in 0.25 mm steps towards the material surface and readings taken at each increment. The material hardness is gauged from the slope of the curve of sensor reading against increment: the steeper the slope, the harder the material. Hardness of samples of wood, foam, and felt were tested; distinctions were made between these materials. The same experiments were conducted for liquid glue, clay and wood, each contained in a plastic bag. The clay demonstrated plasticity characteristics, because it retained its deformed state after the probe was retracted.

Texture detection was carried out by moving the finger slowly along the surface of the material with the sensor just touching the sample. Readings were taken every 5.6 ms as the finger moved along. The fineness of the surface texture was determined by the magnitude of the local variation of the surface, spectral analysis being used to obtain the frequency of each pressure scale. Felt, cloth and a wire grid were tested, and acceptable results were obtained. The felt was classified as smooth because the sensor did not have sufficient spatial resolution to detect the small surface variations.

In further work, Klatzky, Bajscy and Lederman built these active experiments into a formalism based on exploratory procedures by associating specific hand movements with exploratory sensing for three classes of property: surface properties (texture, hardness, etc.), structural properties (shape, volume, etc.), and functional properties (part motion, specific functions) (Klatzky *et al.*, 1987b). Other experiments, combining vision and touch, were conducted at the University of Pennsylvania by Allen (e.g., Allen (1984)), who continued this work for object recognition using passive stereo vision and active tactile sensing (Allen, 1988), and then investigated exploratory sensing using the Utah/MIT dextrous hand at the University of Columbia (Allen & Michelman, 1990).

Current work in haptic sensing is very promising, and tackles the major problems of active tactile sensing. Particularly, object properties such as hardness, elasticity, and thermal characteristics are being investigated that exploit the capabilities of tactile sensors over other sensory modalities. Integration of the manipulator system, through kinesthesia, with conventional tactile sensing is a major feature of haptic systems.

Further explanation of haptic sensing, exploratory procedures and experiments using the Utah/MIT hand are described in Chapters 8 – 11 of this volume, with contributions from Klatzky, Lederman, Allen and colleagues on these topics.

3.6 Conclusion

The survey by Nicholls and Lee (1989) anticipated that tactile sensors will eventually be integrated in factory-based robot systems. This has not yet occurred, although there have been further developments in sensor technology. What is particularly interesting is that tactile domain research has concentrated on systems, rather than individual sensor, issues and on active sensing. Although tactile sensing started off by applying techniques from computer vision, later work has concentrated on active sensing, and on haptic perception in particular. The close integration of sensory and motor subsystems as part of a coherent exploratory strategy is common to several promising systems. More research is needed to establish how and in what contexts active tactile sensing should be employed. These issues are considered in more detail throughout the remainder of this volume.

References

Al-Mohammad, H.A., & Al-Mouhamed, M. 1987. Identification of object faces by a tactile sensor. *Pages 97–108 of: Proc. 8th int. conf. automated inspection and product control (Chicago, IL).* Kempston, UK: IFS (Pubns).

Allen, P. 1984. Surface descriptions from vision and touch. *Pages 394–397 of: Proc. IEEE int. conf. on robotics (Atlanta, GA).* Washington, DC: IEEE Computer Soc. Press.

Allen, P.K. 1986. Sensing and describing 3-D structure. *Pages 126–131 of: Proc. IEEE conf. on robotics and automation (San Francisco, CA).* Washington, DC: IEEE Computer Soc. Press.

Allen, P.K. 1988. Integrating vision and touch for object recognition tasks. *Int. j. robotics research*, **7**(6), 15–33.

Allen, P.K., & Michelman, P. 1990. Acquisition and interpretation of 3-D sensor data from touch. *IEEE trans. on robotics and automation*, **RA-6**(4), 397–404.

Allen, P.K., & Roberts, K.S. 1989. Haptic object recognition using a multi-fingered dextrous hand. *Pages 342–347 of: Proc. IEEE int. conf. on robotics and automation (Scottsdale, AZ)*. Washington, DC: IEEE Computer Soc. Press.

Bajcsy, R. 1985. Shape from touch. *In:* G.N.Saridis (ed), *Advances in automation and robotics*. London: JAI Press Inc.

Bajcsy, R., & Goldberg, K.Y. 1984 (May). *Active touch and robot perception*. Tech. rept. MS-CIS-84-42, GRASPLAB 14. Dept. Computer & Information Science, Moore School, Univ. Pennsylvania, Philadelphia, PA.

Ballard, D.H., & Brown, C.M. 1982. *Computer vision*. Englewood Cliffs, NJ: Prentice Hall.

Basu, P.R., Russell, R.A., & Trott, G. 1988. Extracting 3-dimensional surface features from tactile sensor data. *Pages 908–920 of: Proc. int. symposium on industrial robots (Sydney, Aus.)*. Kempston, UK: IFS (Pubns).

Berger, A.D., & Khosla, P.K. 1991. Using tactile data for real-time feedback. *Int. j. robotics research*, **10**(2), 88–102.

Berkemeier, M.D., & Fearing, R.S. 1990. Determining the axis of a surface of revolution using tactile sensing. *Pages 947–949 of: Proc. IEEE int. conf. on robotics and automation (Cincinnati, OH)*. Los Alamitos, CA: IEEE Computer Soc. Press.

Bicchi, A. 1990. Intrinsic contact sensing for soft fingers. *Pages 968–973 of: Proc. IEEE int. conf. on robotics and automation (Cincinnati, OH)*. Los Alamitos, CA: IEEE Computer Soc. Press.

Bicchi, A., Salisbury, J.K., & Dario, P. 1989. Augmentation of grasp robustness using intrinsic tactile sensing. *Pages 302–307 of: IEEE conf. on robotics and automation (Scottsdale, AZ)*. Washington, DC: IEEE Computer Soc. Press.

Brady, M. 1989. Introduction: The problems of robotics. *Chap. 1, pages 1–35 of:* Brady, M. (ed), *Robotics science*. Cambridge, MA: MIT Press.

Briot, M. 1979. The utilisation of an artificial skin sensor for the identification of solid objects. *Pages 529–547 of: Proc. 9th int. symposium on industrial robotics*. Kempston, UK: IFS(Pubns).

Brown, D.J. 1980 (Dec). *Computer architecture for object recognition and sensing*. Tech. rept. MS-CIS-80-22. Dept. Computer & Information Science, Moore School, Univ. Pennsylvania, Philadelphia, PA.

Buttazzo, G., Bicchi, A., & Dario, P. 1991. Robot tactile perception. *Pages 25–39 of:* Lee, C.S.G. (ed), *Sensor-based robots: Algorithms and architectures (NATO ASI series)*. New York: Springer-Verlag.

Cameron, A. 1989. Optimal tactile sensor placement. *Pages 308–313 of: Proc. IEEE int. conf. on robotics and automation (Scottsdale, AZ)*. Washington, DC: IEEE Computer Soc. Press.

Chalupa, V., Marik, V., & Volf, J. 1982 (May). Tactile matrix for shape recognition. *Pages 339–348 of: Technological & methodological advances in measurement – proc. 9th IMEKO congress of the Int. Measurement Confed. (Berlin)*.

Christ, J.P., & Sanderson, A. 1982 (Sept). *A prototype tactile sensor array*. Tech. rept. CMU-RI-TR-82-14. Robotics Institute, Carnegie-Mellon Univ., Pittsburgh, PA.

Cole, R., & Yap, K.C. 1987. Shape from probing. *J. algorithms*, **8**, 19–38.

Cutkosky, M.R., & Howe, R.D. 1988 (Sept). Dynamic tactile sensing. *In: Proc. ROMANSY 88: 7th CISM-IFToMM symposium on the theory and practice of robots and manipulators (Udine, Italy)*.

Cutkosky, M.R., Jourdain, J.M., & Wright, P.K. 1987. Skin materials for robotic fingers. *Pages 1649–1654 of: Proc. IEEE int. conf. on robotics and automation (Raleigh, NC)*. Washington, DC: IEEE Computer Soc. Press.

Dario, P. 1989. Contact sensing for robot active touch. *Chap. 3, pages 138–163 of:* Brady, M. (ed), *Robotics science*. Cambridge, MA: MIT Press.

De Groot, A.W. 1988. Effect of sensor size in robotic tactile sensor arrays. *Robotica*, **6**(4), 285–288.

Ellis, R.E. 1987. *A tactile sensing strategy for model-based object recognition*. Tech. rept. COINS 87-96. Dept. Computer & Information Science, Univ. Mass. at Amherst.

Engelmore, R., & Morgan, T. 1988. *Blackboard systems*. Reading, MA: Addison-Wesley.

Fearing, R.S., & Hollerbach, J.M. 1985. Basic solid mechanics for tactile sensing. *Int. j. robotics research*, **4**(3), 40–54.

Freeman, H. 1974. Computer processing of line drawing images. *ACM computing surveys*, **6**(1), 57–98.

Gaston, P.C., & Lozano-Pérez, T. 1984. Tactile recognition and localization using objects models: The case of polyhedra on a plane. *IEEE trans. on pattern analysis and machine intelligence*, **PAMI-6**(3), 257–266.

Gonzalez, R.C., & Wintz, P. 1987. *Digital image processing.* 2 edn. Reading, MA: Addison-Wesley.

Grant, E., Wagstaff, N., Rutherford, S., & Mowforth, P.H. 1988. An array tactile sensor and its value in automation. *Pages 1–19 of: Proc. 18th int. symposium on automotive technology & automation (Florence, Italy).* Croydon, UK: Allied Automation.

Grimson, W.E.L. 1986. Sensing strategies for disambiguating among multiple objects in known poses. *IEEE j. robotics and automation,* **2**(4), 196–213.

Grimson, W.E.L., & Lozano-Pérez, T. 1984. Model-based recognition and localization from sparse range or tactile data. *Int. j. robotics research,* **3**(3), 3–35.

Grimson, W.E.L., & Lozano-Pérez, T. 1987. Localizing overlapping parts by searching the interpretation tree. *IEEE trans. on pattern analysis and machine intelligence,* **PAMI-9**(4), 469–482.

Grupen, R.A., Henderson, T.C., & McCammon, I.D. 1989. A survey of general-purpose manipulation. *Int. j. robotics research,* **8**(1), 38–62.

Guiheux, H., Mardapittas, C., Wilson, D.R., & Winter, S.C. 1988. A model based recognition system for tactile data. *Microprocessors and microprogramming,* **24**, 425–432.

Gurfinkel, V.S., Shneyder, A. Yu, Kanayev, Ye M., & Gurfinkel, Ye V. 1974. Tactile sensitizing of manipulators. *Engineering cybernetics,* **12**(6), 47–56.

Harmon, L.D. 1984. Tactile sensing for robots. *Pages 109–158 of:* M. Brady, L. A. Gerhardt, H. F. Davidson (ed), *Robotics and artificial intelligence (NATO ASI series).* New York: Springer-Verlag.

Hillis, W.D. 1982. A high resolution imaging touch sensor. *Int. j. robotics research,* **1**(2), 33–44.

Houshangi, N., & Koivo, A.J. 1989. Tactile sensor control for robotic manipulations. *Pages 1258–1259 of: Proc. IEEE int. conf. on systems, man and cybernetics (Cambridge, MA).* Washington, DC: IEEE Computer Soc. Press.

Howe, R.D., & Cutkosky, M.R. 1989. Sensing skin acceleration for slip and texture perception. *In: Proc. IEEE int. conf. on robotics and automation (Scottsdale, AZ).* Washington, DC: IEEE Computer Soc. Press.

Howe, R.D., Popp, N., Akella, P., Kao, I., & Cutkosky, M.R. 1990. Grasping, manipulation and control with tactile sensing. *Pages 1258–1265 of: Proc. IEEE int. conf. on robotics and automation (Cincinnati, OH).* Los Alamitos, CA: IEEE Computer Soc. Press.

Hu, M. 1962. Visual pattern recognition by moment invariants. *IRE trans. on information theory*, **IT-8**(Feb), 179–187.

Jacobsen, S.C., McCammon, I.D., Biggers, K.B., & Phillips, R.P. 1987. Tactile sensing system design issues for machine manipulation. *Pages 2087–2096 of: Proc. IEEE int. conf. on robotics and automation (Raleigh, NC)*. Washington, DC: IEEE Computer Soc. Press.

Jurczyk, J., & Loparo, K.A. 1989. Mathematical transforms and correlation techniques for object recognition using tactile data. *IEEE trans. on robotics and automation*, **RA-5**(3), 359–362.

Kaneko, M., & Tanie, K. 1990. Contact point detection for grasping of an unknown object using self-posture changeability (SPC). *Pages 864–869 of: Proc. IEEE int. conf. on robotics and automation (Cincinnati, OH)*. Los Alamitos, CA: IEEE Computer Soc. Press.

Klatzky, R., Lederman, S.J., & Reed, C. 1987a (Feb). *There is more to touch than meets the eye: The salience of objects attributes for haptics with and without vision*. Tech. rept. CSTR 8714. Univ. of California, Santa Barbara.

Klatzky, R.L., Bajcsy, R., & Lederman, S.J. 1987b. Object exploration in one and two fingered robots. *Pages 1806–1809 of: Proc. IEEE int. conf. on robotics and automation (Raleigh, NC)*. Washington, DC: IEEE Computer Soc. Press.

Lederman, S.J., & Klatzky, R. 1987. Hand movements: A window into haptic object recognition. *Cognitive psychology*, **19**(3), 342–368.

Luo, R.C., & Loh, H.H. 1987. Tactile array sensor for object identification using complex moments. *Pages 1935–1940 of: Proc. IEEE int. conf. on robotics and automation (Raleigh, NC)*. Washington, DC: IEEE Computer Soc. Press.

Montana, D.J. 1988. The kinematics of contact and grasp. *Int. j. robotics research*, **7**(3), 17–32.

Muthukrishnan, C., Smith, D., & Myers, D. 1987. Edge detection in tactile images. *Pages 1500–1505 of: Proc. IEEE int. conf. on robotics and automation (Raleigh, NC)*. Washington, DC: IEEE Computer Soc. Press.

Nguyen, V-D. 1988. Constructing force-closure grasps. *Int. j. robotics research*, **7**(3), 3–16.

Nicholls, H.R., & Lee, M.H. 1989. A survey of robot tactile sensing technology. *Int. j. robotics research*, **8**(3), 3–30.

Nowlin, W.C. 1991. Experimental results on Bayesian algorithms for interpreting compliant tactile sensing data. *Pages 378–383 of: Proc. IEEE int. conf. on robotics and automation (Sacramento, CA)*. Los Alamitos, CA: IEEE Computer Soc. Press.

Ozaki, H., Waku, S., Mohri, A., & Takata, M. 1982. Pattern recognition of a grasped object by unit-vector distribution. *IEEE trans. on systems, man and cybernetics*, **SMC-12**(3), 315–324.

Printz, H. 1987. Finding the orientation of a cone or cylinder. *In: Proc. IEEE computer soc. workshop on computer vision (Miami Beach, FL)*. Washington, DC: IEEE Computer Soc. Press.

Rebman, J., & Kallhammer, J-E. 1986. A search for precursors of slip in robotic grasp. *Proc. SPIE - Int. Soc. Opt. Eng. (Intelligent robots and computer vision, Cambridge, MA)*, **726**(Oct), 329–337.

Rebman, J., & Morris, K.A. 1986. A tactile sensor with electrooptical transduction. *Pages 145–155 of:* Pugh, A. (ed), *Robot sensors vol.2: Tactile and non-vision*. Kempston, UK: IFS (Pubns).

Roach, J.W., Paripati, P., & Wright, J. 1987. A CAD system based on spherical dual representations. *IEEE computer*, **20**(8), 37–44.

Roach, J.W., Paripati, P.K., & Wade, M. 1989. Model-based object recognition using a large-field passive tactile sensor. *IEEE trans. systems, man and cybernetics*, **SMC-19**(4), 846–853.

Schneiter, J.L., & Sheridan, T.B. 1990. An automated tactile sensing strategy for planar object recognition and localization. *IEEE trans. pattern anal. mach. intell*, **PAMI-12**(8), 775–86.

Seeley, C., & Williams, A.M. 1990. Generic assembly - the InFACT way. *In: Proc. 11th int. conf. on assembly automation (Dearborn, MI)*. Kempston, UK: SME/IFS (Pubns).

Snyder, W.E., & St.Clair, J. 1978. Conductive elastomers as a sensor for industrial parts handling equipment. *IEEE trans. instrumentation and measurement*, **IM-27**(1), 94–99.

Stansfield, S.A. 1991a. A haptic system for a multifingered hand. *Pages 658–664 of: Proc. IEEE int. conf. on robotics and automation (Sacramento, CA)*. Los Alamitos, CA: IEEE Computer Soc. Press.

Stansfield, S.A. 1991b. Robotic grasping of unknown objects: A knowledge-based approach. *Int. j. robotics research*, **10**(4), 314–326.

Tan, M. 1990. CSL: a cost-sensitive learning system for sensing and grasping objects. *Pages 858–863 of: Proc. IEEE int. conf. on robotics and automation (Cincinnati, OH)*. Los Alamitos, CA: IEEE Computer Soc. Press.

Trinkle, J.C., Abel, J.M., & Paul, R.P. 1988. An investigation of frictionless grasping in the plane. *Int. j. robotics research*, **7**(3), 33–51.

Tsujimura, T., & Yabuta, T. 1989. Object detection by tactile sensing method employing force/torque information. *IEEE trans. on robotics and automation*, **RA-5**(4), 444–450.

Vaidyanathan, C.S., & Wood, H.C. 1990. Interpretation of the task status of a gripper from tactile sensor data. *Pages 877–882 of: Proc. IEEE int. conf. on robotics and automation (Cincinnati, OH)*. Los Alamitos, CA: IEEE Computer Soc. Press.

Williams, A.M., & Lill, B.H. 1987 (May). Commercially available flexible assembly cell. *In: Proc. BRA-10, 10th British Robot Assoc. Conf.*

Zhou, Xiju, & El-Zorkany, H. 1987. Automatic location determination using simple touch sensing. *Pages 333–346 of: Advances in CAD/CAM and robotics: NRC contributions*. Report NRCC-27454, National Research Council Canada.

Chapter 4

Planar Elasticity for Tactile Sensing

Ronald S. Fearing

4.1 Introduction

A compliant tactile sensor provides information about the elastic contact between a robot finger and objects in the environment. The indentation and sub-surface deformation are complicated functions of the finger and contact shape, applied loading, and material properties of the sensor and contacted object. Hence, to obtain the most information from a tactile sensor, and to implement successfully a tactile sensor, it is very helpful to examine some of the properties of elasticity. This chapter gives a brief introduction to planar elasticity and contact problems. The planar elasticity model is also used to determine relations between sensing element placement within the tactile sensor and sensor performance.

It is possible to sense the shape of an object using a tactile sensor without a compliant layer. For example, a tactile sensor could be a planar device with needles sticking out which would measure the profile of an object pressed against the surface (Page *et al.*, 1976). This type of sensor is fine if objects always indent normally into the sensor, and do not slide, which could cause the sensing needles to jam or break. Another potential problem with the array of needles is the possibility of a small object feature being between needles, and thus missed.

Most tactile sensing applications will require a compliant skin. One of the main functions of the compliant skin is to increase the size of the contact area to allow sensing of local shape. (A hard finger would have only a point contact, thus local shape could only be determined through repeated contacts.) In robotic grasping applications, the compliance is also necessary to achieve good frictional properties during grasping (Akella & Cutkosky, 1989), for example, to take advantage of the increased grasp stability possible at corners.

The most challenging tactile sensing issues are perhaps in tactile sensing for

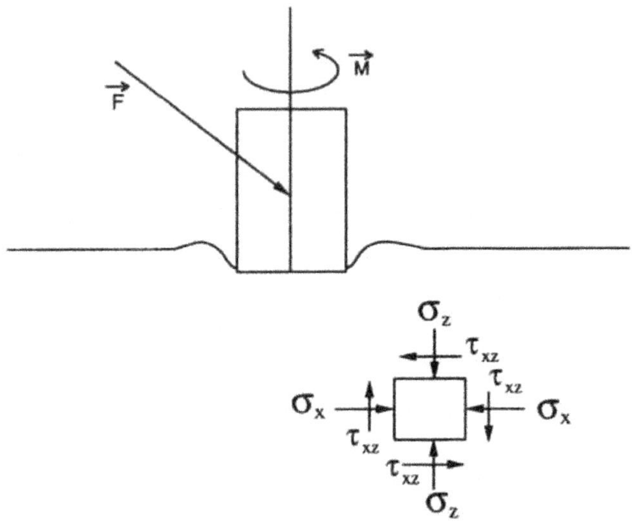

Figure 4.1: *General tactile sensing problem.*

multi-fingered dextrous hands. Unfortunately for the problem of contact interpretation, a tactile sensor for a dextrous hand must be round to allow rolling contacts and motion of the object with respect to the fingers. A flat finger sensor is much easier to analyse than a cylindrical finger, and this chapter examines planar models which can provide a first approximation to the behaviour of a cylindrical sensor for small contacts.

The first problem in tactile sensing is inferring the contact loading characteristics (shape and loading forces) from a limited set of measurements inside the sensor. As we shall see, there is a complicated relationship between the contacting object, the compliant finger, the contact forces, and the sensor placement within the finger, which can be modelled by the theory of elasticity. The reader interested in an in-depth treatment of the problems of elasticity is referred to the books by Johnson (1985), and Gladwell (1980). To give an introduction to the interpretation of contact geometry from tactile sensor measurements, some very simplified mechanical models will be examined. It has been shown that these simplified, planar models make reasonable predictions of the sensor behaviour for various stimuli in some cases (Phillips & Johnson, 1981; Fearing & Binford, 1988). Analytical models are relatively easily obtainable for the 2-dimensional (planar) case. A real sensor would probably need to be modelled using empirical or finite element methods.

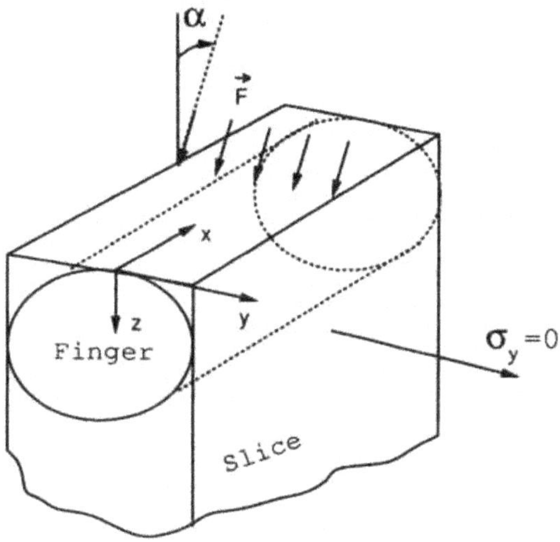

Figure 4.2: *Finger geometry for plane stress assumption.*

4.2 Stresses and Strains in Plane Stress

In its most general form, the tactile sensing problem can be posed as follows. Given a set of sub-surface strain measurements (or displacements) in an elastic body, find the local shape of the contacting body, and the applied forces. Although the net applied forces could be determined using a force sensor, a tactile sensor is required to determine force distributions over multiple contacts. Figure 4.1 shows an object being pressed into a compliant finger with applied force and moment ($\vec{F}$, $\vec{M}$), which cause contact stresses σ_x, σ_y, σ_z, τ_{xy}, τ_{xz}, τ_{yz} distributed over the interface area. The tactile sensor will measure either the displacement of the elastic material (u, v, w) or strains in the material ϵ_x, ϵ_y, ϵ_z, γ_{xy}, γ_{xz}, γ_{yz}. (Stresses, strains and elastic compatibility are defined in Appendix A.)

Figure 4.2 shows the geometrical assumptions used in the planar approximation. Consider a slice of elastic material in the $x - z$ plane with the applied force on top constant in the y direction, and the normal stresses on the face of the slice (σ_y) equal to 0. This is the plane stress assumption. The thickness of the slice roughly corresponds to the diameter of a cylindrical tactile sensor. In the planar analysis, the curvature of a cylindrical finger will be neglected. The contacts are assumed to be small compared with the thickness of the skin so that the infinite half-space assumption is valid.

The material of the finger is assumed to be linear and elastic, although rubber-like materials typically have significant viscoelastic behaviour. The effects of inclusion of sensor elements and the possible harder core of the finger, which violate the assumptions of homogeneous and isotropic material, are neglected. Linear elasticity theory is only valid for small strains (strains can be large with up to 30% strain for very compliant sensors), but provides a reasonable first order approximation.

The stresses and strains inside the elastic slice for any arbitrary loading on the surface can be found by the superposition of responses to a series of line loads, which have constant magnitude across the width of the slice. The normal strain for a line load is derived in Appendix A; the other strains and stresses can be similarly derived. The stresses in the slice for a line load applied normal to the surface are (Johnson, 1985):

$$\sigma_x = \frac{-2F_z z x^2}{\pi r^4}, \quad \sigma_z = \frac{-2F_z z^3}{\pi r^4}, \quad \tau_{xz} = \frac{-2F_z z^2 x}{\pi r^4} \tag{4.1}$$

where F_z is the force per unit thickness of the slice in Nm^{-1}, and $r^2 = x^2 + z^2$. For a concentrated tangential force F_x at the surface:

$$\sigma_x = \frac{-2F_x x^3}{\pi r^4}, \quad \sigma_z = \frac{-2F_x x z^2}{\pi r^4}, \quad \tau_{xz} = \frac{-2F_x z x^2}{\pi r^4}. \tag{4.2}$$

Since $\sigma_y = 0$ in the plane stress approximation, the strains ϵ_x and ϵ_z will be a function of only σ_x and σ_z:

$$\epsilon_z = \frac{1}{E}[\sigma_z - \nu\sigma_x], \quad \epsilon_x = \frac{1}{E}[\sigma_x - \nu\sigma_z] \tag{4.3}$$

where E is the elastic modulus (Nm^{-2}), and ν is Poisson's ratio. Thus for a line load applied at angle α from normal, the strains can be found using the combined stresses from eq. 4.1 and 4.2:

$$\epsilon_z = \frac{-2|\vec{F}|}{E\pi r^4}[z\cos\alpha - x\sin\alpha]\,[z^2 - \nu x^2] \tag{4.4}$$

$$\epsilon_x = \frac{-2|\vec{F}|}{E\pi r^4}[z\cos\alpha - x\sin\alpha]\,[x^2 - \nu z^2] \tag{4.5}$$

$$\gamma_{xz} = \frac{\tau_{xz}}{2\mu} = \frac{-2|\vec{F}|zx}{2\mu\pi r^4}[z\cos\alpha + x\sin\alpha] \tag{4.6}$$

with μ the shear modulus, and the applied line force $\vec{F} = |\vec{F}|(\hat{z}\cos\alpha + \hat{x}\sin\alpha) = F_z\hat{z} + F_x\hat{x}$.

The normal strain ϵ_z has a component $h_z(x, z)$ due to F_z and a component $h_x(x, z)$ due to the tangential force component F_x. The *impulse response* for normal strain (with a unit line load applied at angle α) is defined as:

$$h(x, z) = h_z(x, z)\cos\alpha + h_x(x, z)\sin\alpha. \tag{4.7}$$

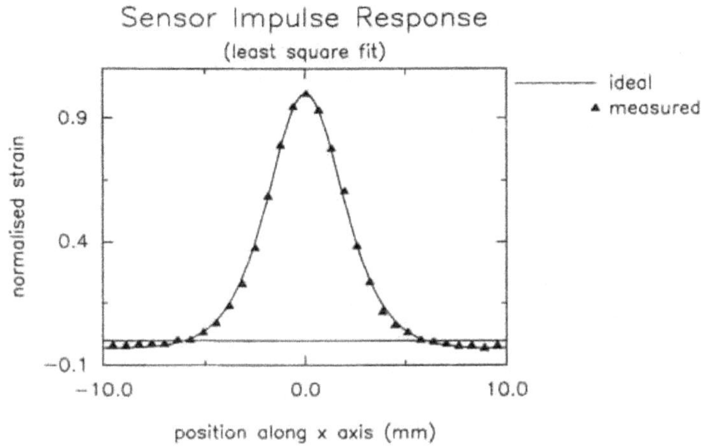

Figure 4.3: *Plane stress model and experimental data.*

The plane stress model can be compared with the response from an actual cylindrical tactile sensor (Fearing, 1990). A knife edge (which approximates a line load) with 0.5 N was applied normally along the finger length at 0.6 mm increments while recording the ϵ_z output from one sensor element. After amplitude normalisation, a least-squares fit to the plane stress model of eq. 4.5 was done by adjusting the depth and Poisson's ratio parameters. The best fitting parameters were a depth of 3.8 mm and $\nu = 0.4$. The theoretical and experimental results are compared in Figure 4.3, showing surprisingly close agreement between experiment and simple mechanical model.

An alternative planar elasticity model can be used under conditions of *plane strain*. An infinitely long line of force applied to the surface of an elastic half-space allows the assumption that $\epsilon_y = 0$ and that $\sigma_y = \nu(\sigma_x + \sigma_z)$, thus:

$$\begin{aligned} \epsilon_z &= \frac{1}{E}[\sigma_z - \nu(\sigma_x + \sigma_y)] \\ &= \frac{1+\nu}{E}[\sigma_z(1-\nu) - \nu\sigma_x] \\ &= \frac{-2F_z z(1+\nu)}{\pi r^4 E}[z^2(1-\nu) - \nu x^2]. \end{aligned} \tag{4.8}$$

It is interesting to compare the normal strains for a line load with the normal strain from a point load on a half-space, the Boussinesq problem. The normal strain

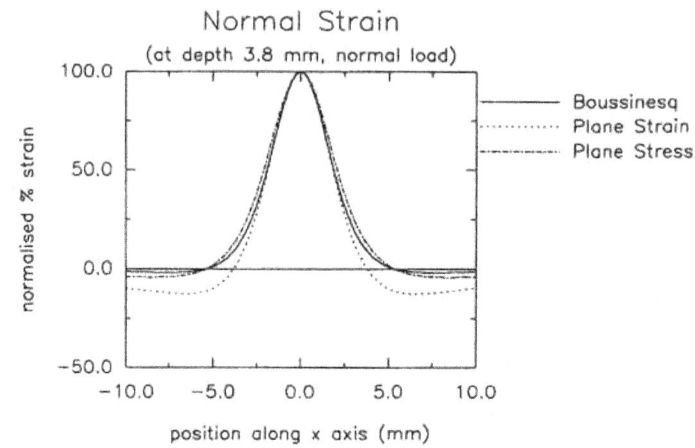

Figure 4.4: *Comparison of impulse responses for normal load.*

for a normal point load is given in Appendix A (eq. A.72) as:

$$\epsilon_z = \frac{F_z z}{4\pi\mu r^3}\left(2\nu - \frac{3z^2}{r^2}\right) \tag{4.9}$$

where here, $r^2 = x^2 + y^2 + z^2$. The normal strains ϵ_z for the Boussinesq, plane stress, and plane strain models are compared in Figure 4.4, for a sensor depth of 3.8 mm and an incompressible material ($\nu = 0.5$). The shape of the impulse-response, i.e., the response to a line or point force, is qualitatively the same for the planar and 3-dimensional models. The normal strain impulse-responses for loads applied 30° from vertical (Figure 4.5) also appear similar under the different models.

The plane stress model assumes that the tactile strain sensors are in an elastic medium of infinite extent. If the tactile finger has a hard core, a more accurate model would be an elastic strip with rigid foundation in two dimensions. Assuming that the strip is rigidly attached to the foundation, the impulse response can be found using potential methods (Fearing, 1990; Domenici *et al.*, 1989; Gladwell, 1980), and is shown in Figure 4.6. The impulse responses for the infinite slice and strip are quite similar in shape and width.

4.2.1 Deflections for Plane Strain

A tactile sensor can measure either sub-surface strains, or surface deflections. From the definitions of strain (see Appendix A) the deflections (for plane strain case) for

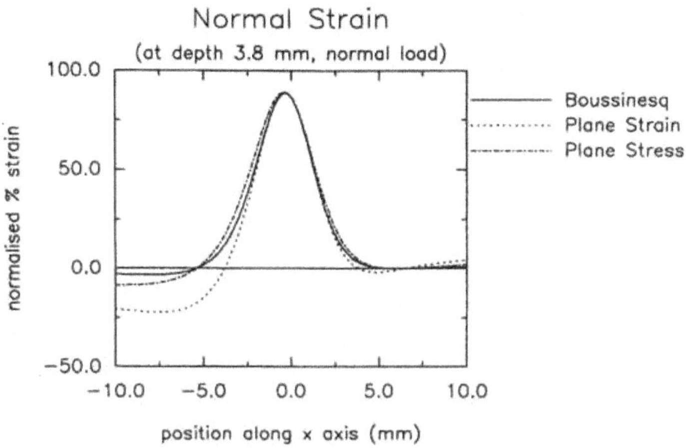

Figure 4.5: *Comparison of impulse responses for inclined load.*

a line load can be determined as (Johnson, 1985):

$$u(x,\ z = 0) = \int_{-\infty}^{x} \frac{\partial u}{\partial x} dx = \int_{-\infty}^{x} \epsilon_x dx \tag{4.10}$$

$$= -\frac{(1 - 2\nu)(1 + \nu)}{2E} F_z - \frac{-2F_x(1 - \nu^2)}{\pi E} \log|x| + C_1$$

$$w(x,\ z = 0) = \int_{-\infty}^{z = 0} \frac{\partial w}{\partial z} dz = \int_{-\infty}^{z = 0} \epsilon_z dz \tag{4.11}$$

$$= \frac{-2F_z(1 - \nu^2)}{\pi E} \log|x| + \frac{(1 + \nu)(1 - 2\nu)}{2E} F_x + C_2.$$

Note the singularity in displacement (due to finite force over zero area) at the origin. Other artifacts of planar elasticity are the constants C_1, C_2, which are chosen to give zero displacement at some reference distance from the line load.

4.2.2 Strain and Deflection Given Surface Traction

Surface traction is defined as the combination of normal and shear stresses applied at a contact. The strains and deflections can be calculated in a straightforward manner from arbitrary surface tractions using the principle of superposition. Unfortunately, the inverse problem of calculating surface tractions from strain or displacement measurements is not as straightforward.

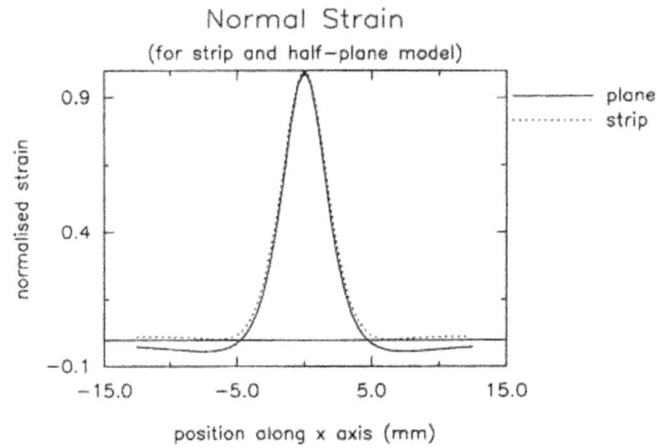

Figure 4.6: *Comparison of impulse response for half space and layer.*

4.2.2.1 Convolution Methods

Now to calculate the normal strain at the sensor at constant depth d, due to pressure distribution $p(x)$ on surface $z = 0$, with $\tau_{xz}(x, z = 0)$, we have by superposition with the impulse response of eq. 4.7:

$$\epsilon_z(x) = \int_{x_o=-\infty}^{\infty} p(x - x_o) h_z(x_o, d) dx_o \tag{4.12}$$

where $\epsilon_z(x)$ is the strain at $(x, z = d)$ due to the pressure on the surface.

Similarly, the normal displacement at the surface can be found from superposition:

$$w(x, z = 0) = \int_{x_o=-\infty}^{\infty} p(x - x_o) k(x_o) dx_o \tag{4.13}$$

where $k(x)$ is $\frac{-2F_z(1 - \nu^2)}{\pi E} \log|x| + C_2$.

4.2.2.2 Frequency Domain Methods

The convolution of eq. 4.12 can be replaced by multiplication of the Fourier Transforms of the pressure distribution $p(x) \rightarrow P(s)$ and the impulse response $h_z(x) \rightarrow H_z(s)$, $h_x(x) \rightarrow H_x(s)$, where s is the spatial frequency, e.g., in cycles mm^{-1}. With the plane stress assumption, the frequency response of $h_z(x)$ (normal strain due to a normal applied line load at surface) at depth $z = d$ is:

$$H_z(s) = (2\pi d|s|(1 + \nu) + 1 - \nu)\frac{e^{-2\pi d|s|}}{E}. \tag{4.14}$$

The frequency response due to the tangential load component is:

$$H_x(s) = \frac{-j}{E}\text{sgn}(s)[2\nu - (1 + \nu)2\pi d|s|]e^{-2\pi d|s|} \quad (4.15)$$

where

$$\text{sgn}(s) = \begin{cases} -1 & \text{if } s < 0 \\ 0 & \text{if } s = 0 \\ 1 & \text{if } s > 0 \end{cases}.$$

If the contact force is not normal to the finger, the assumption can be made that at any point on the surface, the tangential stress is proportional to the normal stress (Smith & Liu, 1953), that is:

$$\sigma_z(x, z = 0) = p(x)\cos\alpha \quad \text{and} \quad \tau_{xz}(x, z = 0) = p(x)\sin\alpha \quad (4.16)$$

where α is less than ϕ_s, the friction angle. With this assumption, the convolution of eq. 4.12 can be replaced by a multiplication in the frequency domain giving:

$$E_z(s) = P(s)\,[H_z(s)\cos\alpha + H_x(s)\sin\alpha] \quad (4.17)$$

where α is the angle of force. The strain can be obtained by taking the inverse Fourier transform of $E_z(s)$.

4.2.3 Determining Surface Tractions from Object Shape and Loading

The previous section has shown how to determine displacements or strains given the surface traction. However, in a typical robotic contact problem, the shape and net force are known, but not the contact tractions. The intuitive, but incorrect, idea is that the surface displacement and object shape are the same. In fact, if the indenter is convex, the displacement $w(x)$ will be equal to the indenter profile in the contact region. However, the width of the contact region will be a function of contact shape, total load, and elastic properties of the sensor skin, and will not necessarily be the same as that of the object. The other boundary condition on the surface is the contact stress $\sigma_z(x, z = 0)$ which is zero outside the contact region. Determining $w(x)$ outside the contact region and $\sigma_z(x, z = 0)$ inside the contact region requires solution of a mixed boundary value problem.

As an example of the relation between loading and surface displacement, consider the normal displacement for a uniform pressure distribution across a region of contact. The displacement (calculated using eq. 4.13) is shown in Figure 4.7. Note that $w(x)$ is non-zero outside the region of pressure application. For plane strain and normal load, the displacement can also be calculated using Fourier transforms (see Appendix A):

$$2\mu W(z, s) = e^{-2\pi|s|z}P(s)\left(z + \frac{2(1 - \nu)}{2\pi|s|}\right). \quad (4.18)$$

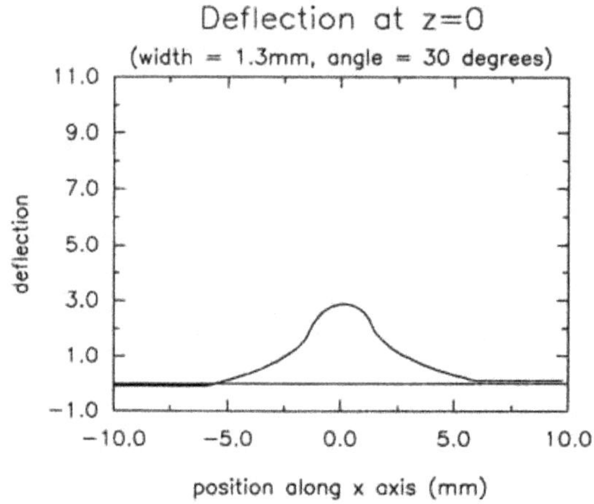

Figure 4.7: *Surface deflection for uniform force distribution at 30°.*

The pressure distribution for frictionless indentation can be found by solving the integral equation:

$$w(x) = \int_a^b p(x - x_o)k(x_o)dx_o \tag{4.19}$$

where $w(x)$, the displacement inside the contact region, is known from the shape of the indenter. The width of the contact (the limits of integration, a and b) is determined from the location and geometry of indenter, and its depth of indentation. (The width of the contact is difficult to calculate, except in the case of a rectangular punch. For a rectangular punch normally indenting a half-plane, the contact width is simply the width of the punch.) The other constraint is that the integral of the pressure distribution must equal the net force on the contact in the z direction:

$$F_z = \int_a^b p(x)dx. \tag{4.20}$$

A survey of pressure distributions for rectangular, cylindrical, and wedge shaped contacts and resulting normal strains is presented by Fearing & Hollerbach (1985), and derivations are provided by Johnson (1985).

4.3 Tactile Signal Processing

So far, we have presented a theoretical model of the forward problem, calculating displacements and strains given contact traction or contact shape and loading. For tactile interpretation, we must solve the inverse problem: solving for shape and force from strains or displacements. By exploiting linear system properties of the planar tactile sensor model, algorithms are derived for localisation, pressure recovery, and determining the angle of force from normal strain measurements.

4.3.1 Deflection vs. Strain Measurements

A continuing unsettled issue in tactile sensing is the choice between measuring surface displacement and measuring strain beneath the skin. Surface displacement sensors (e.g., Nowlin (1991)) appear to offer the ability to measure directly contact shape, but as shown in section 4.2.3, the surface displacement sensor has spreading of the response into adjacent regions as does a sub-surface strain sensor. Determining the extent of contact with a displacement sensor requires finding the surface pressure, perhaps by inverting eq. 4.18. The non-zero pressure region would indicate the extent of contact.

A further difficulty with using surface displacement tactile sensors is that linear elasticity theory presupposes infinitesimal deflections. With compliant sensors, relatively large surface deflections are possible, which could cause surface displacement to disagree with theory.

The main drawback to using sub-surface strain sensors seems to be the loss of resolution due to the low-pass filtering function of the skin layer. The frequency response of $H_z(s)$ (eq. 4.14) is a decaying exponential function of frequency and sensor depth. High spatial frequency detail in the pressure distribution on the surface is dramatically attenuated with increased sensor depth.

4.3.2 Sensor Modelling

To interpret contact shape and forces, accurate models of the sensor behaviour including impulse response, linearity, and elastic modulus are required. While the response of finger-shaped sensors to a line-load agrees well with the planar elasticity predictions, the response of a cylindrical tactile sensor to a point load may not agree well with the 3-dimensional strain profile predicted by the Boussinesq solution (Fearing & Binford, 1988). Sensor models can be developed using either *ad hoc* empirical functions, or finite element modelling. With linear sensor behaviour, it has been shown (Fearing & Binford, 1988) that carefully measured *ad hoc* models can be used to find contact shapes.

4.3.3 Filter Methods for Obtaining Surface Pressure and Deflection

In principle, linear transforms can be used to solve for surface pressure and deflection from sub-surface strain, by inverting the relations in eq. 4.17 and 4.18. In practice, these inverses are ill-conditioned due to the need to high-pass filter noisy sensor data. Regularisation approaches to the inverse problem can be taken, and have worked well in simulation using approximately 40 sensors in a linear array (see for example, De Rossi *et al.* (1991), Pati *et al.* (1988), and Yang (1987)). If the type of contact is known, for example, it is known that the finger is touching a surface of revolution, a parameterised contact model-fitting approach works well. Using this approach, contact location, force, and principal curvature directions were determined in experiments using only a 4×4 set of normal strain measurements (Fearing & Binford, 1988).

Although linear inverse filters for surface pressure estimation are ill-conditioned, it is worth examining them to understand better the problems of contact localisation and magnitude estimation. This section examines 1-dimensional inverse filters for determining contact shape or surface pressure from strain measurements. There is a linear transform K between surface deflection and measured strain, and a linear transform H between surface stress and measured strain that can be inverted using Fourier Transforms. In principle, given adequate sampling density to avoid aliasing, a bandlimited surface stress or deflection profile can be recovered from the discrete strain measurements.

For plane stress, the surface pressure with frictionless indentation is obtained from:

$$P(s) = \frac{-Ee^{2\pi d|s|}}{(2\pi d|s|(1+\nu)+1-\nu)}E_z(s) = H_z^{-1}(s)E_z(s). \tag{4.21}$$

In the plane stress model, the deflection at depth z can be obtained by integrating the equation for normal strain $\epsilon_z = \frac{\partial w}{\partial z}$ from ∞ to z. The surface deflection can then be found as a function of the measured strain at depth d (Gladwell, 1980):

$$W(s) = \frac{2}{2\pi|s|}P(s) = H_z^{-1}(s)E_z(s)\frac{2}{2\pi|s|E} = K^{-1}(s)E_z(s). \tag{4.22}$$

Both $K^{-1}(s)$ and $H^{-1}(s)$ are ill-behaved functions, which grow exponentially with frequency, thus they cannot be used directly as inverse filters without band limiting. (For further discussion of inverse-filtering issues, see Cameron *et al.* (1988).)

The filter methods discussed have assumed that a continuous strain signal was available. However, sampling of the strain function puts an upper limit on the frequency response of the inverse filter. For 3.3 mm sampling along the x direction (the sensor spacing along the axis), the anti-aliasing filter needs a cutoff below 0.15 cycles mm^{-1}. By taking repeated measurements as the finger is scanned along a fixed object, the effective sensor density can be increased to reduce the aliasing effects. Because of undersampling, the inverse filter approach will provide an aliased version of the original input. These sampling effects will be discussed further in section 4.3.5.

4.3.4 Force Angle Estimation

It is possible to estimate the angle of force using strain measurements if the contact pressure distribution is known to be an even function. The measured normal strain is given by eq. 4.17, and since in this equation $P(s)$ and $H_z(s)$ are even and real functions, and $H_x(s)$ is odd and imaginary, we can split the frequency responses into real and imaginary parts:

$$\begin{aligned} Re E_z(s) &= E_{zr}(s) = P(s) H_z(s) \cos\alpha \\ Im E_z(s) &= E_{zi}(s) = -jP(s) H_x(s) \sin\alpha. \end{aligned} \quad (4.23)$$

We can then solve for pressure:

$$P(s) = \sqrt{(\frac{E_{zr}}{H_z})^2 + (\frac{E_{zi}}{-jH_x})^2} \quad (4.24)$$

and for angle of force:

$$\tan\alpha = j\,\frac{H_z(s_o)}{H_x(s_o)}\frac{E_{zi}(s_o)}{E_{zr}(s_o)} \quad (4.25)$$

where s_o is a particular frequency where the ratio is well conditioned.

Another possible approach to recovering the angle of force is suggested by Speeter (1987), who proposes using differences between strain sensors at two depths in the skin. Alternatively, separating the strain response into odd and even components, corresponding to tangential and normal surface force components respectively, can be used to find the angle of force (Fearing, 1990). The easiest approach is to include an intrinsic finger force sensor, such as that developed by Brock & Chiu (1985), to determine force angle, and then use the tactile sensor to determine contact shape and location.

4.3.5 Sampling Density

The filter methods of the previous section assumed that a continuous $\epsilon_z(x)$ was available. Tactile sensors are usually spatially sampled, resulting in aliasing when reconstruction is done by interpolation. One approach to avoid aliasing effects is developed by De Rossi *et al.* (1987), who consider placing sensors close enough such that the stress will be uniform across the sensors, to within the sensor dynamic range. The uniform stress criteria requires sensor spacings less than one tenth the skin thickness. However, reasonable interpolation can be obtained with sensor spacing equal to sensor depth (Fearing, 1990).

This section describes an interpolation method for recovering the location and magnitude of a line load applied normally to the surface of the sensor from discrete measurements, considering effects due to the finite number of strain samples. To illustrate the effects of aliasing, consider a plane stress sensor model with seven normal strain sensors (ϵ_z) uniformly spaced by a distance b apart along the x axis at depth d. Then the discretised strain measurements are given by:

$$\epsilon_z[x] = h_z(x) \sum_{n=-\infty}^{\infty} [\delta(x - nb)\,\mathrm{rect}(\frac{x}{3b})] \quad (4.26)$$

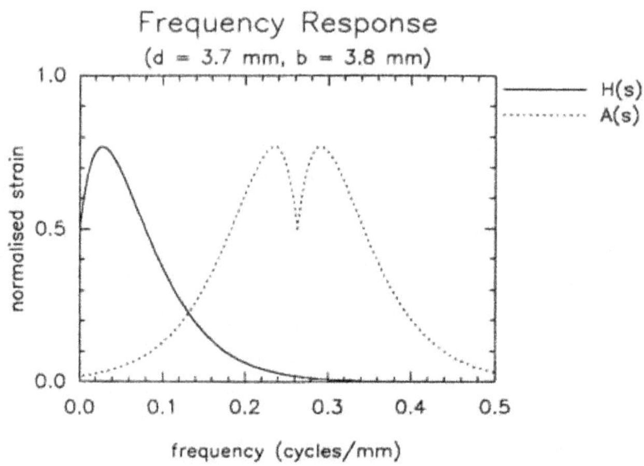

Figure 4.8: *Frequency response $H(s)$ and aliased response $A(s)$.*

where $h_z(x)$ is the normal strain impulse response from eq. 4.5. The rect() function is defined by:

$$\mathrm{rect}(x) = \begin{cases} 1 \text{ if } \mid x \mid \ \leq 1 \\ 0 \text{ if } \mid x \mid \ > 1 \end{cases} . \tag{4.27}$$

In the frequency domain, eq. 4.26 leads to:

$$\tilde{E}_z(s) = H_z(s) * \frac{1}{b} \sum_{n=-\infty}^{\infty} [\delta(s - \frac{n}{b}) * 6b\mathrm{sinc}(6\pi bs)] \tag{4.28}$$

with spatial frequency s in cycles mm^{-1}, and $H_z(s)$ is the Fourier transform of $h_z(x)$. Because of sampling, $\tilde{E}_z(s)$ is periodic in frequency, and the frequency response is widened by the convolution with the sinc($6\pi bs$) term which is due to having only seven samples.

Figure 4.8 shows the frequency response for $d = 3.8$ mm. $A(s)$ is the aliased spectrum, which interferes with recovery of the original spectrum of $H(s)$. If the sensor width is finite, and integration of strain occurs along the sensor element, the low pass properties of $H(s)$ may be slightly improved, reducing overlap with $A(s)$. In this example, the aliasing energy is significant, but would be reduced in practice by low-pass contact pressure distributions. The continuous strain signal can in principle be recovered by low pass filtering of the periodic spectrum $E_z(s)$, but this process is limited by aliasing, and the finite extent of $\epsilon_z[x]$.

The relative error due to aliasing can be estimated from the ratio of energy in the reconstruction error ($E_o - E_R$) to the energy in the original signal (E_o) (Pratt,

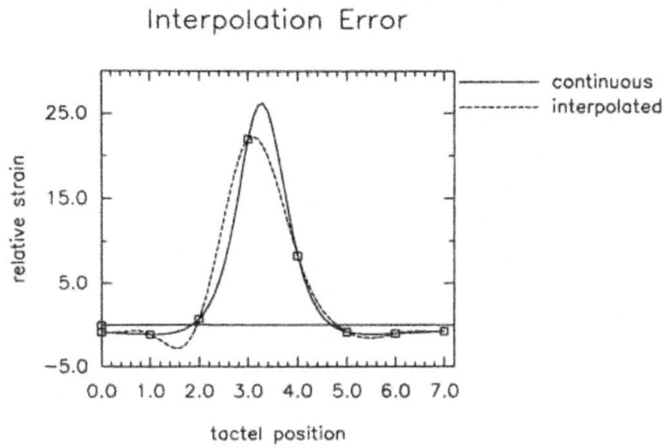

Figure 4.9: *Reconstruction error due to aliasing (b= 3.8 mm).*

1978). Assuming an ideal low pass filter, the error is given by:

$$\xi \;=\; \frac{E_o \;-\; E_R}{E_o} \;=\; \frac{\int_{-\infty}^{\infty} \mid H_z(s) \mid^2 ds \;-\; \int_{-1/2b}^{1/2b} \mid H_z(s) \mid^2 ds}{\int_{-\infty}^{\infty} \mid H_z(s) \mid^2 ds}. \tag{4.29}$$

This expression includes reconstruction errors due to the cutoff of the high frequencies by $H_z(s)$. Not included are the effects of non-ideal low-pass filtering. Evaluated numerically for equal sensor depth and spacing, we get $\xi \approx 0.20$ which implies up to 20% error for localisation and amplitude in the interpolated function $\epsilon_x(y)$. If $P(s)$ is sufficiently bandlimited, eq. 4.29 will be an upper bound on the expected error.

The effects of aliasing on localisation and amplitude estimation will now be shown for seven equally spaced sensors at depth d. The normal-strain impulse-response estimate using a windowed sinc interpolation is:

$$\hat{h}(x_o) \;=\; d \sum_{n=-3}^{3} \frac{\sin \pi(x_o - nb)}{\pi(x_o - nb)} \; w_{Han}(\frac{x_o}{b} - n) \; h[nb] \tag{4.30}$$

with w_{Han} the Hanning window (Oppenheim & Schafer, 1975) which is:

$$w_{Han}(n) \;=\; \frac{1}{2}[1 \;-\; \cos(\frac{2\pi n}{N-1} + \pi)], \quad \text{for} \quad |n| \leq \frac{N-1}{2} \tag{4.31}$$

and N is the total number of samples. The window makes the low pass filter finite in extent, and reduces ringing. Figure 4.9 shows the peak displacement and amplitude error for the interpolated response to a point force in the plane stress model when compared to the continuous strain response.

For a fixed sensor spacing, greater sensor depth will reduce aliasing, by increasing impulse-response overlap between elements. This overlap has sometimes been referred to as mechanical crosstalk between elements, and attempts have been made to reduce it. However, as we have seen, this impulse-response overlap is necessary for good localisation. Ultimately, localisation is limited by the sensor elements' noise and linearity. In experiments using a tactile sensor with equal sensor spacing and depth, localisation to within 12% of sensor spacing was obtained (Fearing, 1990), with aliasing being the dominant error source.

4.3.6 Finger Softness

Perhaps surprisingly, the elastic modulus of the finger only affects the amplitude, but not the spatial frequency response of a tactile sensor. This can be seen from the expression for $H_z(s)$ (eq. 4.14) where the elastic modulus E is only a multiplicative factor. According to the planar models developed in this chapter, soft fingers do not have wider impulse-responses than hard fingers.

However, the softness of a sensor is an important factor in determining the size of the contact on the sensor. Intuitively, to be able to estimate the shape of a contact, the contact area should be significantly larger than the sensor spacing. To illustrate the dependence of contact area on force and finger elasticity, we will examine the classical Hertz contact problem, with contact between a cylindrical finger of radius R_f and a hard cylinder of radius R_B, with an angle ψ between the two cylinder axes. The pressure distribution corresponding to a frictionless contact between two cylinders is ellipsoidal (Gladwell, 1980):

$$p(x, y) = \frac{3F}{2\pi ab} \sqrt{1 - \frac{x^2}{a^2} - \frac{y^2}{b^2}} \tag{4.32}$$

where a and b are the major and minor axes respectively of the ellipsoid at $z = 0$, F is total force, and $p(x, y) = 0$ outside the contact area. The length of the major axis of the contact ellipse (Timoshenko & Goodier, 1951) is:

$$a = f(R_f, R_B, \psi) \left(\frac{F}{E_f}\right)^{\frac{1}{3}} \tag{4.33}$$

where E_f is the elastic modulus of the finger, and it is assumed that the contacting body is very rigid compared with a soft finger. The function $f(R_f, R_B, \psi)$ is determined from the solution of the elastic deflection equation (Gladwell, 1980) which gives the contacting ellipse size. The solution uses the complete elliptic integrals of the first and second kind, and is tabulated by Cooper (1969) as a function of eccentricity of the contact ellipse. As can be seen from eq. 4.33, the contact dimensions are proportional to the cube root of the ratio of contact force to finger stiffness. To double the contact length requires eight times the force.

4.4 Summary

This chapter has used planar elasticity theory to model the transformation from contact to measured strain or displacement. Although the linear elastic model is simplistic, and not physically realistic for actual sensor geometries, it is useful in practice. The simple, analytic expressions are useful for predicting the influence of sensor design parameters such as Poisson's ratio, elastic modulus, sensor depth, and sensor spacing on the sensor response. In spite of the simplifying assumptions, model predictions for aliasing effects due to discrete sensors, and the indentation of an inclined line-load, agree well with experiments on a real tactile sensor (Fearing, 1990). The simple models can also be used to show trade-offs between the measurement of normal strain, shear strain, or displacement.

Another important use of the planar model is understanding the inverse contact-shape-from-strain problem. Even with a near-ideal sensor, the exponential decaying frequency response of the strain measurements prevents the use of a simple linear inverse filter because of noise amplification at high frequencies. However, if the contact type is known *a priori*, and the sensor response has been accurately modelled, it is possible to use a matched filter approach to determine contact size, load, and orientation (Fearing & Binford, 1988).

The ability to interpret contacts from tactile sensor data depends not only on the numerical accuracy of the individual strain or displacement sensors, but on the accuracy of the sensor impulse-response model and contact model. Although a tactile sensor may not behave as nicely as the ideal planar elastic model, the sensor must have linearity and obey the principle of superposition to make solutions tractable. If necessary, good empirical models of the tactile sensor impulse-responses can be derived using careful measurements. (For example, the impulse-response can be mapped out by probing the sensor surface.) The quality of these models will determine overall tactile sensing system performance.

References

Akella, P., & Cutkosky, M. 1989. Manipulating with soft fingers: Modelling contacts and dynamics. *Pages 764–769 of: Proc. IEEE int. conf. on robotics and automation (Scottsdale, AZ)*. Washington, DC: IEEE Computer Soc. Press.

Brock, D., & Chiu, S. 1985. Environment perception of an articulated robot hand using contact sensors. *Pages 89–96 of: ASME Pubn. PED-Vol. 15, Robotics and flexible automation*. ASME.

Cameron, A., Daniel, R., & Durrant-Whyte, H. 1988. Touch and motion. *Pages 1062–1067 of: Proc. IEEE int. conf. on robotics and automation (Philadelphia, PA)*. Washington, DC: IEEE Computer Soc. Press.

Cooper, D.H. 1969. Hertzian contact-stress deformation coefficients. *J. applied mechanics*, **36**(2), 296–303.

De Rossi, D., Nannini, A., & Domenici, C. 1987. Biomimetic tactile sensor with stress-component discrimination capability. *J. molecular electronics*, **3**, 173–181.

De Rossi, D., Caiti, A., Bianchi, R., & Canepa, G. 1991. Fine-form tactile discrimination through inversion of data from a skin-like sensor. *Pages 398–403 of: Proc. IEEE int. conf. on robotics and automation (Sacramento, CA)*. Los Alamitos, CA: IEEE Computer Soc. Press.

Domenici, C., De Rossi, D., Bacci, A., & Bennati, S. 1989. Shear stress detection in an elastic layer by a piezoelectric polymer tactile sensor. *IEEE trans. electr. insul*, **EI-24**(6), 1077–1081.

Fearing, R.S. 1990. Tactile sensing mechanisms. *Int. j. robotics research*, **9**(3), 3–23.

Fearing, R.S., & Binford, T.O. 1988. Using a cylindrical tactile sensor for determining curvature. *Pages 765–771 of: Proc. IEEE int. conf. on robotics and automation (Philadelphia, PA)*. Washington, DC: IEEE Computer Soc. Press.

Fearing, R.S., & Hollerbach, J.M. 1985. Basic solid mechanics for tactile sensing. *Int. j. robotics research*, **4**(3), 40–54.

Gladwell, G.M.L. 1980. *Contact problems in the classical theory of elasticity*. Netherlands: Sijthoff and Nordhoff.

Johnson, K.L. 1985. *Contact mechanics*. Cambridge, UK: Cambridge University Press.

Nowlin, W.C. 1991. Experimental results on Bayesian algorithms for interpreting compliant tactile sensing data. *Pages 378–383 of: Proc. IEEE int. conf. on robotics and automation (Sacramento, CA)*. Los Alamitos, CA: IEEE Computer Soc. Press.

Oppenheim, A.V., & Schafer, R.W. 1975. *Digital signal processing*. Englewood Cliffs, NJ: Prentice-Hall.

Page, C.J., Pugh, A., & Heginbotham, W. 1976. Novel techniques for sensing in a 3D environment. *In: Proc. 6th int. symposium on industrial robots*. Kempston, UK: IFS(Pubns).

Pati, Y.C., Friedman, D., Krishnaprasad, P.S., Yao, C.T., Peckerar, M.C., Yang, R., & Marrian, C.R.K. 1988. Neural networks for tactile perception. *Pages 134–139 of: Proc. IEEE int. conf. on robotics and automation (Philadelphia, PA)*. Washington, DC: IEEE Computer Soc. Press.

Phillips, J.R., & Johnson, K.O. 1981. Tactile spatial resolution III. A continuum mechanics model of skin predicting mechanoreceptor responses to bars, edges, and gratings. *J. neurophysiology*, **46**(6), 1204–1225.

Pratt, W.K. 1978. *Digital image processing*. New York: John Wiley.

Smith, J.O., & Liu, C.K. 1953. Stresses due to the tangential and normal loads on an elastic solid with application to some contact stress problems. *J. applied mechanics, trans. ASME*, **75**(June), 157–166.

Speeter, T.H. 1987 (Jan). *Analysis and control of robotic manipulation*. Ph.D. thesis, Dept. of Biomed. Eng., Case Western Reserve Univ.

Timoshenko, S., & Goodier, J.N. 1951. *Theory of elasticity*. New York: McGraw-Hill.

Yang, R. 1987. *Tactile perception for multifingered hands*. Tech. rept. SRC TR 87-162. Systems Research Center, Univ. Maryland.

Chapter 5

Integrating Tactile Sensors – ESPRIT 278

Zenon G. Rzepczynski

5.1 Introduction

ESPRIT Project 278 brought together major European industrial partners and leading academic institutions in an attempt to solve some of the considerable problems associated with sensor integration. Particular emphasis was placed on the development of a gripper-mounted tactile image sensing system, and investigation of the role of such a sensor when deployed in co-operation with an advanced vision system and other sensors (Ghani, 1987).

The automation problem to be addressed by the project was that of workpiece handling: the grasping and orientating of parts within a manufacturing task (Raymond *et al.*, 1983). As part of an incremental progression towards the final application goal, "The Bin-picking Task", the project application task was constrained to objects lying in a single plane, with the intention of progressing this to accommodate a degree of component overlap (up to 50%) within the duration of the project.

This chapter will take the case study of Project 278 and review its approach and accomplishments, with particular consideration being given to the integration and deployment of a tactile imaging sensor.

5.2 Project Structure

During the first year, a major thrust of the project was to set up two experimental test sites, the primary one at Fraunhofer-Institut fuer Productiontechnik und Automatisierung (IPA), Germany, and a subsidiary one at Joyce Loebl, UK. Each test site started off with standard commercially available equipment: a Kuka 160/15 robot controlled by a Bosch Rho2 controller interfaced to a Joyce Loebl Magiscan 2A vision system. In parallel, research and development work began on the requisite non-standard components; the first prototypes appeared during the second year. The task of integrating and testing these prototypes then began in earnest.

The technical approach adopted by the project was, where possible, to take existing components and subsystems and then to adapt and enhance them in order to achieve the overall system capabilities required. Where this could not be achieved, such as with commercially available tactile sensors and gripper systems, then the project undertook the development of these components.

Having identified the various developments required to achieve the defined objectives, the task of integrating them into a coherent piece of automation equipment was addressed. This required not only enhancements to the existing hardware and software of the robot controller, but also a complete evaluation of the most appropriate control schemes including where, within the control architecture of the system, data processing and decision making should take place.

The two test sites, set up at an early stage of the project, played an invaluable role in supporting not only the early experimentation work, but also the iterative development and integration cycles adopted by the project.

The partners in this collaborative project and their roles were as follows:

- MARI Computer Systems Limited, Gateshead, UK (prime contractor, and tactile system developer);
- Vickers Plc, Joyce Loebl Division, Gateshead, UK (vision system developer);
- Fraunhofer-Institut fuer Productiontechnik und Automatisierung, Stuttgart, Germany (gripper developer, systems integration and application engineers);
- Robert Bosch GmbH, Erbach, Germany (robot controller system developer);
- University of Newcastle upon Tyne, UK (specialist support in sensors, VLSI and control);
- National Technical University of Athens, Greece (advanced control techniques);
- New University of Lisbon, Portugal (investigation of CAD integration issues).

5.3 The Manufacturing Requirement

Most manufacturing processes require the presentation of correctly oriented workpieces. Unfortunately it is rare for a workpiece to arrive at a workplace in a known orientation and be positioned exactly. When produced volume is high, specially engineered solutions can be justified, for example, for preserving orientation through workpiece-specific carriers, or for restoring orientation through specialised devices such as vibratory bowl feeders.

With the trend towards flexible manufacturing, implying frequent changes of workpiece, the specially engineered solution is less viable. Equally, the alternative of using manual labour for orientation and feeding is not attractive in this context (Birk *et al.*, 1981).

There is therefore a need for a means of acquiring unoriented workpieces from a general-purpose carrier (pallet, belt, bin) in a manner that is flexible, reliable, easily adaptable to new workpieces, and capable of cycling fast enough. This is the prime industrial need which was to be investigated by ESPRIT Project 278 as reported in this case study.

5.4 The Sensing Problem

The basic problem is to determine the position and orientation of a (known) workpiece with sufficient accuracy to allow proper grasping, transfer and placement. This apparently simple problem can easily be complicated in a number of ways:

- there may be more than one possible type of workpiece at any time;
- there may be more than one workpiece in the field of operation;
- components may be touching;
- objects may be overlapping;
- objects might not be lying on a horizontal plane;
- there may be objects with no stable states.

The worst case is obviously that of a bin holding a variety of jumbled parts (Ghani & Rzepczynski, 1986). To provide a more realistic target for a three year timescale, the project concentrated on the simpler aspects of the problem. It is considered however that dealing with these complicating factors in practice will require significant advances in the application of knowledge-based techniques to the problem.

In the simple case, then, we have a known object on its own, lying on a horizontal plane, in one of a number of known stable states. Taking a pragmatic approach, the problem then reduces to one of determining: (a) which stable state the object is in, (b) the rotational angle, from some fixed axis, of that stable state in the horizontal plane, and (c) the position on the horizontal plane of a point on the object. From these, and knowledge of the object, it is possible to compute the complete positional and orientation data for the robot to grasp the object.

Using a vision system with a camera mounted above the plane we can begin to determine these data. The problems that arise here include calibration, optical distortion, sensitivity to lighting, image skewness, object rotation, and ambiguity. These problems may be tackled by various techniques (Robinson & Hage, 1986), but in the final analysis, there will often be some degree of uncertainty as to the output of a vision system.

Take, for example, a rectangular valve block lying on its side. It may be impossible to determine from a view from above which of the two possible orientations and which two possible stable states are correct. Further, the position of the block as computed by the vision system may have a degree of error associated with it resulting from some of the above mentioned problems.

The role of a tactile sensor becomes clearer here. Incorporated into the gripping surfaces of the robot's end effector, a pair of tactile sensor pads should provide further information about the object during the process of grasping. (The gripper is assumed to be a two-finger device with parallel motion.) Assuming that distinguishing features are present on the gripping surfaces, the tactile image data should help resolve orientation ambiguity. Furthermore, as the gripper position is known, and the tactile data is derived from direct contact between object and gripper, the accuracy of any positional information that can be determined from the tactile data is limited only by the resolution of the tactile sensor pad, and the accuracy of the robot's positional sensing mechanisms.

Thus, while the vision system provides a 2-dimensional view of the entire object, though with some uncertainty, the tactile sensor should provide a 2-dimensional contact image of subsections of the object's surface, with well-defined accuracy.

The problem then is how to integrate these two complementary types of data (as well as other sensory inputs) into the robot control scheme such that the system can be adapted to a new part with a minimum of time/cost/effort/specialist skills, and can carry out the task reliably, with as low a cycle time as possible.

5.5 An Integrated Approach

5.5.1 System Architecture

In essence, the system developed consisted of a mainly conventional robot controller linked to three intelligent subsystems via a specially developed interface controller.

Starting from its Rho2 robot controller with its native BAPS programming language, Bosch Gmbh's approach was to make enhancements to the hardware, software, and communications abilities of the system towards the needs of sensor integration. In approaching these, the main achievements were:

- enhancement of the BAPS programming language by the addition of a set of commands which gives the programmer access to the functions within the tactile, vision and gripper subsystems;

- the definition of a system-wide communications protocol based on the EIA RS511 (1393A) draft standard;

- the development of a specialised sensor supervisor (intelligent interface con-

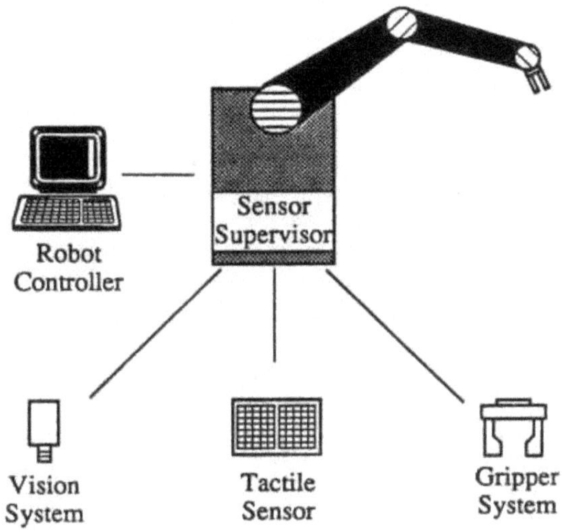

Figure 5.1: *Logical system configuration.*

troller) as an add-on component to the existing controller to handle sensor system communications and interpretation of the new sensor-specific commands.

The three subsystems linked to the robot controller were those concerned with tactile and vision sensing and the control of the two axis gripper. The logical configuration of the components of the system is shown in Figure 5.1.

5.5.2 Tactile and Gripper System

5.5.2.1 Tactile Sensor System

There were two tactile sensing techniques developed within the project (piezoresistive and optical) both of which were integrated into the gripper. The details of these techniques (Ghani, 1988) are of less relevance here than how the information produced by them was analysed and used in the control scheme. A more detailed description of the sensors and their transduction methods is presented Chapter 2.

The MARI piezoresistive sensor is a 32×32 sensor array on 1.27 mm pitch on a pad 15 mm thick incorporating preprocessing electronics to produce a binary thresholded image. Two such sensors were mounted on the rotating pads of the parallel fingers of the gripper and connected to the tactile processor subsystem.

5.5.2.2 Gripper System

The gripper system developed by IPA was designed to meet the needs of flexibility in workpiece handling and of sensor integration. Flexibility was achieved by a highly modular design which allows for easily exchanged finger and sensor modules. Manipulative flexibility was increased by including rotating pads at the gripping points of a pair of fingers. This was found to be of significant advantage in reducing cycle times for workpiece orientation.

Integration of the two types of tactile sensor was achieved in different ways. The MARI sensor pad was easily mounted on all the finger types including those with the rotating axis. The optical sensor from Newcastle University was designed as an integral part of a finger with a freewheeling rotating pad, to be used in opposition to one with a driven pad.

The gripper is controlled by a Motorola M68000-based system which communicates serially with the robot controller and interprets commands from it. The gripper controller also relays commands from the robot to the tactile processor.

5.5.2.3 The Preprocessor

The function of the preprocessor subsystem is to scan the sensor array, threshold the analogue signals and transmit the digital data to the processor.

Because the array is a matrix of conducting rows and columns, in order to isolate a sensor point from the effects of all other points, a voltage balancing technique is employed. This involves holding all non-selected columns at the same potential as the sensor point being interrogated, ensuring the measured current flow is dependent on its resistance alone.

To remove as much of an overhead as possible from the software, the analogue data from each sensor point is thresholded by the hardware at eight levels. This provides the processing software with eight binary images of each sensor, each corresponding to a different force range.

In order to allow the system to be mounted on the end of a robot arm, interconnections are kept to a minimum. Data from each sensor point is transmitted serially at a rate controlled by an external clock line. The present system operated at 1 Mbit/s per sensor pad, allowing a pair of images to be captured in just over a millisecond.

The tactile processor developed is a Motorola M68000 system augmented by a Texas TMS-32010 digital signal processor. This system implements the image processing and analysis algorithms that act on the binary images obtained from the sensors in order to confirm the identification of the object grasped. The matching technique is based on reducing the features in each pair of tactile images to a set of typed vector points, and minimising the search space during matching by exploiting geometric constraints.

5.5.2.4 Tactile Image Processing

In order to maximise the number of features available for matching and measurement, each pair of images from the two sensor arrays are treated as a unit. Features from both images are regarded as being on one common plane to facilitate spatial measurements on the sensor plane. It is of course necessary to remember for each feature, which image it appeared on.

Whether the system is in a recognition or a learning mode, the binary images undergo basically the same processing. Firstly, a contour following algorithm is used to produce chain codes which identify each contour (e.g., identifying a hole and the shape containing the hole). A second algorithm then traverses each contour, monitoring the gradient and identifying very slight changes in gradient (straight lines) and very sharp changes in gradient (corners). Thus, every tactile image pair is reduced to an image descriptor (ID) which consists of two lists: a list of feature points and a list of lines.

A line is represented by the coordinates of the two points on a contour which delimit a straight-line segment. A feature point is a point on the sensor plane associated with a specific feature (either a corner or a shape), and is represented by a 5-tuple (TYPE, POS, SIDE, RI, RD) defined as follows:

TYPE identifies one of the following:
: **corner:** the contour changes direction by more than 70° over 10 adjacent pixels;
: **shape:** the start and end points of the contour are the same, but the contour does not touch the sensor boundary.

POS represents the (x, y) coordinates, relative to the sensor frame, of a specific point associated with the feature as follows:
: **corner:** the apex of the corner;
: **shape:** the centre of gravity of the pixels contained within the contour.

SIDE indicates on which of the two sensor pads the feature appeared.

RI is a feature-dependent rotation-invariant parameter as follows:
: **corner:** magnitude of the corner angle;
: **shape:** area (the number of pixels in the shape).

RD is a feature-dependent rotation-dependent parameter as follows:
: **corner:** slope of the line bisecting the corner angle;
: **shape:** slope of the line of maximum radius (if present).

5.5.2.5 Image Recognition

The matching algorithm takes as its input two IDs, and outputs a correction vector (known as the Correction Matrix). Matching is a process of searching for pairs of

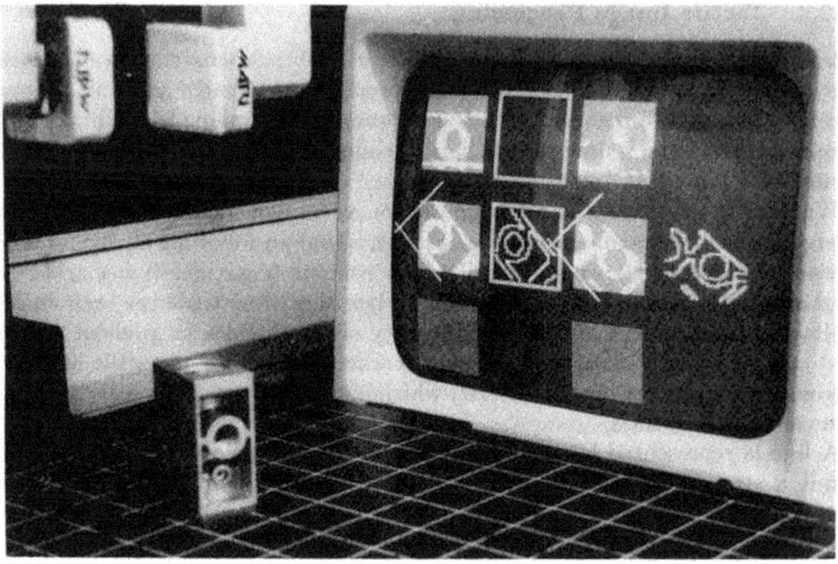

Figure 5.2: *The gripper mounted tactile sensors and example images obtained from the valve-block shown.*

matched feature points from the two IDs (identical TYPE and SIDE, and close-valued RI) and checking that the spatial relationships between the matched feature points from one ID are close to the feature points in the other ID. Spatial relationships are computed using the POS and, where appropriate, the RD components as well.

During matching, the list of lines in each ID (if present) is used to generate additional feature points. Each line is used to compute the position of a point which is the intersection of that line (or its extension) with a perpendicular drawn from the position of the first feature point to be matched. This technique allows lines to be used for matching even if the visible segments vary considerably due to the windowing effect.

The speed of the algorithm is highly dependent on the order in which searching is carried out. The preferred order is to match on shapes first, followed by corners and derived points. Corners are further ordered by distance from the centre of the image. Once an initial matching pair of features is found, the subsequent search tree is effectively pruned by applying the spatial relationship test on each following potential matched pair. The processing time is between one and four seconds, depending on the degree of accuracy required.

The screen image shown in Figure 5.2 shows the tactile images captured by a pair of MARI sensors from a rectangular valve block. The top pair of images is a reference pair, captured during a learning phase. The pair of axes superimposed on the images on the middle line represents the spatial offset between these images and those learnt previously. This offset data (the Correction Matrix, CM) is transmitted by the tactile processor to the robot controller.

The format of the Correction Matrix CM is:

$$CM = [x, y, \theta, ConfidenceLevel] \tag{5.1}$$

For example,

$$CM = [12, 8, 46, 9]$$

where:

12 is the offset in mm in the x direction of the gripper coordinate frame,
8 is the offset in mm in the y direction of the gripper coordinate frame,
46 is the rotation in degrees around the z axis of the gripper coordinate frame,
9 is the confidence level (an integer in the range 1 – 9).

5.5.3 Vision System

Although there are many vision systems available, their use for robot guidance is still uncommon. This is mainly due to high costs, insufficient performance and flexibility, and the effort needed in developing algorithms and achieving integration with the robot.

The approach taken by Joyce Loebl in this development was to base its initial work on its existing Magiscan II vision system, and to develop algorithms, software and hardware towards overcoming these problems. It was supported by Newcastle University in the area of VLSI.

5.5.4 Communications Infrastructure

The definition of a system-wide communications protocol based on the EIA RS511 (1393A) draft standard proved a vital step in achieving system integration. Although the implementation using V24 (RS232C) might be considered a potential limitation to the performance of the overall system, this was never in fact the case. Through an appropriate allocation of processing responsibility, and definition of the system interfaces, the requirements on communications bandwidth were kept to a minimum. The choice of V24 also greatly eased the integration with the existing components of the system.

The approach adopted by the project was based on a single-master many-slave star topology. The robot controller is the only system able to issue commands and request responses (e.g., status information and data) from the subsystems. None of the subsystems (tactile, gripper and vision) are able to initiate a communication transaction.

The implemented protocol includes a level of error checking, based on a CRC polynomial and automatic retransmission of erroneously received data.

Message Name	Command (C)/ Request (R)	Description
Vision System		
INIT_VISION	C	Initialise vision system
CALI_VISION	C	Calibrate vision system
VISION_CALI	R	Request calibration data
VISION_STAT	R	Request for status message
CLAS_VISION	C	Library selection
TRAI_VISION	C	Training of a new workpiece
IDEN_VISION	C	Command for object information
VISION_IDEN	R	Request for object data
Tactile System		
INIT_TACTILE	C	Initialise tactile sensor system
TACTILE_STAT	R	Request for status message
TRAI_TACTILE	C	Training of a new grasp position
IDEN_TACTILE	C	Command for object identification
TACTILE_IDEN	R	Request for object data
Gripper System		
INIT_GRIPPER	C	Initialise gripper system
VELO_GRIPPER	C	Select finger speed
GRIPPER_VELO	R	Request actual finger speed
GRIPPER_STAT	R	Request gripper status
POSI_GRIPPER	C	Command for finger positioning
GRIPPER_POSI	R	Request actual finger position
GRIPPER_FORC	R	Request actual grasp force
ROTA_GRIPPER	C	Command to rotate finger pad
GRIPPER_ROTA	R	Request actual pad angular position

Table 5.1: *Command set for vision, tactile and gripper systems.*

Table 5.1 lists the various subsystem commands which may be invoked via this protocol.

5.5.5 System Operation

In the system concept adopted by the project, the robot uses a plan view camera to locate workpieces in the work area, and then moves in to take a closer view of an individual workpiece using an arm-mounted camera in order to recognise the part and measure its position and orientation as accurately as possible. The recognition process is based on comparison with the results of an earlier learning phase during which the various possible views of the object are presented to the vision system.

The results of the visual recognition process, a predefined object face identifier, are sent to the robot controller which then constructs the appropriate recognition command for the tactile subsystem. This recognition command will include as an argument the predefined identifier of the expected tactile images.

As explained earlier, the tactile processor compares the active images obtained from the object being grasped with those captured and processed during the learning phase. The feature based matching algorithm confirms the accuracy of the match through the calculation of a confidence level, in the range 1 to 10. Most importantly for the control scheme, a correction matrix is calculated.

The correction matrix describes how the current images differ spatially from those learnt previously in terms of offsets in the x and y axis of the gripper coordinate frame and an angular rotation about the origin of the axis.

All the above information is transmitted back to the robot controller where it is used to calculate how the rotating finger pads and the end effector should be moved to allow accurate positioning of the workpiece. For example, if the correction matrix reports an angular displacement of the object of 10° then the pads on the fingers of the gripper (and the object gripped between them) will be rotated by this amount in order to arrive at the desired object orientation. Similarly the x and y offsets will used by the robot controller to calculate the required position for final placing of the object.

If the tactile system cannot confirm the object recognition performed by the vision system, then an appropriate recovery strategy is employed which might involve another attempt by the vision system to identify another alternative.

The practical implementation of the above scenario required the definition of a command set of messages for the transfer of information between the robot controller and the other subsystems. Table 5.1 lists these messages and their function.

5.6 Conclusion

The project described here was one of the first to integrate successfully a tactile imaging sensor together with commercially available automation equipment and sensors to produce a truly flexible workpiece handling system.

The technical approach adopted illustrated how intelligent sensing subsystems could be incorporated into a control scheme using a relatively simple communications infrastructure and command set. As more flexible communication interfaces become more readily available for automation equipment then implementations using, for example Fieldbus, should be even simpler to achieve.

The author believes that the requirement for tactile sensing in an industrial environment is still very real. The historic impediments to its adoption, i.e., resolution, ruggedness, cost, etc. are constantly the subject of pure and applied research and will be overcome in the near future. Perhaps the area of research that still presents the greatest obstacle and greatest promise is that concerned with the true integration of sensing and control functions.

Rather than a continual striving for greater and greater spatial resolution (the

MARI sensor is already comparable with that of human skin) it is the way in which the sensing capabilities are used that is important. It is the combination of movement and sensing (i.e., active sensing) which holds the promise of significant increases in "perceived" resolution. When we have mastered the integration of the techniques of adaptive control and tactile sensing, then we shall see tactile sensing predominating in our "lights out" factories of the future.

Acknowledgements

The author is indebted to the many people who have worked on Project 278. Particular mention must be made of Din Ghani, Steve Bell, Steve Redman, Peter Hage, Hans Lindner, Werner Glaess, Wolfgang Gerke, Charles Allen, John Finch, and Spyros Tzafestas.

The project described in this paper was partly funded by the European Community under the ESPRIT Programme. Particular mention should be made of Mrs. P. MacConnaill, head of ESPRIT CIM, who was instrumental in bringing together the ingredients of the project's success. The consortium enjoyed the support of all of her team particularly that of Dr. A. Ikonomopoulos.

References

Birk, J.R., *et al.* 1981. An orientating robot for feeding workpieces stored in bins. *IEEE trans. systems, man and cybernetics*, **SMC-11**(2).

Ghani, N. 1987. Tactile with vision: An integrated approach to robot sensing. *In:* Bernold, T. (ed), *Artificial intelligence in manufacturing*. Amsterdam: Elsevier.

Ghani, N. 1988. Visual and tactile senses in collaboration. *Sensor review*, **8**(4), 210–215.

Ghani, N., & Rzepczynski, Z.G. 1986 (Dec). A tactile sensing system for robotics. *In: Proc. int. conf. on intelligent autonomous systems (Amsterdam)*.

Raymond, C., Donath, M., & Olson, W.R. 1983. Problems of vision directed robots in an unstructured parts handling environment. *Pages 14/1–14/18 of: Proc. 13th int. symposium on industrial robots (Chicago, IL)*. Dearborn, MI: Robotics Inst. of SME.

Robinson, A., & Hage, P. 1986. Co-ordinating vision and tactile sensing with robotics. *Sensor review*, **6**(3), 130–132.

Chapter 6

Distributed Touch Sensing

Howard R. Nicholls and Nigel W. Hardy

6.1 An Alternative To Regular Array Sensing

Much of this book has been concerned with the processing of data obtained from regular arrays of sensing elements. Some of these arrays are organised as planar devices, whilst others are on curved surfaces, for example moulded into a finger shape. A common feature of such sensors is that they are mounted on robot fingers, and some may enable dextrous manipulation. This chapter presents an alternative approach to tactile sensing, through the use of a set of simple binary touch probes distributed around a flexible assembly machine.

The main issues of transduction and fabrication of tactile sensor arrays have already been discussed. The complexity of some sensor designs and their interfacing requirements dictates that the economic cost of a particular tactile sensor may be a significant proportion of the hardware costs of a robotic system. There may also be considerable processing requirements for the analysis and use of tactile data. Thus, it is worth considering approaches to solving contact sensing problems that may provide economic benefits through simplicity of physical design combined with suitable control software.

Many of the sensing scenarios described in the previous chapters involve mounting a tactile sensor array on a finger of a robot hand, so that the sensor is taken to the object to be sensed. This may be appropriate for certain tasks, but inappropriate or unnecessary for others. A European project to design and produce a prototype machine for flexible assembly has taken a radically different approach to tactile sensing from those presented so far. Although having a compliant end effector with multiple grippers and changeable fingers, the sensing in the parallel-jaw grippers is very basic, but can include the detection of the presence of parts. Grasped parts are constrained by specially shaped fingers, whilst positional uncertainties are dealt with by six degree of freedom compliance.

Some uncertainties in the assembly machine's environment are resolved through

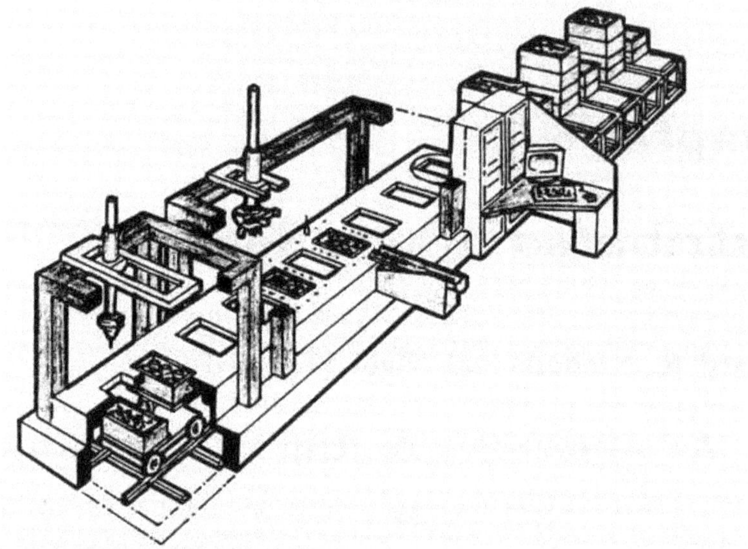

Figure 6.1: *The InFACT machine.*

tactile sensing, but in this scenario the objects to be sensed are taken to the sensor, the converse of many of the examples studied so far. Essentially, a set of binary touch probes are distributed around the machine in strategic places. Each probe can be read as a simple binary device, indicating whether or not there is something touching the probe. The set of probes form a non-orthogonal array of binary tactels, and can be repositioned by the manipulators in the machine. By monitoring state changes in the probes, and combining this with other sensory information and task-specific knowledge, a wide range of information can be extracted, from simple confirmation of parts feeding to object orientation error detection.

The remainder of this chapter describes the project and its use of touch probes in an automated flexible assembly environment.

6.2 The InFACT Project: Flexible Assembly

The InFACT Project (EU321: Integrated Flexible Assembly Cell Technology) was funded under the EUREKA/FAMOS initiative in flexible automated assembly systems. The project's nine partners come from the UK, France, Italy, and Austria.

The project consortium designed and built a flexible assembly machine (Figure 6.1) capable of putting together a wide range of products. The machine is particularly suited for automated assembly of small and medium sized light electromechanical goods, with minimal changeover effort required between different product assemblies (Loughlin, 1992). Examples of such assemblies include automotive light clusters, electrical wall sockets, and electrical meters.

The constructed machine (Figure 6.2) is a near-market prototype incorporating a number of novel techniques; consequently both research issues and economic considerations were encompassed in design decisions.

The InFACT machine is intended to provide users with versatility of operation at an economic cost, thus, many of the elements of the machine had to be designed to be generic across assembly tasks and product ranges to minimise product-specific tooling and maximise versatility. Another design consideration was that the generic components themselves had to be relatively cheap to manufacture, making the complete machine cost-competitive. The idea for the machine arose from work on the Windsor flexible assembly workcell (Williams & Lill, 1987); the technical director of that project, A. M. Williams, has managed the technical development of the InFACT machine and is responsible for many of the concepts embodied in the current machine.

The design and perceived role of the machine was influenced by two key factors particular to automated assembly (Seeley & Williams, 1990):

Assembly is generic: in many cases, assembly is a generic process. The way in which component parts fit together is a function of how certain key features mate, and is independent of the size, shape and function of the component parts. Consequently, most assembly operations can be described using a small number of generic processes.

Assembly machinery should be flexible: there is an increasing requirement for product variation within manufacturing industry. Manufacturers are now seeking automated assembly systems which can cope with frequent product modifications, handle similar product items, and which minimise retooling for product changeovers. This indicates a trend away from dedicated or semi-flexible automated assembly systems towards fully flexible assembly systems.

These factors influenced actuation, sensing, and control. Mechanical actuators were developed that were able to handle a wide range of assembly operations. The combination of a high level machine supervision system, a subservient sensor integration system, and versatile actuators enable the machine to deal efficiently with the frequent product changes becoming characteristic of manufacturing industry.

6.3 The InFACT Machine Configuration

The InFACT machine is divided into three zones: supply, assembly, and power operations. Small parts such as nuts and screws are presented to the assembly and power zones by linear vibratory feeders, whilst larger parts or subassemblies are pre-loaded into pallets in the supply zone and brought into the working area

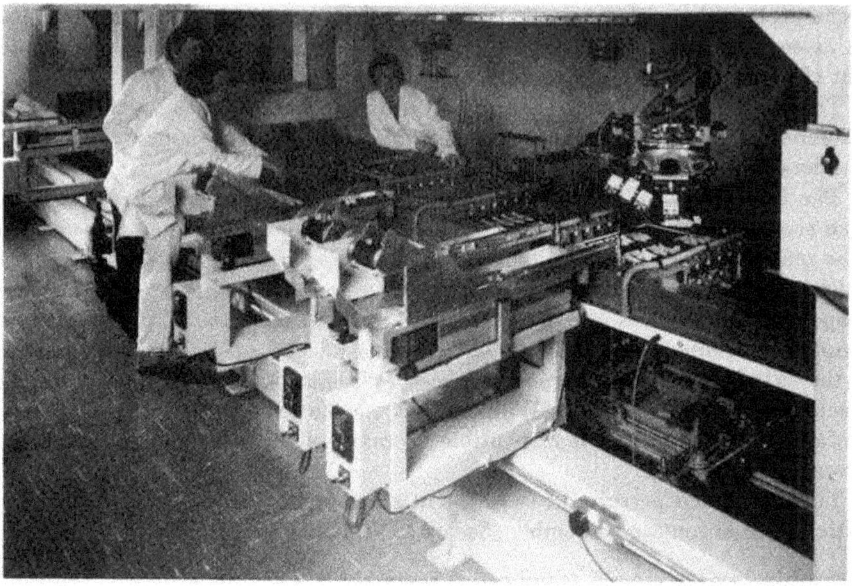

Figure 6.2: *General view of the InFACT machine manipulators.*

on a track-based shuttle system. The shuttle is also used to transport completed assemblies to an output stack of pallets.

Assembly operations are carried out by two gantry-based Cartesian motion manipulators, known as the main and the power operations manipulators (Figure 6.2). The four degrees of freedom (x, y, z, θ_z) main manipulator is situated in the central assembly zone and is designed to carry an end effector with a turret capable of holding four grippers. The end effector and gripper mechanism have been designed and are under development. Grippers are usually pneumatically actuated parallel jaw devices with interchangeable fingers; both grippers and fingers can be changed automatically during an assembly sequence. The end effector functions as a remote centre compliance device, whose stiffness can be altered at any time during an assembly cycle.

Many of the uncertainties about parts and sub-assemblies are engineered out through suitable cost effective fixturing, finger design, and pre-assembly inspection. Sensing on the end effector and on the grippers mainly relates to compliance settings, gripper types, and the state of internal counters. A variety of simple binary switches is available to determine the success or failure of grasp and release operations.

The second gantry-based manipulator has three degrees of freedom (x, y, z) and is used for heavier assembly operations requiring the use of force, such as press fits and screw driving. Different power tools are attached according to the type of power operation required, and tools may be changed during an assembly cycle. Again, the sensors in the power operations zone are fairly simple devices, put to maximum effect through a hierarchical control system.

The InFACT machine has a distributed control architecture, running on a network of INMOS transputers (INMOS, 1987). The machine is controlled by a software supervision system, which maintains and coordinates overall machine activity, and a sensor integration system, which manages all the sensor data. Low level control and sensory data gathering are dealt with by processes called action modules. Both actuation control and sensing control is highly modular, each action module presenting a simple command-level interface to the supervisor and to the sensing system. Control algorithms for specific devices, such as manipulators, the end effector, the materials handling system, etc. are each implemented on a transputer.

The supervisor is under the control of the human-computer interface (HCI) which runs under the Unix operating system on a PC. This computer also connects to factory communications facilities, enabling the integration of the machine into a factory-wide production system.

The simple, general purpose, low-cost sensing approach contrasts with that of some of the other research described in this book. Successors of the prototype machine are intended to be in place in factories from 1993. The design reflects near-market automated assembly solutions, rather than indicating what may be feasible in factories in 5 - 10 years time, when the requirements of flexible manufacturing and assembly systems will be more demanding of manipulative tasks.

However, the InFACT machine will still have to handle for example, feeding errors, positional uncertainties, and orientation errors. These are resolved through use of the binary touch probes and coordinated sensing operations, which are described in detail in section 6.6.

6.4 Sensing Philosophy and Sensor Integration

The sensing philosophy in the InFACT machine involves having simple, cheap physical sensors which are controlled in a uniform way from a coordinating subsystem, called the sensor integration system. The sensor integration system performs various operations upon sensor data, presenting sensory information to the supervisor in a suitable format when it requires it. This approach has several benefits, since it relieves the supervisor of the burden of directly managing a large volume of sensory information, and the use of cheap physical sensors helps to keep down the overall cost of the machine.

The sensor integration system accesses the physical sensors in the machine via the action modules. The raw data are processed and transformed by the sensor integration system before it is passed on to the supervisor. Such a strategy enables abstractions of sensor data, presenting sensors as *virtual* entities whose physical sensing details are hidden from the higher levels of the control system (Rowland &

Nicholls, 1992). Sensor values can be combined, and also coordinated temporally.

The system is designed to accommodate five classes of sensing facilities, providing a rich source of information about the machine's activities (Nicholls & Rowland, 1992). The most primitive sensor class, *state*, allows the current value of a virtual sensor to be obtained, whilst the *event* class can provide notification that a change of value of a state sensor has occurred. A higher level sensor class, *trigger*, enables the measurement of the current value of a sensor when another sensor changes its value, i.e., a temporal combination of *state* and *event* sensing. The design supports the logging of sensor data over time, and the production of simple statistics, known as *trend* sensing. A fifth sensor class, *drift*, is used for detecting when trend statistics cross a threshold. This can be used to indicate that the machine is trending towards an undesirable condition, perhaps due to tool wear. There is also a facility for checking the integrity of the sensing system.

The supervisor obtains all sensory information from the sensor integration system via a command-response protocol (Hardy *et al.*, 1992), and employs the sensed data in assembly task generic procedures where uncertainties need to be resolved. Sensor use strategies are related to situations where there will be uncertainties in specific machine environment parameters; typically, this involves resolving parts feeding problems, but sensing is also used to confirm the correct completion of assembly sub-tasks. The supervisor combines sensory processing, performed by the sensor integration system, with actuation using task-specific knowledge. It is the integration of information from virtual sensors with task-specific knowledge that provides the versatility of control necessary for the fulfillment of the machine's functional requirements.

6.5 The Touch Probes

Certain problems, mostly associated with component feeding, were identified that required sensing to resolve uncertainty. Various sensing solutions to these problems could have been adopted; for example, a vision system could have been employed to resolve orientation uncertainties. Since design requirements included versatility and low cost, such solutions were not adopted. Instead, these problems are solved by using a combination of low-cost contact position sensors, appropriate commands from the sensor integration system, and task-specific knowledge in the supervisor.

The contact position sensors are binary touch probes (Figure 6.3). When an object comes into contact with the probe stylus, displacing it by more than 0.1 mm, the binary sensor output indicates contact and continues to do so as long as the contact is maintained. When the object is no longer touching the stylus, the spring-loaded stylus returns to its upright sentinel position and the output changes to indicate there is no contact. The probe is sensitive in 2.5 dimensions: x, y, and $-z$; i.e., it can detect being knocked sideways and also that it is being pushed vertically downwards.

Figure 6.3: *A sentinel touch probe from the InFACT machine.*

The principle of operation is that displacement of the stylus increases the air gap between the two halves of a ferrite pot core, causing a change in inductance of a coil within the core. Conditioning electronics within the probe detect this change and generate a binary output signal but no measurement of magnitude of deflection or direction of movement is made.

The touch probes can be located in any of 32 sockets distributed around the InFACT machine. Each probe is mounted on a jack plug and has a grasp pick-up area, enabling the main manipulator to grasp the probe, unplug it from its socket and discard it or reposition the probe in another location. The touch probes located at strategic sites around the machine can be considered to be a non-regular array of binary tactels, whose spatial distribution can be altered during the machine's assembly cycle, or more usually, between batches of different products to be assembled. Probes are usually located in front of the linear feeders, to enable verification of small parts feeding, and several rows of four probes are normally sited in the assembly zone, enabling location of key features on parts, angular orientation measurement, etc.

The sentinel probe concept was originated by Thorn-EMI in the Windsor workcell (Williams & Lill, 1987), and has been refined and packaged by MARI Advanced Microelectronics, of Newcastle, UK, as part of the InFACT project. The versatility of the relocatable touch probe concept fits well with the requirements of the InFACT project. The next section describes how the simple binary output is used to resolve uncertainties in assembly operations through the use of sensing functions provided by the sensor integration system and task-specific knowledge available to the supervisor.

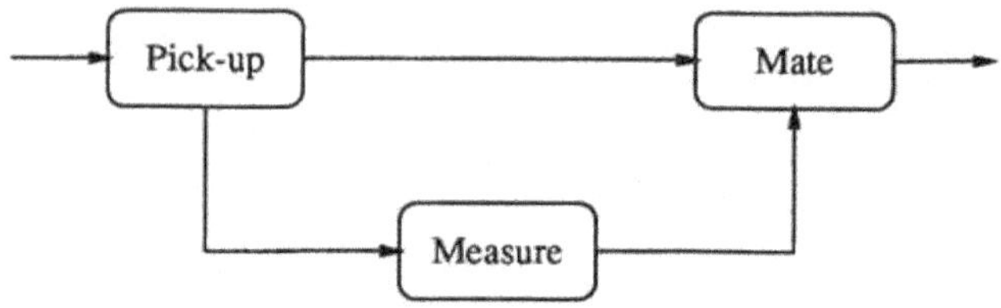

Figure 6.4: *A simplified railroad diagram of task classes.*

6.6 Assembly Sequence Specification

Following both previous work (Liebermann & Wesley, 1977; Barnes *et al.*, 1983; Williams & Lill, 1987) and the InFACT philosophy that assembly is generic, the applications programming or assembly sequence specification on the machine is accomplished by the use of so-called task generics. These can be considered as parameterised subroutines (to follow a traditional procedural programming analogy), as task templates, or (conveniently) as forms to be filled in at the user interface level. At run time, execution of a task generic is in many ways similar to a call to a subroutine with appropriate parameters. For the purposes of off-line program verification it is important that the pre- and post-conditions for the execution of tasks be clearly established and enforced. The characterisation of a task generic as a template is valuable here. The whole of the assembly sequence specification activity is presented at the HCI as a form filling exercise and part of this entails the selection of suitable task generics and the provision of specific data for each invocation.

The design of the task generics incorporates the following points:

- The InFACT design team's expertise in assembly automation is built into the generics, both in the selection of an appropriate and acceptably comprehensive set of generics, and in the internal implementation of the generics such that they will be effective and reliable.

- Consideration was given to the design criteria of the whole machine such that effective and rational use of the machine is encouraged or enforced.

Task generics are grouped into classes according to their high level functionality. To create an assembly program, the user must (among other things) specify a sequence of task generics. The program verification performed by the supervisor places constraints on the legal sequencing, not at the level of individual generics but at the class level. The legal sequences may be represented in "railroad" diagrams and Figure 6.4 shows a much simplified example.

It indicates that a legal task sequence must begin with a task from the class `Pick-up`, that it must conclude with a task from the class `Mate` and that optionally a task from the class `Measure` can occur between them. Classes contain a range of tasks to cover both the mechanical capabilities of the InFACT machine and the variations in activity which the machine will need to handle. The mate class contains, for example, generics which operate with the main manipulator and ones which operate with the power operations manipulator. For each, there are variations which cope with sensed and unsensed parts mating, with single point insertions and with double point insertions, etc.

6.6.1 Measurement Tasks

The function of a measurement task is to remove an error resulting from locational uncertainty about a fed component. This error may be in both position and orientation. It should be noted that in tasks of any class, sensing may be used for automatic verification of the success of actions or for quality assurance purposes. The simple sensors in the end effector (section 6.3) are particularly important here.

Consider a task sequence based on the simple situation in Figure 6.4. Here, any deviation from the nominal position of the fed part will result, after execution of the chosen `Pick-up` task, in a spatial relationship between the component and the gripper which is not the nominal or expected one on which the correct operation of the `Mate` relies. If variation in the feed location is unavoidable, a `Measure` task can be included in the sequence. This will determine the deviation from the nominal spatial relationship for each individual component, thereby enabling calculation of a correction value which can be applied to all subsequent locations of the arm.

For example, consider the mating of a control knob onto a spindle where a key on the knob and a keyway in the spindle must engage (Figure 6.5). We will assume that the initial orientation of the spindle is known and reliable but that the initial orientation of the knob cannot be guaranteed. The orientation of the gripper during the `Mate` task must be specified assuming some particular relationship between the knob and gripper. If the feed orientation were consistent, the orientation of the gripper during `Pick-up` could be chosen such that this relationship would be achieved. If the orientation is not consistent and a `Measure` task must be used, the `Mate` is specified based on a nominal feed orientation. The `Measure` task measures the deviation from this and can apply its inverse as a correction to the location for the `Mate`.

More formally and generally, the framework used can be described with homogeneous transformations ($\mathbf{T}$) to represent the spatial relationships of coordinate systems set up to denote features in the system. The location of a feature is the position and orientation of its coordinate system relative to some other stated system.

Relating all locations to the "world" coordinate system of a manipulator, we can express the location ($\mathbf{L}_{c1}$) of the component to be picked up (from a pallet in this

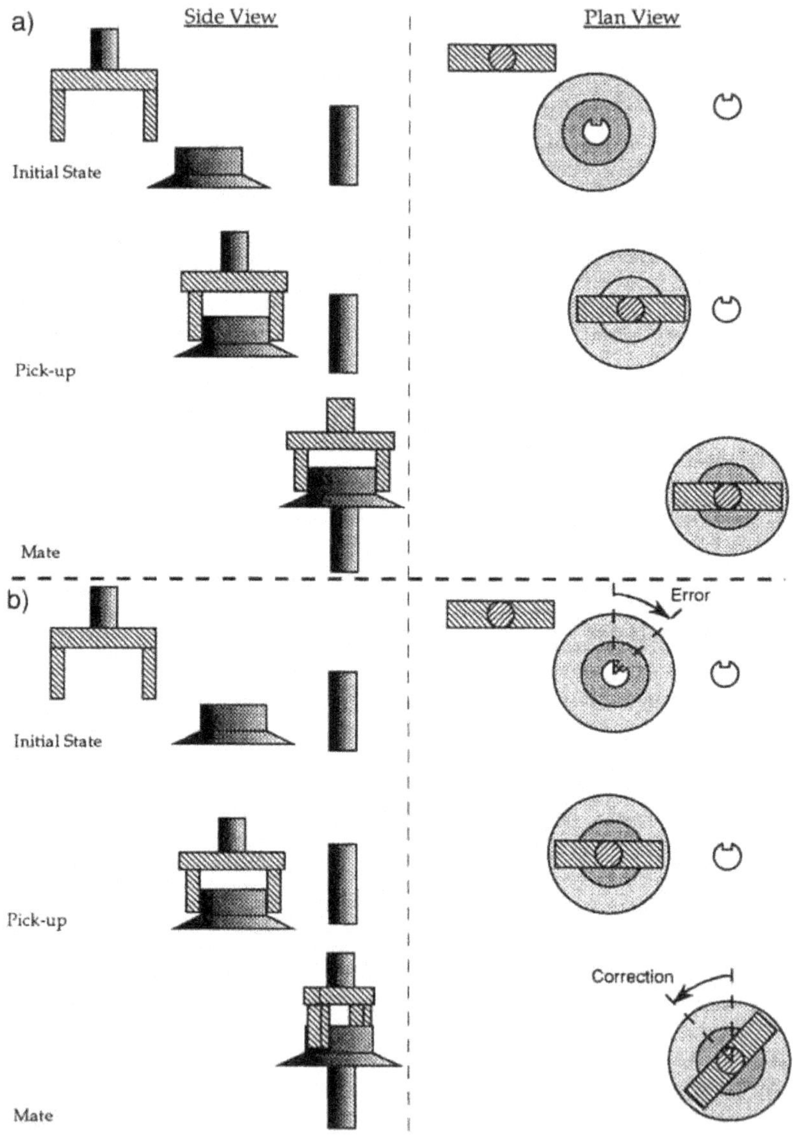

Figure 6.5: *Side and plan views of the knob assembly where: a) the knob is fed at the nominal orientation, and b) the error in feed orientation is corrected during the mate.*

example; an analogous system applies to components fed by linear feeders) as:

$$\mathbf{L}_{c1} = {}_w\mathbf{T}_p \; {}_p\mathbf{T}_o \tag{6.1}$$

and the location ($\mathbf{L}_t$) of the tool point as:

$$\mathbf{L}_t = {}_w\mathbf{T}_a \; {}_a\mathbf{T}_g \tag{6.2}$$

where the transformations are between coordinate systems at:

w – the manipulator world origin;
p – the component's pallet origin;
o – the component origin;
a – the end of manipulator arm;
g – the tool point.

It should be noted that:

1. it is assumed that the tool point is to be brought into coincidence with the component origin for pick-up (a component grasp point, defined relative to this origin could be used with appropriate enhancements to the method), and

2. ${}_a\mathbf{T}_g$ could be calculated from a variety of transforms allowing for exchangeable grippers, fingers, special tools, etc.

All transformations in these expressions are provided by the user with the exception of ${}_w\mathbf{T}_a$. The supervisor can control ${}_w\mathbf{T}_a$ when commanding the manipulator and thereby make $\mathbf{L}_t = \mathbf{L}_{c1}$, achieving pick-up. Uncertainty in the location of feeding (likely to be in ${}_p\mathbf{T}_o$) will result in the tool point and the object origin not being coincident after the pick-up, i.e., ${}_g\mathbf{T}_o$ is not the identity transform.

All measurement tasks require that an alignment feature on the object be brought into coincidence with a coordinate system established at the point of a probe sensor. The probe frame location (${}_w\mathbf{T}_s$) is known to the supervisor through a combination of initial machine calibration (which establishes the precise location of all probe sockets) and knowledge of the size of the probe. The location ($\mathbf{L}_{c2}$) of the alignment feature at this time is expressed by:

$$\mathbf{L}_{c2} = {}_w\mathbf{T}_a \; {}_a\mathbf{T}_g \; {}_g\mathbf{T}_o \; {}_o\mathbf{T}_f \tag{6.3}$$

where f is the coordinate system at the alignment feature and ${}_o\mathbf{T}_f$ is specified by the user. ${}_w\mathbf{T}_a$ is reported by a trigger sensor (section 6.4) based on a manipulator position state sensor and a probe contact event sensor. The pick-up error (${}_g\mathbf{T}_o$) is therefore the only unknown and can be calculated by:

$${}_g\mathbf{T}_o = {}_o\mathbf{T}_f^{-1} \; {}_w\mathbf{T}_s \; {}_w\mathbf{T}_g^{-1} \quad . \tag{6.4}$$

In all manipulations subsequent to the measurement task, the grasp error transformation can be taken into account. As part of a mate task, for example, where a coordinate system (*m1*) relative to the grasped component origin is to be made coincident with a coordinate system (*m2*) relative to the coordinate system (*pa*) of a partial assembly on a pallet to achieve a desired spatial relationship, the expression for the fixed part is:

$$\mathbf{L}_{c3} = {}_w\mathbf{T}_p \; {}_p\mathbf{T}_{pa} \; {}_{pa}\mathbf{T}_{m2} \tag{6.5}$$

and this must be made equal to that for the mobile part, which is:

$$\mathbf{L}_{c4} = {}_w\mathbf{T}_a \; {}_a\mathbf{T}_g \; {}_g\mathbf{T}_o \; {}_o\mathbf{T}_{m1} \tag{6.6}$$

where again the only variable is ${}_w\mathbf{T}_a$, the manipulator position, which can be controlled.

6.6.2 Measurement Task Design

Using the framework above, the following specific points must be considered in the design of an individual `Measure` task:

1. The characterisation of the feed position uncertainty.
2. The definition of an alignment point; the components of the coordinate system at this point may have special significance, for example, the orientation may be important in the execution of the task.
3. The definition of an appropriate trajectory to ensure collision with the probe or probes; it should be noted that the alignment point need not be the point at which a probe collision occurs and that more than one collision can occur. The alignment point's position must be established relative to the fixed position of the probe. Therefore, a number of collisions may be used to estimate the alignment point position, or a number of probes may be used for collisions.
4. The specification of the parameterisation data required from the user; the position of the alignment point relative to the component origin will always be required, although constraints may be placed on its value. The trajectory for some tasks may require additional information.

6.6.3 Uncertainties

The selection and design of the measurement tasks is based on the belief that, for a well designed assembly application on an integrated assembly machine, the positional uncertainty associated with a component when it is fed will be well understood. An analysis of the current machine configuration and feeding mechanisms and of common manufacturing procedures leads to the conclusion that the following types of positional error may exist in fed components:

1. A 180° rotation about the vertical axis; this situation arises from the use of linear vibratory feeders in the machine (Redford, 1989). These devices are extremely efficient and cost effective at presenting small parts, sometimes of considerable intricacy, in a desired orientation. It is well understood that they often cannot discriminate between alternative orientations of objects which are nearly rotationally symmetric. A policy is therefore adopted which selects one orientation as "correct" and the other as "incorrect". Rejection of incorrectly fed parts is unnecessary since the main manipulator can easily correct for the error without loss of speed. Distinguishing the orientations requires that there be an asymmetric feature which can be detected. Detection can be rapid.

2. An unknown rotation about the vertical axis; this situation can arise with parts fed by linear vibratory feeders (where no orientation feature is usable by them) or with pallet fed parts which cannot be or are not reliably orientated when fed. Measurement of the rotation from some nominal orientation requires that an asymmetric feature be present which can be detected. It is likely that searching for this feature will be a relatively slow process.

3. A translation along either the x or y axis; this translation is constrained to be along one of the horizontal axes of the component. This is done for simplicity but represents no major loss of applicability since the user has control over the positioning of the components axis system. This uncertainty is only likely to exist with pallet fed parts. An edge, of reasonable length, roughly perpendicular to the axis along which uncertainty exists is desirable for this measurement.

4. x and y translations and rotation about the vertical axis; here, both translational and rotational certainty about the location of a component lying on a surface is absent. Two edges are required for this measurement.

5. The (x, y) position of the centre of a vertical hole; this is used in "hole-on-peg" applications where accurate placement of a hole in a component carried by the manipulator is required. It is assumed that if accuracy is crucial in mating, the diameter of the hole will be known. Measurement is likely to be relatively slow.

It is important to note that in many cases there will be "no" positional uncertainty. That is, the uncertainty will be within the tolerance of the machine and the application software. This is particularly true when end of arm compliance and well designed fingers or other capture mechanisms and good capture strategies are used. Such acceptable levels of uncertainty are expected to be the norm in applications and the use of measurement tasks will be kept to a minimum due to the time penalties most of them impose.

6.7 Two Example Measure Tasks

The definition of the alignment point, the trajectory, and the data required from the user will be examined for two of the types of uncertainty described above. The trajectories are described in terms of the points through which the nominal alignment point must pass. The data considered are only those essential to the algorithms. Additional information, such as arm speeds, which are common to all classes of task would also be required.

6.7.1 The "Translate Until" Task

For the correction of a translational error along the x or y axis, the so-called "translate until" task is used. Essentially, the task causes the component to be translated in an appropriate direction until a single probe collision is detected.

The determination of the alignment point ${}_o\mathbf{T}_f$ has two aspects. The translational components must be provided by the user and should be chosen to ensure an unambiguous collision at a suitable point. The rotational components are used to determine the angle at which the component will be presented to the probe. Only rotation about the vertical axis is permitted and can be made analogous to the four cardinal points of the compass.

The translational components of the probe position ${}_w\mathbf{T}_s$ are known and fixed. For simplicity, the rotational component about the vertical axis is used to control the direction from which the (suitably oriented) component will approach. Again, the rotation can be considered in terms of the four compass points. It is now possible to define the two points in the trajectory. They must be on either side of the nominal collision point, sufficiently far away to allow for the maximum permissible translational error (set to ± 7.5 mm in the InFACT machine) and are therefore simple offsets from ${}_w\mathbf{T}_s$.

To parameterise this task the user must provide:

1. the translational offsets from the component grasp position to the alignment point (for most regular components this will be in a single dimension),
2. the orientation of the part with respect to the probe, and
3. the direction of approach to the probe.

In operation the arm will move to position the component at the first trajectory point, the trigger sensor will be armed, the component will be moved in a straight line to the second trajectory point, and then continue on its way while the latched arm position at collision is obtained.

6.7.2 The "Find Hole Centre" Task

In this task the alignment point is the centre of the hole. Clearly, no collision with the alignment point can be contrived.

The orientation component of the alignment point has no special significance and is taken to be the same as that of the grasp point. The orientation of the probe point is again fixed to reflect the approach direction of the component to the probe.

The trajectory requires six points. The first is a vertical offset from coincidence of the probe and alignment points. That is, the component is held over the probe. The second point brings the probe and nominal alignment points into coincidence, i.e., the component is placed over the probe. The next three points are calculated to achieve three widely spaced collisions with the hole circumference. The hole radius is required for their calculation and it is used, along with the latched arm positions at the the collisions to calculate the observed centre of the hole and thus establish the grasp error. The final point on the trajectory lifts the component back off the probe from where it can be taken on towards the mate position.

To parameterise this task the user must provide:

1. the hole centre position with respect to the component grasp point,
2. the direction of approach to the probe,
3. the radius of the hole, and
4. a height for the first and last trajectory positions.

6.8 Conclusion

The correction of grasp errors caused by unavoidable locational errors in feeding is an important requirement in the InFACT machine. Sensing is obviously part of the solution since automatic mechanical corrections are not always possible or time efficient. It was however important, for economic reasons, that sensing strategies imposed as little time penalty as possible on the assembly cycle and that the sensors be moderately cheap and robust.

The strategy adopted reflects the fact that in automatic industrial assembly uncertainties will not be great. An assembly cell is a relatively highly constrained world where a machine does not have to accommodate unknown objects, objects in entirely unknown positions, objects of different types mixed in a single feed area, or objects behaving in novel ways. Moreover, if some anomaly is detected, it is not necessary to investigate and understand it, since economic considerations dictate that abandoning that individual assembly (possibly putting it aside for human attention) is the correct response. The machine does not therefore require the sophisticated and expensive sensing which an autonomous rover vehicle or a remote manipulator hand working in an unknown environment might have. Simple, reliable and cheap sensing can be used in an industrial situation, together with existing data about the world (in which we can have a high degree of confidence) to solve problems in an entirely appropriate way.

References

Barnes, D.P., Lee, M.H., & Hardy, N.W. 1983. A control and monitoring system for multiple-sensor industrial robots. *Pages 471–480 of: Proc. 3rd int. conf. on robot vision and sensory controls (Cambridge, Mass.)*. Kempston, UK: IFS (Pubns).

Hardy, N.W., Nicholls, H.R., & Rowland, J.J. 1992 (Nov). The design of sensing commands in the InFACT assembly machine. *In: Proc. 23rd int. symp. industrial robots.*

INMOS. 1987. *The transputer architecture.* INMOS, Bristol, UK.

Liebermann, L.I., & Wesley, M.A. 1977. Autopass: an automatic system for computer controlled mechanical assembly. *IBM j. research and development*, **21**(4), 321–333.

Loughlin, C. 1992. *InFACT: Project, concepts, machine.* Bradford, UK: MCB University Press.

Nicholls, H.R., & Rowland, J.J. 1992. Sensor integration in automated assembly. *Pages 24.5–24.8 of: Proc. Canadian conf. on industrial automation (Montreal).* Montreal, Canada: IAIA.

Redford, A.H. 1989. A materials handling system for flexible assembly. *In: Proc. int. conf. on manufacturing science and technology (Sweden).*

Rowland, J.J., & Nicholls, H.R. 1992. Virtual sensors in flexible assembly. *Pages 241–246 of: Proc. SICICA '92 – IFAC symposium on intelligent components and instruments for control applications (Malaga, spain).* Malaga: IFAC.

Seeley, C., & Williams, A.M. 1990. Generic assembly – the InFACT way. *In: Proc. 11th int. conf. on assembly automation (Dearborn, MI).* Kempston, UK: SME/IFS (Pubns).

Williams, A.M., & Lill, B.H. 1987 (May). Commercially available flexible assembly cell. *In: Proc. BRA-10, 10th British Robot Assoc. Conf.*

Chapter 7

The Human Tactile System

Letty Moss-Salentijn

7.1 Introduction

The human upper extremity presents an exquisite example of a structure that is capable both of delicate tactile sensing and of precisely adjusted, coordinated movements. It is probably not necessary or even useful in the design of robotic arms to duplicate all biological structures that are responsible for the efficiency of the human hand and arm. Nevertheless, some of the functional components of this successful model, if properly understood and used, may be of advantage in robotic design. This chapter describes in some detail the various sensory components of a human extremity.

The human extremity in its simplest form may be considered as a series of skeletal leverarms to which the muscles responsible for their movement are attached. The musculoskeletal elements are surrounded by connective tissue, which provides mechanical and metabolic (vascular) support. The extremity is covered with an outermost layer of skin.

Several "nerves of distribution" run in the connective tissue. They are part of the complex neural circuitry of the body and have both efferent (motor) and afferent (sensory) components. On the following pages, one part of the circuitry will be described: the general somatic innervation which conveys different sensory stimuli from the body surface and from the deeper tissues to the central nervous system.

A continuous, coordinated inflow of sensory information is essential for the normal function of an extremity. To this end sensory receptors are located both in or near the outermost layer of skin (exteroceptors), as well as deep within the muscular and connective tissues (proprioceptors). The purpose of this chapter is to provide a description of the exteroceptors in the hairless (glabrous) skin and of the proprioceptors in muscles, tendons, and connective tissues around the joints, with an emphasis on their structure, distribution and function. This description, in addition to whatever inherent value it may possess, will serve as a structural and functional basis for Chapter 8 by Lederman and Pawluk, which considers lessons from biological touch for robot tactile sensing.

7.2 The Peripheral Nerve

Sensory end organs, as their name implies, are specialised terminals of peripheral nerves. Within the nerves, long cellular extensions, or axons, of first-order receptor neurons are bundled together. The cell bodies of these neurons are located alongside the spinal cord, in dorsal root ganglia. Information from the end organs is conducted via the axons of the first-order neurons to second-order neurons in the central nervous system for further processing.

7.2.1 Organisation of a Peripheral Nerve

The structure of a peripheral nerve is complex, as it contains the axons of numerous neurons that have different tasks and/or target tissues. In order to accommodate these differences, groups of axons travel in separate bundles or fascicles within the peripheral nerve. The organisation into fascicles appears to provide a near-complete segregation of motor axons and sensory axons. Thus, it has been possible, by intraneural microstimulation of a selected peripheral nerve in the upper extremity, to demonstrate that 20 - 40% of the fascicles in that nerve carry general somatic afferents and project to the skin (Schady *et al.*, 1983).

The fascicles are separated from their neighbours by connective tissue elements: collagen fibres, cellular elements and small blood vessels. The collagen fibres are arranged in a well-organised framework that provides mechanical support for the nerve (Ushiki & Ide, 1990). In the more proximal segments of larger nerves, connective tissue may comprise as much as 85% of the cross-sectional area of the nerve (Slingluff *et al.*, 1987; Sunderland & Bradley, 1949).

Each fascicle is surrounded by a perineurium: a sheath of several layers of flattened cells of connective tissue origin, alternating with thin layers of collagen fibrils, that surround the fascicle like a support stocking. The perineural cells are attached to each other by tight junctions and form a fairly impenetrable cellular insulation layer ("blood-nerve barrier"). Thus, in addition to providing mechanical support, the perineurium effectively protects and regulates the internal milieu of the fascicle. Within the fascicle are the nerve fibres, separated from each other by minimal quantities of connective tissue elements: collagen fibres, fibroblasts and small blood vessels (Figure 7.1).

A nerve fibre is an axon supported by Schwann cells, that are lined up along the length of the axon in single file. The Schwann cells are surrounded by a basement membrane, a delicate sheath of collagen and macromolecules. Depending on the characteristics of the axon (see below), a nerve fibre is either myelinated or unmyelinated. In the former case, the Schwann cells may be wrapped around a single axon, forming a lipid myelin sheath which is composed of many concentric

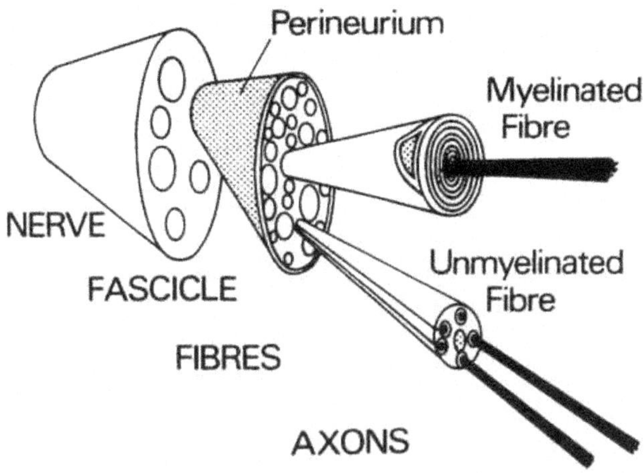

Figure 7.1: *Organisation of a peripheral nerve. Each subunit is further magnified to show additional detail.*

layers of Schwann cell membrane. The myelin sheath acts as a high-resistance, low-capacitance insulator (Andres & v.Düring, 1973; Krstić, 1978; Terzis & Smith, 1990).

Many Schwann cells are present along the length of a nerve fibre. Where two Schwann cells meet, the myelin sheath is interrupted. The site of interruption is a node of Ranvier. The interval between adjacent nodes of Ranvier is an internode, which ranges in length from 0.18 – 1.83 mm (Carlsen & Behse, 1980; Behse, 1990; Barchi, 1980). The length of an internode, and thus the speed of impulse conduction, are directly related to the diameter of the nerve fibre (which includes the myelin sheath): the greater the fibre diameter, the faster the speed of conduction.

In unmyelinated nerve fibres, several small-diameter axons are supported in infoldings of the cell membranes of a single file of Schwann cells. While the speed of impulse conduction in unmyelinated nerves is considerably slower than in myelinated nerves, the speed increases as a function of the square of the axons' diameter (Barchi, 1980).

7.2.2 Classification of Afferent Nerves

The functional and morphological differences of nerve fibres historically have led to various attempts at classification of those fibres. Since the early observations that afferent fibres to muscles are larger in diameter than those to the skin, several classifications have been proposed based on fibre diameter (see: Terzis & Smith (1990) for a review).

The classification that will be used in this chapter was developed by Erlanger and Gasser (1937; Gasser & Grundfest, 1939). It applies to all fibres in the peripheral nerves (somatic afferent and efferent, as well as autonomics), and divides them on the

Class	Myelin Sheath	Diameter (μm)	Conduction Speed (ms^{-1})	End Organ
Aα	+	12 - 20	70 - 120	Muscle spindles (primary endings) Golgi tendon organ
β	+	6 - 12	35 - 75	Mechanoreceptors
(γ)	+			(muscle spindles - motor axons)
δ	+	1 - 5	5 - 30	Thermoreceptors (cold) Nociceptors
C	-	0.2 - 1.5	0.5 - 2.0	Thermoreceptors (warm) Nociceptors (polymodal)

Table 7.1: *Classification of afferent nerve fibres serving sensory end organs.*

basis of the related parameters of fibre diameter and speed of conduction. The fibres are classified in groups A, B and C. The fibres in groups A and B are myelinated, those in group C unmyelinated. A-fibres have the greatest diameters (the myelin sheath is included in the diameter) and highest conduction speed, while C-fibres have the smallest diameters and lowest conduction speed (Table 7.1) (Daniel & Terzis, 1977; Darian-Smith, 1984; Hensel, 1973; Martin, 1991).

7.2.3 Afferent Fibre Distribution in a Peripheral Nerve

The information that is available on the actual frequencies of the different types of afferent fibres is very limited and based largely on single fibre analysis by microelectrode recording of somatic afferent nerves. It can be shown that a somatic afferent fibre responds selectively to changes in its environment: mechanical deformation, changes in temperature, or potentially injurious stimuli. Recordings of a single fibre's discharge patterns and measurement of the conduction speed of the action potentials provide reliable estimates of that fibre's diameter (Darian-Smith, 1984).

An example of such a study, conducted on one of the purely sensory nerves in the lower extremity of a young boy, has produced the typical frequency histogram of afferent fibres in a peripheral nerve (Boyd & Davey, 1968; Ochoa & Mair, 1969) and is illustrated in Figure 7.2.

7.2.4 Nerve Conduction

Information is conveyed from the periphery to the central nervous system by a rapidly propagated travelling wave of electrical excitation ("nerve impulse") along the cell membrane of an individual axon.

The resting neuron normally maintains a "resting potential", an electrical gradient of about 70 mV. This condition is maintained importantly by gated sodium

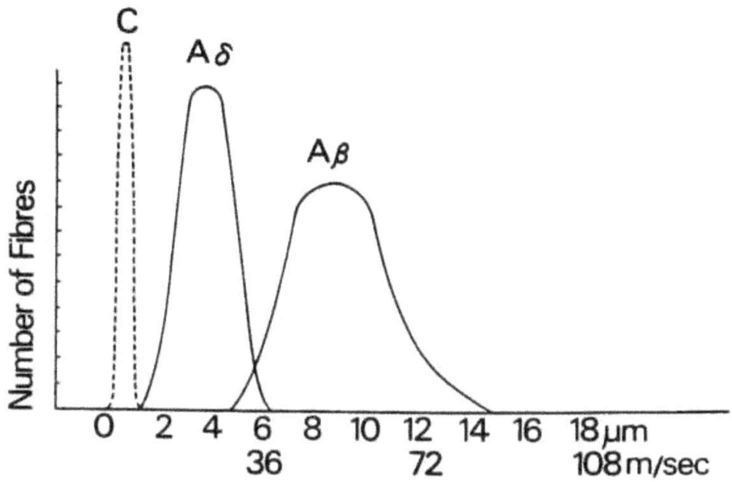

Figure 7.2: *Relative frequency of different classes of afferent fibres in a purely sensory nerve (adapted from Boyd & Davey, 1968).*

channels in the cell membrane, which actively prevent the sodium ions that are present in higher concentrations outside the membrane from re-entering the cell.

The peripheral sensory ending of a receptor neuron is excited when modality-specific (e.g., chemical, pressure or stretch) gated sodium and potassium channels open, allowing an influx of sodium ions and some efflux of potassium ions. This results in a temporary depolarisation of the membrane and the production of receptor potentials. The magnitude of the receptor potentials and the subsequent discharge frequency of the nerve fibre increases with the increasing intensity of the peripheral stimulus. When the receptor potentials reach the first node of Ranvier, an integrated potential is generated. As soon as the integrated potential reaches a required threshold, the spike of an action potential is produced. The action potential travels without decrement along the axon membrane as it activates sequentially the voltage-gated sodium and potassium channels of the axon, thus altering the local cell membrane permeability and leading to a reversal of membrane polarity. Following the passage of the impulse, the membrane repolarises during a brief refractory period of a few milliseconds and returns to resting potential values.

In unmyelinated fibres the action potential is propagated along the entire axon membrane surface. Each depolarised patch of membrane produces an action potential that in turn triggers depolarisation in an adjacent area. Thus the action potential travels along smoothly, at a constant, but relatively slow speed.

In contrast, the action potential in myelinated fibres is propagated in a discontinuous fashion, jumping from one node of Ranvier to the next. Voltage-gated sodium

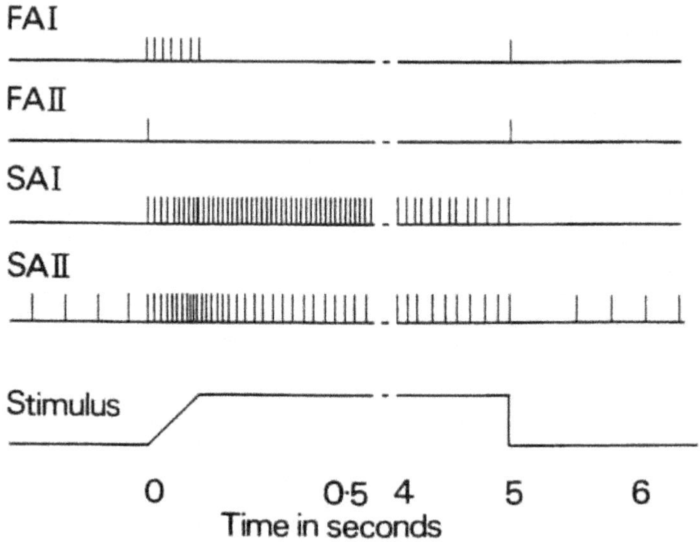

Figure 7.3: *Four different patterns of discharge in nerve fibres in response to peripheral stimulation (mechanical displacement). Two patterns of fast adapting (FA) and two of slowly adapting (SA) mechanoreceptors are illustrated (adapted from Malinovský, 1986; Terzis & Smith, 1990).*

and potassium channels are present in relatively high concentrations in the exposed axon membrane at the nodes of Ranvier. The myelinated internodes (Schwann cells) are passive conductors. The speed of conduction of the action potential is positively correlated with the thickness of the myelin sheath and the length of the internodes (Koester, 1991a;b; Martin, 1991; Noback *et al.*, 1991; Terzis & Smith, 1990).

7.2.5 Fast Adapting and Slowly Adapting Receptors

The term "adaptation" refers to the output of impulses by the receptor. A fast adapting receptor will produce impulses that will rapidly return to zero or a baseline value, when the intensity of the stimulus does not vary. Another impulse discharge follows only in response to a changing (increasing or decreasing) stimulus.

A slowly adapting receptor will respond to a stimulus by producing impulses for the duration of the stimulus. When there is a change (increase or decrease) in the intensity of the stimulus, the slowly adapting receptor will change the frequency of its output (Figure 7.3).

The pattern of adaptation often results from characteristic properties of the excitable membrane of the receptor neuron, but it is also possible for non-neural components of a particular receptor to determine its adaptation pattern. For example, slippage between the connective tissue-derived capsular layers of the fast adapting Pacinian corpuscle may be responsible for the absorption of the static component of an effective stimulus that reaches the encapsulated axon of the receptor neuron (Loewenstein & Mendelson, 1965; Malinovský, 1986).

7.3 Nerve Terminals in the Skin

The sensory end organs that provide information about conditions at the body surface are located at various levels, and in different layers, of the skin. It is therefore important to consider not only the structural characteristics of the end organs themselves, but also the physical composition of the surrounding tissues and their interactions with the end organs.

7.3.1 Structure of the Skin

The skin is a layered organ. It consists of two interdigitating layers: a superficial, relatively thin layer of epidermis (mean value: 0.1 mm) and a deep, considerably thicker layer of dermis (mean value: 1.0 mm).

The epidermis is an epithelium, composed of several cell layers, of which the most superficial are keratinized. The thickest keratin layers are found in the non-hairy skin of palms and soles (0.3 - 1.5 mm respectively) (Leeson & Leeson, 1979; Millington & Wilkinson, 1983). The brick-and-mortar structure of keratinized cells and lipid intercellular material in the superficial layers renders the epidermis mechanically strong and impermeable to water-soluble substances (Landmann, 1988).

The dermis determines most of the mechanical characteristics of the skin. There are two layers of dermis. One layer is thick and contains both densely packed, stretch-resistant collagen fibres, that run parallel to the skin surface, and a smaller population of elastic fibres. A second, much thinner layer is the papillary layer, located at the interface with the epidermis. This layer contains most of the capillary and nerve supply of the dermis, as well as some of the sensory end organs. The dermal papillae interdigitate with deeply protruding "rete pegs" of epithelium.

Underneath the dermis lies the hypodermis which is not considered part of the skin. This is a loose connective tissue layer, containing blood vessels, nerves and fat cells. The mobility of the skin relative to the underlying musculoskeletal tissues is a function of the density and structure of the hypodermis.

7.3.2 Dermal-Epidermal Environment

A basement membrane is present at the interface between epidermis and dermis. This structure separates the epithelial and connective tissue compartments and serves as a selective barrier for the passage of certain macromolecules. More importantly, for purposes of this discussion, the basement membrane provides an attachment surface for both epidermis and dermis. Therefore, by means of this intermediary element, both tissues are firmly tied together.

One of the major components of the basement membrane is a special type of collagen (Type IV). This is a non-fibrillar collagen, the molecules of which are capable of forming a sheet of interconnecting tetrameres. Associated with this sheet are proteoglycans as well as laminin and fibronectin, glycoproteins that serve as adhesive macromolecules for epithelial cells and connective tissue elements respectively (Krieg & Timpl, 1986; LeBlond & Inoue, 1989; Sams, 1990).

The epithelial cells of the epidermis attach to each other with desmosomes and to the basement membrane with hemidesmosomes. These attachments are cell membrane specialisations that allow adhesion not only externally, but also on the internal surface of the membrane, where the intracellular skeleton of tonofibrils is anchored into the desmosomal attachment plaque. Thus, the tonofibrillar skeleton of an individual cell becomes part of a tissue-wide apparatus, firmly attached to the basement membrane (Sams, 1990; Skerrow, 1986). While its mechanical role is not completely understood at this time, it is generally assumed that this apparatus provides mechanical stability to the epithelium.

On the dermal side, anchoring fibrils are attached to the basement membrane that extend into and are connected with the collagen meshwork that lies deep to it (Millington & Wilkinson, 1983).

The surface area of the basement membrane, and possibly its efficiency as an attachment structure, is increased at the tissue level by the interdigitation of dermal papillae and epithelial rete pegs. This interdigitation is also present at the cellular level. The basal cell membranes of the non-dividing epithelial cells that lie adjacent to the basement membrane, interdigitate with the undulating surface of that membrane. The dividing cells in the basal cell layer of the epithelial rete pegs generally have a smoother basal cell membrane (Bale & White, 1982; Briggaman & Wheeler, 1975; Zampetti *et al.*, 1989).

7.3.3 Distinction Between Hairy Skin and Glabrous Skin

The human skin has several appendages: hairs, nails, sebaceous glands and sweat glands. While the presence of nails is limited to the terminal phalanges of the digits, the other three types of appendages are typically found in hairy skin. A hair follicle is a complex structure. Sebaceous glands are usually part of this complex, as are the small muscles that are capable of erecting the hair follicles. Hair follicles are richly innervated with a variety of sensory end organs, allowing hair to play an important role in mechanoreception. However, the so-called piloneural complex is beyond the scope of this chapter and will not be discussed further (see: Munger & Ide (1988), for review).

The non-hairy or glabrous skin lacks hair and the associated sebaceous glands. Sweat glands are present in large numbers in the glabrous skin of palms and soles, as may be noted at times of emotional stress (Millington & Wilkinson, 1983).

7.3.4 Glabrous Skin

The skin surface of the palms and soles is characterised by the presence of an individual pattern of ridges. Along the summit of these ridges, at fairly regular intervals, the ductal openings of sweat glands may be found. The ridges, which develop early in the foetal period, overlie large, protruding ridges of dermis, the dermal papillae.

The dermal papillae are separated from each other by relatively thin projections of epithelium. The epithelial projections are part of a 3-dimensional net-like structure (rete), resembling a honeycomb. The individual projections that are seen in sectioned skin are commonly known as rete pegs. The largest of these project from the summits of the skin ridges and subdivide the large dermal papillae lengthwise. The epithelial ducts of the sweat glands are continuous with the epithelium at the bases of these large "glandular" or "intermediate" epithelial ridges. Somewhat shorter rete pegs separate adjacent dermal papillae. They run parallel to the intermediate epithelial ridges and are variously named "adhesive" or "limiting" ridges (Figure 7.4). At regular intervals epithelial cross ridges are present, running between the intermediate and the limiting ridges (Halata, 1975; Lederman & Browse, 1988).

7.4 Nerve Terminals in Glabrous Skin

Different types of sensory end organs in the glabrous skin appear to be responsible for the detection of different stimuli: mechanoreceptors can detect various mechanical deformations; thermoreceptors perceive changes in temperature, while nociceptors respond to potentially injurious stimuli.

Single fibre recordings in neurophysiologic studies show that specific discharge patterns are evoked by the various stimuli and that specific somatic afferent nerve fibres are responsible for the conduction of the different sensory modalities (Table 7.1). The results of morphologic studies are not quite so clear cut. Early morphologic descriptions of sensory end organs tend to be rather confusing in their complexity. In the past two decades simpler classifications have been proposed: one was based on ultrastructural descriptions of the end organs (Halata, 1975); a second on the simple distinction of encapsulated and unencapsulated terminals (Chouchkov, 1978). The latter is generally preferred. The sensory end organs of the glabrous skin may be classified as shown in Table 7.2.

An unencapsulated sensory terminal consists of an axon with a slightly bulbous ending. The terminal has no myelin sheath. If the terminal is located in the dermis, an incomplete Schwann cell cover and a basement membrane may separate it from the connective tissue elements. In the epidermis the axon terminal is not covered by any structures but is surrounded completely by epithelial cells.

In all encapsulated sensory terminals, the Schwann cells and the perineural cells have developed into a specialised capsule that surrounds the terminal(s) of the axon.

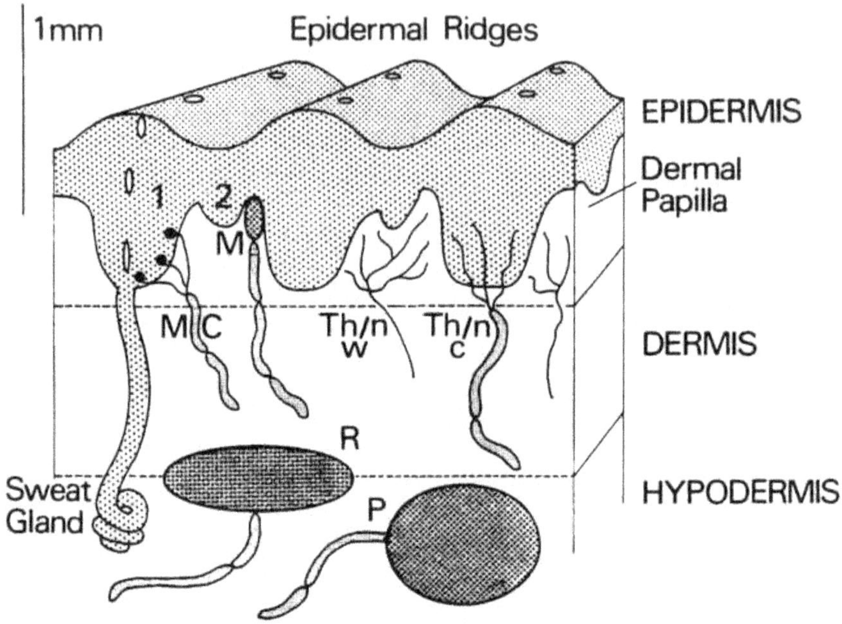

Figure 7.4: *Diagram of a section of skin and hypodermis. The anatomical locations of the various sensory terminals are indicated.* **MC**: *Merkel cell-related endings;* **M**: *Meissner's corpuscle;* **P**: *Pacinian corpuscle;* **R**: *Ruffini corpuscle;* **Th/n c**: *thermoreceptor (cold)/nociceptor;* **Th/n w**: *thermoreceptor (warm)/nociceptor. The last terminal is the only one that is served by an unmyelinated (C) fibre.* **1** *and* **2** *are the glandular and adhesive ridges respectively.*

While there are many good morphologic descriptions of the sensory end organs, the match of the different terminals with the neurophysiologically defined afferent fibre types remains somewhat tenuous (Darian-Smith, 1984). A definite correlation between structure and function has been possible only for the Pacinian corpuscle, which is large enough to be grossly visible, allowing for direct experimentation (Sato, 1961).

Unencapsulated ("free" nerve endings)	epidermal endings dermal endings Merkel cell-related nerve endings
Encapsulated (corpuscles)	Meissner's corpuscle Pacinian corpuscle Ruffini corpuscle

Table 7.2: *Morphologic classification of sensory terminals in glabrous skin.*

7.4.1 Thermoreceptors and Nociceptors

Thermoreceptors and nociceptors are classified as unencapsulated ("free") nerve endings, based almost entirely on circumstantial evidence. They respond either to small temperature changes or to potentially injurious stimuli.

Thermoreceptors resemble mechanoreceptors in that they are low threshold sensory units, responding to small, appropriate stimuli. Thermoreceptors are excited only or preferably by temperature stimuli; they have a very high threshold for mechanical stimuli. These receptors are slowly adaptive. At a constant temperature they have a static discharge and they respond dynamically to temperature changes. Two separate classes of thermoreceptors, cold and warm receptors, are distinguished based on their opposite dynamic responses to changes in temperature (Burgess & Perl, 1973; Hensel, 1973; Terzis & Smith, 1990).

In human skin, separate cold and warm spots may be demonstrated, each approximately 1 mm in diameter. The density of cold spots is higher than that of warm spots. In the skin of the palm there are 1.5 cold spots and 0.4 warm spots per cm^2; in the glabrous skin of the finger the density of warm spots is slightly higher: 1.6 per cm^2 (Rein, 1925; Strughold & Porz, 1931).

In most cases these spots are innervated by separate nerve fibres, although in primates a single fibre may be responsible for the innervation of up to eight cold spots. Based on the findings in electrophysiological studies, it would appear that the nerves that serve cold receptors most likely are myelinated $A\delta$-fibres, while warm receptors usually are served by the unmyelinated C-fibres (Hensel, 1969; Hensel & Boman, 1960; Hensel & Iggo, 1971). However, there has been some suggestion that some myelinated warm receptive afferent fibres exist as well (Darian-Smith & Dykes, 1971).

Studies that employed physiological methods suggest that the cold receptors are located somewhat more superficially in the skin than the warm receptors: 0.15 - 0.22 mm below the skin surface versus 0.3 - 0.6 mm for the warm receptors. This would place the cold receptors in or immediately below the epidermis, while the warm receptors appear to be located in the superficial layers of the dermis (Andres & v.Düring, 1973; Bazett & McGlone, 1930; Bazett *et al.*, 1930).

Actual morphologic observation of these receptors is difficult. Their identification usually rests on criteria such as the presence of a terminal bulboid swelling of the axon, and the presence in the axon of many mitochondria as well as clear core vesicles and axoplasmic vesicles. While cold receptors that terminate in the epidermis have been described, the difficulty in distinguishing them from the Merkel-cell related nerve endings is generally recognised (Hensel, 1973).

The nociceptors, like the thermoreceptors, are unencapsulated and slowly adapting, but unlike the latter they are high threshold receptors that do not respond to less than noxious stimuli, i.e., stimuli in an extreme enough range to be able to cause tissue damage. Nociceptors in the skin either respond to mechanical stimuli only, or to a variety of stimuli: noxious cold or noxious heat in addition to mechanical stimuli, or they may be polymodal in nature responding to mechanical, thermal and chemical irritants.

These receptors are poorly defined morphologically. Similarly, it is not known whether they respond to noxious stimuli directly, or indirectly, through one or more chemical intermediaries, released from traumatised tissues (Martin, 1991).

There are two classes of nociceptive fibres: Aδ and C. The former carry "fast pain", a sharp, pricking pain as well as the vague sensation of tickle. The latter carry, at a slower conduction speed, a sickening, burning pain, as well as the vague sensation of itch (Burgess *et al.*, 1968; Burgess & Perl, 1973; McIntyre, 1980; Ruch, 1979; Dykes, 1975; Martin & Jessell, 1991; Terzis & Smith, 1990).

7.4.2 Mechanoreceptors

Mechanoreceptors are low-threshold sensory nerve terminals, that are involved in the transduction of mechanical to electrical energy. Most are highly specialised and encapsulated structures. Their dynamic responses to stimuli are determined not only by the properties of their receptor membranes (mechanically regulated membrane channels), but also by the properties of their capsules and/or their spatial relationships to the surrounding tissues.

In the human glabrous skin four structurally well-defined types of mechanoreceptors are present. Two of these, the Merkel cell-related endings and Meissner's corpuscles are located superficially at the interface between epidermis and dermis. They are most numerous. The two others are located deeper, in the dermis and in the hypodermis. These are the Ruffini and Pacinian corpuscles.

Neurophysiologically, four patterns of activity have been identified in individual afferent fibres in humans in response to peripheral stimulation. The four patterns of response tentatively have been matched with the four identified types of mechanoreceptors, based on circumstantial evidence. While further confirmation is required, it is now generally assumed that the Meissner and Pacinian corpuscles are fast adapting (FA) mechanoreceptors. Differences in their sensitivity to stimuli of different intensity and frequency has led to a further classification into FAI and FAII respectively (RA or QA and P in an older terminology). Merkel cell-related endings and Ruffini corpuscles are slowly adapting receptors: SAI and SAII respectively (see Figure 7.3). It should be noted that this classification applies to human mechanoreceptors. It is not always possible to distinguish between SAI and SAII units in subhuman primates (Martin & Jessell, 1991; Lederman & Browse, 1988; Terzis & Smith, 1990).

Quantitative data concerning the distribution and densities of the various mechanoreceptors of the skin are very limited. Morphologists must resort to time-consuming studies of specially stained serial sections. Such studies have been done in foetal hands for the distribution of Pacinian corpuscles. For the remainder of the mechanoreceptors the data on density are derived from studies in which electrodes were inserted through the skin and readings were recorded for single nerve fibres of

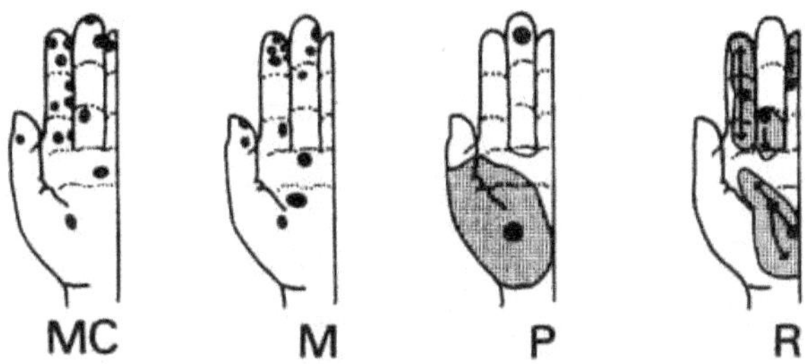

Figure 7.5: *Relative frequencies and sizes of the four types of mechanoreceptors in that part of the glabrous skin of the hand that is served by the median nerve. Abbreviations as in Figure 7.4. The arrows in the diagram of the Ruffini corpuscles indicate the direction of stretch (adapted from Johansson & Vallbo, 1979).*

the median nerve (Figure 7.5) (Darian-Smith & Kenins, 1980; Johansson & Vallbo, 1979; 1983).

In this limited sample the highest densities for all mechanoreceptors were found in the fingertips. 70 Merkel cell-related endings, 140 Meissner's corpuscles, 21 Pacinian corpuscles and 49 Ruffini corpuscles were present per cm^2. In comparison, the density in the plan of the hand was 8, 25, 9, and 16 respectively.

7.4.2.1 Merkel Cell-Related Endings

These endings were first noted and implicated in touch by Friedrich Merkel in 1875. They are located in clusters in the basal epithelial layer of the glandular ridges (Halata, 1975). A single, afferent fibre serves several (4 – 7) Merkel cells.

The receptive fields of these clusters of endings are small (2 – 4 mm in diameter) and sharply bounded. Merkel cell-related endings have the highest density in the glabrous skin of the fingertips (Figure 7.5) (Darian-Smith, 1984; Johansson & Vallbo, 1979; 1983). Morphologically, Merkel cell-related endings are the only mechanoreceptors that are unencapsulated. The afferent fibre loses both its myelin sheath and its Schwann cell cover as it penetrates the basement membrane at the epidermal-dermal interface. The free axon terminates in a disc-like expansion adjacent to a Merkel cell. A synapse is present between the unencapsulated nerve ending and the Merkel cell (Figure 7.6). The Merkel cell is characterised by the presence of neuropeptide-containing granules in its cytoplasm.

There is general agreement that Merkel cell-related endings are SAI mechanoreceptors that respond to sustained indentation, but some doubt continues to be expressed about the role of the Merkel cells in these endings. Developmentally, Merkel cells are present in the epithelium of a ten week-old foetus, well before the presence of any papillary ridges or intra-epithelial nerves (Brenowitz *et al.*, 1980; Lyne & Hollis, 1971). Mearow and Diamond (1988) suggest, on the basis of denervation experiments, that the Merkel cells simply may be targets for growing nerves, or specialised epithelial cells that merely serve to enhance the excitability of

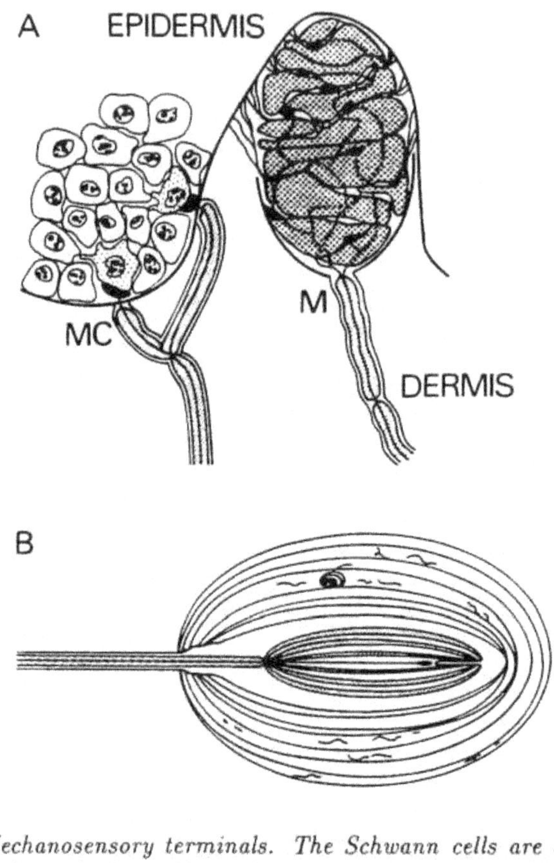

Figure 7.6: *Mechanosensory terminals. The Schwann cells are darkly shaded; the nerve axons and terminals are black.* **A**: *Meissner's corpuscles (***M***) and Merkel cell-related endings (***MC***). The Merkel cells are lightly shaded.* **B**: *Pacinian corpuscle.*

the unencapsulated intra-epidermal nerve terminal.

Even if one accepts that Merkel cells are an essential part of a specialised receptor, there is no general agreement as to which is the prime transducer element: Merkel cell or axon. In favour of a primary role for the former the following points may be noted. The Merkel cell has a number of microvillous processes that extend in the spaces between surrounding epithelial cells. The processes may serve as sensors, enabling the cell to perceive mechanical deformations in the epidermis, that could be relayed to the axon via the synapse. In addition, there are voltage-gated cation channels in the cell membranes of Merkel cells. If these channels are permeable to calcium ions, a mechanism would be available for the discharge of the dense

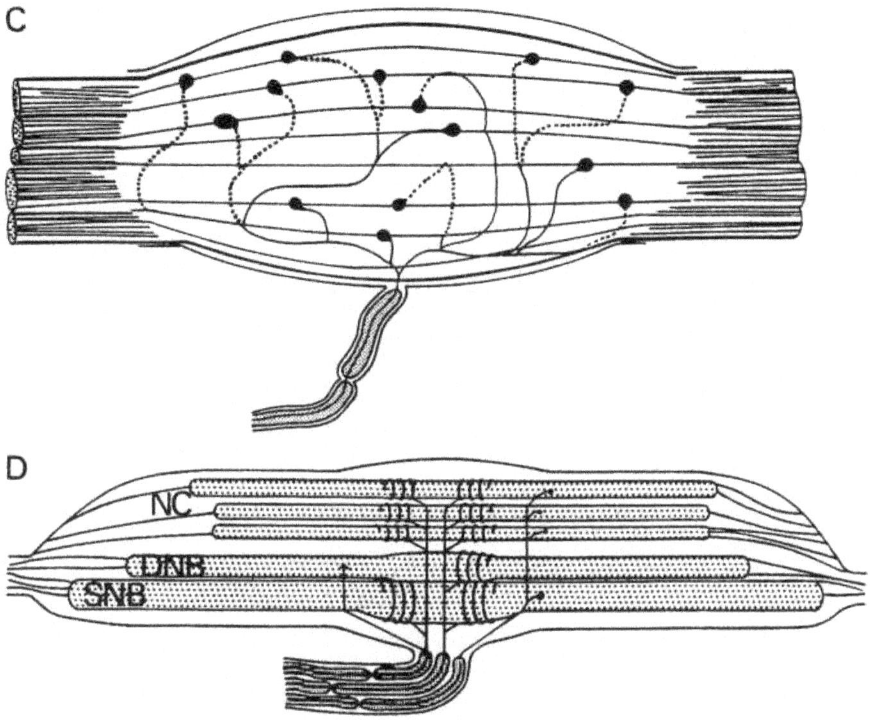

Figure 7.7: *Mechanosensory terminals.* **C**: *Ruffini corpuscle.* **D**: *Muscle spindle. The muscle fibres are lightly shaded.* (**NC**: *nuclear chain fibres;* **DNB**: *dynamic nuclear bag fibre;* **SNB**: *static nuclear bag fibre.)*

core vesicles. Finally, there is evidence that discharge of the neuropeptides in the cytoplasmic granules may lead to a discharge of action potentials in SAI fibres (Iggo *et al.*, 1988; Nurse & Cooper, 1988; Hartschuh *et al.*, 1986).

In favour of a primary role for the axon is the morphologic appearance of the axon terminal, which resembles that of other unencapsulated nerve endings that serve as prime receptors. In addition, the receptor delay of the Merkel cell-related endings is very short and signal transmission via the synapse might be too slow (Gottschaldt & Vahle-Hinz, 1981; Hartschuh *et al.*, 1986). The actual roles of Merkel cells and axons in the normal function of this mechanoreceptor complex have not yet been established.

7.4.2.2 Meissner's Corpuscles

These corpuscles were first described by G. Meissner in 1853. They are located quite superficially in the papillary dermis, between the epithelial rete ridges. Meissner's corpuscles are found almost exclusively in the glabrous skin of digits, where they are most numerous, and in the glabrous skin of palms and soles. They are seldom found in hairy skin (Darian-Smith, 1984; Johansson & Vallbo, 1979; 1983).

A single myelinated fibre serves 12 – 17 evenly distributed corpuscles. The receptive fields of these groups are sharply bounded and relatively small, between 2 and 4 mm in diameter (Figures 7.5 and 7.6).

Meissner's corpuscle is an ovoid structure: approximately 0.1 mm in length and 0.05 mm wide. Its long axis is oriented normal to the surface of the skin. Two to six axons, branches of the single fibre that serves the corpuscle, enter the structure at its lower pole. Inside the corpuscle they run spirally, like ribbon candy, sending out numerous terminals between approximately ten, irregularly stacked, transversely oriented lamellae of Schwann cells. An incomplete perineural capsule tightly surrounds the base of the corpuscle, like an egg cup.

Collagen fibrils run from the corpuscle to the basement membrane which partially surrounds it. As a result, the top of the corpuscle is tied directly to the basement membrane and indirectly, via hemidesmosomes to the tonofibrillar skeleton of epidermis. In this way any epidermal deformation is transmitted directly to the corpuscle.

Meissner's corpuscles are FAI mechanoreceptors subserving the perception of "moving touch". They respond to stepwise indentation and are particularly sensitive to low frequency (50 Hz) sinusoidal mechanical stimuli.

These mechanoreceptors develop at the time of birth. With age they become progressively smaller and substantially reduced in number, while the size and number of their axons decreases (Castano & Ventura, 1978; 1979; Cauna, 1965; Darian-Smith & Kenins, 1980; Halata, 1975; Johansson & Vallbo, 1979; 1983; Malinovský, 1986).

7.4.2.3 Pacinian Corpuscles

These, the largest and least numerous of all mechanoreceptors, were first described by A. Vater in 1741 and later by F. Pacini in 1840. They are customarily named after the latter.

Pacinian corpuscles are located in the deep dermis and hypodermal layers (subcutaneous fat) of the palm, particularly on the thenar (medial) and hypothenar (lateral) eminences. They are also found in the periosteum, between the flexor tendons and around joints.

Only one Pacinian corpuscle is served by any FAII nerve fibre. The single fibre may be responsible for the sensitivity over a large area of the skin: a whole digit or part of the palmar skin. The receptive field is not sharply bounded (Figures 7.5 and 7.6).

Pacinian corpuscles are ovoid and large enough to be seen with the unaided eye. They may be up to 3 mm in length and up to 2 mm in width. The single nerve fibre enters at one of the poles and continues to be myelinated for about a quarter of the length of the corpuscle. Usually the axon remains unbranched and is located centrally, surrounded by an inner core formed by 30 – 60 concentric layers of closely packed cellular lamellae of modified Schwann cells. The inner core is bisected by two symmetrical radial clefts. Occasionally there are two or more axon branches in a Pacinian corpuscle, each with a separate inner core. Between the inner core(s) and the outer capsule a small subcapsular space is present. The outer, fluid-filled capsule is composed of 20 - 70 concentric cellular lamellae, that are derived from modified perineural cells. These cells are far less tightly packed than the cellular lamellae of the inner core. The fluid-filled spaces between the lamellae contain collagen fibrils and even an occasional capillary (Andres & v.Düring, 1973; Darian-Smith, 1984; Krstić, 1978; Malinovský, 1986; Munger & Ide, 1988).

Pacinian corpuscles are FAII receptors, giving a maximal response to 300 Hz vibratory stimuli. They also exhibit directional sensitivity. It has been suggested that there may be two functionally distinct populations of corpuscles, one more sensitive to the stimulus of compression and the other to the withdrawal of that stimulus. This directionality of sensitivity may be due to the orientation of the elliptically shaped, unmyelinated portion of the axon in the hemilamellar clefts of the inner core. A membrane-stretching model has been proposed as a possible explanation of the transduction mechanism in Pacinian corpuscles (Bolanowski, 1988; Martin & Jessell, 1991).

Pacinian corpuscles develop during the foetal period. The actual incidence of the corpuscles has been studied in the hand and in the index finger of human foetuses at the age of seven months (Buschong, 1963; Cauna & Mannan, 1959). The findings of this painstaking research compare well with the data based on single fibre readings in the human median nerve (Johansson & Vallbo, 1979; 1983). With age the corpuscles become more convoluted.

7.4.2.4 Ruffini Corpuscles

These corpuscles were first described by A. Ruffini in 1894. While they appear to be slightly more prevalent than Pacinian corpuscles, they are not as easily identified microscopically and their morphology is not as well understood.

Ruffini corpuscles are spindle-shaped. Their length varies from 0.5 - 1.0 mm and their maximal width is 0.2 mm. They usually are oriented with their long axes parallel to the skin surface. They are present in the deepest layers of the dermis in the skin of the fingers, in the dense fibrous aponeurosis of the palm (below the hypodermis), and in fibrous tendon sheaths, retinacula and ligaments of joints. Their receptive field is large and not sharply bounded (Figures 7.5 and 7.7).

Morphologically there is little ultrastructural information on human Ruffini corpuscles anywhere and none on those in the human hand. In recent years the ultrastructural and functional resemblance of Ruffini corpuscles and Golgi tendon organs has been recognised and some investigators consider them to be similar, except for their location. Excellent ultrastructural studies have provided some insight in the morphology of non-human Ruffini corpuscles (Darian-Smith, 1984; Kannari, 1990; Munger & Ide, 1988).

Only one Ruffini corpuscle is served by any SAII nerve fibre. The corpuscle incorporates the multiple terminal axon branches of the nerve fibre as well as segments of the collagen fibres of the surrounding fibrous connective tissue. A 4 – 5 layer capsule of lamellar, modified perineural cells surrounds the corpuscle. In places the capsule may be thin or even incomplete. Inside, numerous unmyelinated axons of up to 0.2 mm long branch off the single axon, that enters at the equator of the corpuscle. The branches further ramify among the enclosed collagen fibres in the core of the corpuscle producing a typical "bushy" or spray structure. The multiple axon endings are plate- or knob-shaped. The relationship between the axon endings and the accompanying Schwann cells is complicated. More than two Schwann cells may be associated with each ending and the Schwann cell bodies are located at some distance from the axons, somewhat like astrocytes in the central nervous system (Kannari, 1990). The axon endings are partially exposed, although still covered by a basement membrane. They are closely associated with the collagen fibres.

Ruffini corpuscles are SAII receptors. They have a low background discharge in the absence of stimuli. They exhibit directional sensitivity, responding specifically to lateral stretch in the skin. It is thought that the stretch of the collagen fibres in the surrounding tissue is mechanically transmitted to the enclosed collagen fibres and that this will cause deformation (and opening of cation channels in the mechanoreceptive cell membrane) of the axon terminals (Bolanowski, 1988; Iggo, 1974).

Between the capsule and the core of the corpuscle a fluid-filled subcapsular space is present. The core of the corpuscle is further subdivided lengthwise in compartments by "septal cells" (fibroblasts). Andres and v.Düring (1973) have suggested that these fluid-filled spaces may insulate the axon terminals from unspecific stimuli.

7.5 Limb Proprioception

In order for the limb to be used effectively, limb proprioception, that is the sense of its position and of its movements (kinesthesia) is essential. This perception is mediated by an array of proprioceptive mechanoreceptors in and around the joints and in the muscles, as well as the previously described exteroceptive mechanoreceptors of the skin (Martin & Jessell, 1991; Millar, 1979; v.d.Wal *et al.*, 1988).

As in the case of the mechanoreceptors of the skin, it is difficult to obtain reliable quantitative data on the incidence of the various mechanoreceptors around the joints and in the muscles. No organised set of data on the density of these structures can be presented here.

7.5.1 Afferents of the Joints

Articulations between the various bones of the limb are examples of diarthroses: joints with an intervening, lubricant-filled joint space. The joint is surrounded by a joint capsule, which is composed of an innermost layer of loose connective tissue and an outer layer of dense fibrous collagenous tissue. This outer fibrous layer fits like a sleeve around the joint and is continuous at both ends with the fibrous periosteum that covers each of the two articulating bones.

Muscles of the limb either attach directly into the fibrous periosteum or into a separate, dense fibrous structure (tendon or aponeurosis) that in turn is attached to the periosteum.

In and around the joint four types of sensory nerve endings typically are identified. Three of these are mechanoreceptors: Ruffini-type endings, modified Pacinian endings and Golgi tendon organ-like endings. The fourth type consists of a group of unencapsulated endings which are widely distributed through most articular tissues and are part of the nociceptive system (Johansson *et al.*, 1991; Newton, 1982; Wyke, 1981; Zimny, 1988a;b).

The **Ruffini-type corpuscles** are low-threshold, slowly adaptive mechanoreceptors. They are not very numerous and are found in the fibrous layer of the joint capsule, particularly at the flexor side, and also in several of the intrinsic ligaments of the joint. These corpuscles, which in joint tissues may occur in clusters of up to six, served by a single myelinated nerve fibre, function as both static and dynamic mechanoreceptors and provide information about joint angles and/or limb movements (Johansson *et al.*, 1991; Strasmann & Halata, 1988).

The modified **Pacinian corpuscles** are low threshold, fast adapting mechanoreceptors. They are thickly encapsulated, smaller and simpler than those that are found in the skin. They are found in the fibrous layer of the capsule, in the loose connective tissue that is peripheral to the joint capsule, as well as the loose connective tissue between the joint capsule and the aponeuroses into which the muscles attach (the "shearing spectrum" (Strasmann *et al.*, 1990)). Pacinian corpuscles are present also in the aponeuroses themselves and in selected parts of the periosteum, along the long axis of the bone. These corpuscles serve solely as dynamic mechanoreceptors, becoming active at acceleration and deceleration (Ide, 1988; Johansson *et al.*, 1991; Strasmann & Halata, 1988; Strasmann *et al.*, 1990; v.d.Wal *et al.*, 1988).

The presence of **Golgi tendon organ-like** endings in joint tissues may be debatable. They have been identified by some in collateral ligaments and intrinsic ligaments. They are thought to be high threshold, slowly adaptive mechanoreceptors that are completely inactive in immobile joints and most efficient when joints are in extreme positions, providing information about the tension in the ligaments (Johansson *et al.*, 1991; Zimny, 1988a;b). Others have noted Golgi tendon organ-like endings only at the boundaries between muscle fibres and tendinous tissues, where one would ordinarily expect their presence (Strasmann *et al.*, 1990). Their structure will be described below.

The afferents of the joints appear to be capable of detecting noxious stimuli and of providing information about speed, acceleration, position and direction of movements. Nevertheless, the greatest density of the mechanoreceptors of the joints is in areas related to extremes of movement. They seem to be most sensitive to the extremes of joint angles and do not appear to play a dominant role in providing information about the limb at rest (Burgess *et al.*, 1982; Johansson *et al.*, 1991; Martin & Jessell, 1991; Strasmann *et al.*, 1990).

7.5.2 Afferents of the Muscles

There is growing evidence that an exclusive role for joint capsule receptors in kinesthesia is doubtful and that mechanoreceptors in muscle tissue and even in the skin play a much more substantial role than was previously thought. The joint angle, in particular, is judged predominantly from information about muscle length. This information is provided by muscle spindle receptors. These slowly adapting mechanoreceptors are spindle-shaped and are located in the fleshy portion of the skeletal muscles, with their long axes parallel to the skeletal muscle fibres (Burgess *et al.*, 1982; McCloskey, 1978).

A second type of slowly adapting mechanoreceptor is found at the junction between groups of skeletal muscle fibres and the collagen fibres of tendon or aponeurosis. This type of mechanoreceptor is the Golgi tendon organ. It is also spindle-shaped and is lined up in series with the skeletal muscle fibres. Both types of mechanoreceptors are encapsulated stretch receptors. They are distributed extensively throughout most skeletal muscles (Gordon & Ghez, 1991; Martin & Jessell, 1991; v.d.Wal *et al.*, 1988).

Muscle spindles are 4 – 10 mm long. Their perineural capsule is multilayered, composed of an outer and an inner cellular layer. This capsule forms a permeability barrier, that is impervious at the equator of the spindle, but quite leaky at the poles (Ovalle & Dow, 1988). Muscle spindles contain three different types of highly specialised muscle fibres: usually one static nuclear bag fibre, one dynamic nuclear bag fibre and approximately five nuclear chain fibres. In the centre of the spindle these fibres are surrounded by fluid to facilitate sliding. While these fibres contract, they do not contribute significantly to muscle contraction.

In the nuclear bag fibres, the nuclei are clustered in the centre of the fibres. Nuclear bag fibres are thicker and longer than nuclear chain fibres. The latter contain nuclei that are arranged in a single file in the centre of the fibres. The innervation of a muscle spindle is quite complex and is composed of two types of sensory endings and a motor ending (Figure 7.7).

A large, myelinated Aα fibre comprises the first type of sensory ending (primary ending). Branches of this axon encircle each of the muscle fibres in the spindle. Branches of a smaller, myelinated Aβ fibre terminate on all of the spindle's muscle fibres, except the dynamic nuclear bag fibre. Finally, motor axons, branches of small Aγ fibres, innervate the polar regions of the spindle's muscle fibres.

When the spindle's fibres are stretched, the stretch-sensitive channels in the sensory axons are activated, leading to an increase in the firing rate. The primary endings are particularly sensitive to the rate of change in the length of the fibres, dynamic stretch and vibratory stimuli. The secondary endings are not much affected during the dynamic phase, but have a regular discharge during the static phase of muscle activity. The motor axons cause the polar regions of the spindle's muscle fibres to contract, and in doing so these fibres lengthen centrally. This allows for increased sensitivity of the spindle (Gordon & Ghez, 1991; Martin & Jessell, 1991; Terzis & Smith, 1990).

Golgi tendon organs are slender, encapsulated structures of about 1.0 mm long and 0.1 mm in diameter. Like Ruffini corpuscles, the Golgi organs are spindle-shaped and their fine structure resembles that of Ruffini endings. Indeed, the classification of these two different types of spray ending may be more or less arbitrary (Johansson *et al.*, 1991).

These receptors are oriented parallel with the tendon fibres, i.e., in the direction of stretch. Unlike the Ruffini corpuscles, they have a distinct perineural capsule and only a minute space below the capsule. The collagen fibres in their inner core are arranged in parallel. Golgi tendon organs are served by a single large, myelinated $A\alpha$ axon (Andres & v.Düring, 1973; Gordon & Ghez, 1991; Martin & Jessell, 1991; Terzis & Smith, 1990; v.d.Wal *et al.*, 1988).

7.6 Conclusion

The glabrous skin of the primate hand is a most efficient tactile instrument with an array of exteroceptors. The richest sensory innervation is found in the tips of the fingers. In addition to superficially placed, slowly adapting sensors for temperature and pain, the skin contains a fairly well defined group of mechanoreceptors. The mechanoreceptors appear to act both in parallel and in serial fashion, as they provide the rich input for tactile sensing. The slowly adapting Merkel cell-related complexes and fast adapting Meissner's corpuscles are located superficially and have sharply bounded receptive fields. Fast adapting Pacinian corpuscles and slowly adapting Ruffini corpuscles are located deeper in the skin and have large, not very sharply bounded receptive fields.

The proprioceptors around the joints have been less well defined to date and can be described as Ruffini-type corpuscles, modified Pacinian corpuscles and Golgi tendon organ-like endings. Their principal function appears to be the detection of extremes of joint movement.

The proprioceptors of the muscles play a major role in kinesthesia. The muscle spindles are arranged in parallel with the muscle fibres and therefore monitor the early stretch of these fibres. Golgi tendon organs are arranged in series with the muscle fibres and thus provide a later response to stretch, since the tendon is much stiffer than the more compliant muscle fibres.

While great strides have been made in recent decades in the study of the morphology and function of the different sensory receptors in the limb, the picture remains incomplete at this time. Much remains to be done in mapping the many terminals and in gaining an understanding of their role as some of them act in parallel and others in a serial fashion. Redundancy is not unknown in biological systems, but at this juncture it is not possible to decide whether any part of this fairly elegant system of biological sensing is redundant or whether all components perform unique and essential functions. A formidable task remains to be done.

Acknowledgements

I thank my colleagues of the Somatosensory Study Group at Columbia University for a congenial and stimulating exchange of information at our 7:30 a.m. meetings.

I am particularly grateful to Mr. Manjit Basra for his invaluable assistance in preparing this manuscript.

References

This list is not an exhaustive bibliography, but represents a selection of, mostly recent, articles, reviews and textbooks that should serve as a resource for additional references.

Andres, K.H., & v.Düring, M. 1973. Morphology of cutaneous receptors. In: (Iggo, A., ed.) *Handbook of Sensory Physiology.2.* Springer Verlag.

Bale, E., & White, F.H. 1982. Quantitative light and electron microscopical studies of the epithelial-connective tissue junction in intraoral mucosae. J. Microscopy, 128:69.

Barchi, R.L. 1980. Excitation and conduction in nerve. In: (Sumner, A.J., ed.) *The Physiology of Peripheral Nerve Disease.* W.B.Saunders.

Bazett, H.C., & McGlone, B. 1930. Experiments on the mechanism of stimulation of end-organs for cold. Am. J. Physiol., 93:632.

Bazett, H.C., McGlone, B., & Brocklehurst, R.J. 1930. The temperatures in the tissues which accompany temperature sensations. J. Physiol. London, 69:88.

Behse, F. 1990. Morphometric studies on the human sural nerve. Acta Neurol. Scand., 82 (Suppl.132):1.

Bolanowski, S.J.Jr. 1988. Transduction mechanisms in Pacinian corpuscles. In: (Hník, P., Soukup, T., Vejsada, R., & Zelená, J., eds.) *Mechanoreceptors. Development, Structure and Function.* Plenum Press.

Boyd, I. A., & Davey, M. R. 1968. *Composition of Peripheral Nerves.* E.&.S. Livingstone Ltd.

Brenowitz, G. L., Tweedle, C. D., & Johnson, J. I. 1980. The development of receptors in the glabrous forepaw skin of pouch young opossums. Neuroscience, 5:1303.

Briggaman, R. A., & Wheeler, C. E. 1975. The epidermal-dermal junction. J. Invest. Dermatol., 65:71.

Burgess, P. R., & Perl, E. R. 1973. Cutaneous mechanoreceptors and nociceptors. In: (Iggo, A., ed.) *Handbook of Sensory Physiology. 2.* Springer Verlag.

Burgess, P. R., Petit, D., & Warren, R. M. 1968. Receptor types in cat hairy skin supplied by myelinated fibers. J. Neurophysiol., 31:833.

Burgess, P. R., Wei, J. Y., Clark, F. J., & Simon, J. 1982. Signaling of kinesthetic information by peripheral sensory receptors. Ann. Rev. Neurosci., 5:171.

Buschong, J. 1963. *Report of the Frequency and Distribution of the Pacinian Corpuscle in a Palm of a Seven Month Human Fetus.* M. A. Thesis, Johns Hopkins University. (Cited by Darian-Smith, 1984.)

Carlsen, F., & Behse, F. 1980. Three-dimensional analysis of Schwann cells associated with unmyelinated nerve fibres in human sural nerve. J. Anat., 130:545.

Castano, P., & Ventura, R. G. 1978. The Meissner's corpuscle of green monkey (*Cercopithecus aethiops* L.) I. The organization of nervous component. J. Submicr. Cytol., 10:327.

Castano, P., & Ventura, R. G. 1979. The Meissner's corpuscle of green monkey (*Cercopithecus aethiops* L.) II. The connective tissue components. Some considerations from a functional standpoint. J. Submicr. Cytol., 11:185.

Cauna, N. 1965. The effects of aging on the receptor organs of the human dermis. In: (Montagna, W., ed.) *Advances in Biology of Skin. Aging.* Pergamon.

Cauna N., & Mannan, G. 1959. Development and postnatal changes of digital Pacinian corpuscles (Corpuscula lamellosa) in the human hand. Anat. Rec., 93:271.

Chouchkov, C. 1978. Cutaneous receptors. Adv. Anat. Embryol. Cell Biol., 54(3):1.

Daniel, R. K., & Terzis, J. K. 1977. Structure and function of the peripheral nerves. In: (Daniel, R. K., & Terzis, J. K., eds.) *Reconstructive Microsurgery.* Little, Brown & Co.

Darian-Smith, I. 1984. The sense of touch: performance and peripheral neural processes. In: (Darian-Smith, I., ed.) *Handbook of Physiology. Section 1: The Nervous System. III. Sensory Processes 2.* Am. Physiological Society.

Darian-Smith, I., & Dykes, R. W. 1971. Peripheral neural mechanisms of thermal sensation. In: (Duner, R., & Kawamura, Y., eds.) *Oral-Facial Sensory and Motor Mechanisms*. Appleton-Century-Crofts.

Darian-Smith, I., & Kenins, P. 1980. Innervation density of mechanoreceptive fibres supplying glabrous skin of the monkey's index finger. J. Physiol. London, 309:147.

Dykes, R. W. 1975. Nociception. Brain Res., 99:229.

Erlanger, J., & Gasser, H. S. 1937. *Electrical Signs of Nervous Activity*. Univ. Pennsylvania Press.

Gasser, H. S., & Grundfest, H. 1939. Axon characteristics in relation to spike dimensions and conduction velocity in mammalian A-fibers. Am. J. Physiol., 127:393.

Gordon, J., & Ghez, C. 1991. Muscle receptors and spinal reflexes: the stretch reflex. In: (Kandel, E. R., Schwartz, J. H., & Jessell, T. M., eds.) *Principles of Neural Science. Ed.3.* Elsevier.

Gottschaldt, K. M., & Vahle-Hinz, C. H. 1981. Merkel cell receptors: structure and transducer function. Science, 214:183.

Halata, Z. 1975. The mechanoreceptors of the mammalian skin. Ultrastructure and morphological classification. Adv. Anat. Embryol. Cell Biol., 50(5):7.

Hartschuh, W., Weihe, E., & Reinecke, M. 1986. The Merkel cell. In: (Bereiter-Hahn, J., Matoltsy, A. G., & Richards, K. S., eds.) *Biology of the Integument. 2. Vertebrates.* Springer Verlag.

Hensel, H. 1969. Cutane Wärmereceptoren bei Primaten. Pflügers Arch., 313:150.

Hensel, H. 1973. Cutaneous thermoreceptors. In: (Iggo, A., ed.) *Handbook of Sensory Physiology. 2.* Springer Verlag.

Hensel, H., & Boman, K. A. 1960. Afferent impulses in cutaneous sensory nerves in human subjects. J. Neurophysiol., 23:564.

Hensel, H., & Iggo, A. 1971. Analysis of cutaneous warm and cold fibres in primates. Pflügers Arch., 329:1.

Ide, C. 1988. Significance of the extracellular matrix for the regeneration of sensory corpuscles with special reference to Pacinian corpuscles. In: (Hník, P., Soukup, T., Vejsada, R., & Zelená, J., eds.) *Mechanoreceptors. Development, Structure and Function.* Plenum Press.

Iggo, A. 1974. Cutaneous receptors. In: (Hubbard, S. J., ed.) *The Peripheral Nervous System*. Pergamon, 208:351.

Iggo, A., Pacitti, E.G., & Findlater, G. S. 1988. Dense cored vesicles in SAI Merkel cells and their role in mechanoelectric transduction. In: (Hník, P., Soukup, T., Vejsada, R., and Zelená, J., eds.) *Mechanoreceptors. Development, Structure and Function*. Plenum Press.

Johansson, H., Sjölander, P., & Sojka, P. 1991. Receptors in the knee joint ligaments and their role in the biomechanics of the joint. CRC Crit.Rev. Biomed. Eng., 18:341.

Johansson, R. S., & Vallbo, A. B. 1979. Tactile sensibility in the human hand: relative and absolute densities of four types of mechanoreceptive units in glabrous skin. J.Physiol. London, 286:283.

Johansson, R. S., & Vallbo, A. B. 1983. Tactile sensory coding in the glabrous skin of the human hand. Trends Neurosci., 6:27.

Kannari, K. 1990. Sensory receptors in the periodontal ligament of hamster incisors with special reference to the distribution, ultrastructure and three-dimensional reconstruction of Ruffini endings. Arch. Histol. Cytol., 53:559.

Koester, J. 1991a. Membrane potential. In: (Kandel, E. R., Schwartz, J.H., & Jessell, T.M., eds.) *Principles of Neural Science. Ed.3*. Elsevier.

Koester, J. 1991b. Voltage-gated ion channels and the generation of the action potential. In: (Kandel, E.R., Schwartz, J. H., & Jessell, T. M., eds.) *Principles of Neural Science. Ed.3*. Elsevier.

Krieg, T., & Timpl, R. 1986. Protein components of the epidermal basement membrane. In: (Bereiter-Hahn, J., Matoltsy, A. G., & Richards, K. S., eds.) *Biology of the Integument. 2. Vertebrates*. Springer Verlag.

Krstić, R. V. 1978. *Die Gewebe des Menschen und der Säugetiere. Ein Atlas zum Studium fur Mediziner und Biologen*. Springer Verlag.

Landmann, L. 1988. The epidermal permeability barrier. Anat. Embryol., 178:1.

LeBlond, C. P., & Inoue, S. 1989. Structure, composition and assembly of basement membrane. Am. J. Anat., 185:367.

Lederman, S. J., & Browse, R. A. 1988. The physiology and psychophysics of touch. In: (Dario, P., ed.) *Sensors and Sensory Systems for Advanced Robots*. NATO ASI Series, Vol. F43. Springer Verlag.

Leeson, T. S., & Leeson, C. R. 1979. *A Brief Atlas of Histology*. W. B. Saunders.

Loewenstein, W. R., & Mendelson, M. 1965. Components of receptor adaptation in a Pacinian corpuscle. J. Physiol. London, 177:377.

Lyne, A. G., & Hollis, D. E. 1971. Merkel cells in sheep epidermis during fetal development. J.Ultrastructure Res., 34:464.

Malinovský, L. 1986. Mechanoreceptors and free nerve endings. In: (Bereiter-Hahn, J., Matoltsy, A. G., & Richards, K. S., eds.) *Biology of the Integument. 2. Vertebrates.* Springer Verlag.

Martin, J. H. 1991. Coding and processing of sensory information. In: (Kandel, E. R., Schwartz, J. H., & Jessell, T. M., eds.) *Principles of Neural Science. Ed.3.* Elsevier.

Martin, J. H., & Jessell, T. M. 1991. Modality coding in the somatic sensory system. In: (Kandel, E. R., Schwartz, J. H., & Jessell, T. M., eds.) *Principles of Neural Science. Ed.3.* Elsevier.

McCloskey, D. I. 1978. Kinesthetic sensibility. Physiol. Rev., 58:763.

McIntyre, A. K. 1980. Cutaneous receptors. In: (Sumner, A. J. ed.) *The Physiology of Peripheral Nerve Disease.* W. B. Saunders.

Mearow, K. M., & Diamond, J. 1988. Merkel cells and the mechanosensitivity of normal and regenerating nerves in Xenopus skin. Neuroscience, 26:695.

Millar, J. 1979. Convergence of joint, cutaneous and muscle afferents into cuneate neurones in the cat. Brain Res., 175: 347.

Millington, P. F., & Wilkinson, R. 1983. *Skin.* Vol.9 in Series Biological Structure and Function. Cambridge University Press.

Munger, B. L., & Ide, C. 1988. The structure and function of cutaneous sensory receptors. Arch. Histol. Cytol. 51:1.

Newton, R. A. 1982. Joint receptor contributions to reflexive and kinesthetic responses. Phys. Ther., 62:22.

Noback, C. R., Strominger, N. L., & Demarest, R. J. 1991. *The Human Nervous System. Introduction and Review. Ed.4.* Lea & Febiger.

Nurse, C., & Cooper, E. 1988. Electrophysiological studies on Merkel cells isolated from rat vibrissal mechanoreceptors. In: (Hnik, P., Soukup, T., Vejsada, R., & Zelená, J., eds.) *Mechanoreceptors. Development, Structure and Function.* Plenum Press.

Ochoa, J., & Mair, G. P. 1969. The normal sural nerve in man. I. Ultrastructure and numbers of fibres and cells. Acta Neuropathol., 13:197.

Ovalle, W. K., & Dow, P. R. 1988. The capsular sleeve of muscle spindles in mouse and man with special reference to the cytoskeleton. In: (Hník, P., Soukup, T., Vejsada, R., & Zelená, J., eds.) *Mechanoreceptors. Development, Structure and Function.* Plenum Press.

Rein, H. 1925. Ueber die Topographie der Warmempfindung. Beziehungen zwischen Innervation und receptorischen Endorganen. Z.Biol., 82:513.

Ruch, T. C. 1979. Somatic sensation: receptors and their axons. In: (Ruch, T., & Patton, H. O., eds.) *Physiology and Biophysics. The Brain and Neural Function.* W. B. Saunders.

Sams, W. M. 1990. Structure and function of the skin. In: (Sams, W. M., & Lynch, P. J., eds.) *Principles and Practice of Dermatology.* Churchill Livingstone.

Sato, M. 1961. Response of Pacinian corpuscles to sinusoidal vibration. J. Physiol. London, 159:391.

Schady, W., Ochoa, J. L., Torbjork, H. E., & Chen, L. S. 1983. Peripheral projections of fascicles in median nerve. Brain, 106:745.

Skerrow, C. J. 1986. Desmosomal proteins. In: (Bereiter- Hahn, J., Matoltsy, A. G., & Richards, K. S., eds.) *Biology of the Integument. 2. Vertebrates.* Springer Verlag.

Slingluff, C. L., Terzis, J. K., & Edgerton, M. T. 1987. The quantitative microanatomy of the brachial plexus in man: reconstructive relevance. In: (Terzis, J. K., ed.) *Microreconstruction of Nerve Injuries.* W. B. Saunders.

Strasmann, T., & Halata, Z. 1988. Topography of mechanoreceptors in the connective tissue of the elbow joint region in Monodelphis domestica, a laboratory marsupial. In: (Hník, P., Soukup, T., Vejsada, R., & Zelená, J., eds.) *Mechanoreceptors. Development, Structure and Function.* Plenum Press.

Strasmann, T., v. d. Wal, J. C., Halata, Z., & Drukker, J. 1990. Functional topography and ultrastructure of periarticular mechanoreceptors in the lateral elbow region of the rat. Acta Anat., 138:1.

Strughold, H., & Porz, R. 1931. Die Dichte der Kaltpunkte auf der Haut des menschlichen Körpers. Z. Biol., 91:563.

Sunderland, S., & Bradley, K. C. 1949. The cross-sectional area of peripheral nerve trunks devoted to nerve fiber. Brain, 72:428.

Terzis, J. K., & Smith, K. L. 1990. *The Peripheral Nerve. Structure, Function and Reconstruction.* Raven Press.

Ushiki, T., & Ide, C. 1990. Three-dimensional organization of the collagen fibrils in the rat sciatic nerve as revealed by transmission and scanning electron microscopy. Cell Tissue Res., 260:175.

Wal, J. C. v. d., Strasmann, T., Drukker, J., & Halata, Z. 1988. Sensory nerve endings in the deep lateral cubital region: a topographical and ultrastructural study in the rat. In: (Hník, P., Soukup, T., Vejsada, R., & Zelená, J., eds.) *Mechanoreceptors. Development, Structure and Function.* Plenum Press.

Wyke, B. 1981. The neurology of joints: a review of general principles. Clin. Rheum.Dis., 7:223.

Zampetti, M., Fattorossi, A., Grieco, T., & Calvieri, S., 1989. Basal keratinocyte subsets: ultrastructural and morphometric features. Acta Derm.Venereol., 69:59.

Zimny, M. L. 1988a. Mechanoreceptors in articular tissues. Am. J. Anat., 182:16.

Zimny, M. L. 1988b. Mechanoreceptors in human intra- articular tissues. In: (Hník, P., Soukup, T., Vejsada, R., & Zelená, J., eds.) *Mechanoreceptors. Development, Structure and Function.* Plenum Press.

Chapter 8

Lessons From the Study of Biological Touch for Robotic Tactile Sensing

Susan J. Lederman and Dianne T. Pawluk

8.1 Introduction

The current introduction should also be read in conjunction with the accompanying chapter in this book by Lederman, Klatzky, & Pawluk, "Lessons From Biological Touch For Robotic Haptic Sensing".

8.1.1 What Is Touch Good For?

The sense of touch may be used to sense, perceive, recognise, grasp and manipulate solid (and virtual) objects in the environment. Typically, it has been treated as little more than an inferior form of vision, emulating the emphasis placed on geometry (e.g., shape and size) by its mainstream cousin. The research in our two chapters should help to dispel this narrow perspective.

To be sure, touch can and must be used as a less precise substitute for vision, especially when the exploring arm/hand occludes one's sight of the target object, or when the illumination is too low to permit focal vision. However, it can provide information about surfaces not available from a given visual perspective, such as the hidden back or sides of an object. Also, given the anatomical and mechanical structure of the hand, it can provide 3-dimensional information about both the object's external and internal structure (e.g., holes). Our chapters also emphasise the complementary functions of vision and touch. Vision is indeed exceptionally effective at obtaining structural information by the extraction of contours. In contrast, the sense of touch demonstrates a particular competence for extracting the material characteristics of objects (e.g., surface texture; strain-related properties such as compliance, brittleness, elasticity; thermal properties). The special ability of touch to sense vibration and static deformation can be used to detect the occurrence of an event, both at a distance or at a site directly on the skin. Its ability to detect slip is critical for maintaining a stable precision grasp between the thumb and index finger. Finally, we will suggest that many of the cues mentioned above may further prove important in achieving the fine, post-grasp manipulatory skills that living organisms demonstrate within highly unstructured environments (see our companion chapter in this book).

8.1.2 Definitions: Tactile, Kinesthetic, Haptic

In keeping with the review by Loomis and Lederman (1986), the term "tactile" is used to refer to processing of solely cutaneous information. The term "kinesthetic" is used to refer to processing of inputs from receptors embedded in muscles, tendons, and joints. The term "haptic" describes the processing of both cutaneous and kinesthetic inputs for perceiving and guiding actions within the real world. While we do not directly discuss kinesthetic perception, the interested reader is referred to Clark and Horch (1986).

8.1.3 Focus and Chapter Outline

Our two chapters in this book discuss selected aspects of biological tactile and haptic sensing, respectively, which we judge to be most relevant to robotics and telerobotics. The chapters focus primarily on perceptual processing by humans and monkeys, since most of the biological research has been performed with these organisms. We discuss both "psychophysical" (as well as other behavioural) and "neurophysiological" experimental approaches. The former is a non-invasive technique that studies the response of the whole living organism to the controlled variation of physical stimulus parameters. The latter is more invasive, since it directly examines the function of an organism's nervous system and its various subsystems. Collectively, these methods allow investigators to determine precisely which stimulus parameters (for example, any of those suggested by the first principles of physics) are critical to the performance of the system under question, be it the response of a single neuron, a network of neurons or that of the entire organism. These methods can therefore be used in conjunction with analytical methods to determine the most relevant properties of biological systems (e.g., insects, birds, mammals, primates, etc.) known to function effectively in tasks similar to those that autonomous and/or telerobotic systems might be designed to perform.

The remainder of this introduction considers the potential benefits from studying biological touch. It also includes a mini-tutorial on the basic principles of the scientific method, with particular emphasis on the experimental method. The tutorial is intended to help roboticists understand the general perspective and methodological approaches that they will encounter when consulting the biological literature.

Section 8.2 considers skin mechanics, and its implications for cutaneous processing. Section 8.3 describes cutaneous sensitivity, addressing the cutaneous input-output relation and its limits of resolution. In sections 8.4 and 8.5, respectively, we discuss the cutaneous processing of fine surface textures and micropatterns, where the constraints imposed by the cutaneous input-output relation become critical. In section 8.6, thermal sensing is also briefly considered, given its potential significance in robotic applications. Section 8.7 raises a number of critical issues pertaining to the integration of multiple spatially distributed cutaneous inputs. The relevance of biological cutaneous systems for the design of effective robotic tactile sensing is considered throughout the chapter.

8.1.4 What Can the Study of Biological Touch Contribute to Robotics?

It is possible to describe a continuum along which various approaches to robotic design would fall. At one extreme exists the unquestioning attempt to mimic living systems; at the other end prevails an equally blind rejection of this anthropomorphic approach. There is, however, another way for roboticists to approach the study of biological systems, one that attempts to learn generally from new conceptualisations and methodologies, as well as from specific empirical findings, about how living systems approach tasks that the roboticists have not yet solved.

We would agree substantially with Whitney (1986) that when the environment can be precisely controlled or modified, as with many automated industrial environments, there may be little value in copying human processing. Indeed, a robot might perform such tasks better using some entirely different processing method. Yet, as we note below in greater detail, many of the conceptual and methodological issues we treat in the remainder of this section would still be highly relevant. There are also many robotic environments over which humans can exert little, if any, control. Consider, for example, the highly unstructured environments encountered by autonomous or telerobotic systems servicing a space station, exploring an unknown planet, recovering aviation and ship wreckage from deep or murky waters, or applications such as agriculture, radioactive waste disposal, and microrobotic surgery. Here, we argue, it should prove valuable to learn, in addition, how complex living systems specifically process information to achieve relatively flexible and efficient performance. The specific empirical details of biological tactile processing are also the subject of our two chapters.

8.1.4.1 Overlapping Problem Domains

The scientific study of biological systems has for quite some time addressed many of the same sensing problems that now challenge roboticists. Consider, for example, sensor performance, sensor fusion, selection of primitives for scene segmentation and object recognition, feature extraction algorithms, active versus passive perception, selective attention, control and planning, grasping and manipulation, and sensory-guided motor control. For both the biological and robotic domains, such topics are enormously complex. Both domains must address critical hardware considerations to understand fully the nature of the constraints imposed, as well as their effect on system performance. Our chapters highlight the case with biological touch, and further suggest ways in which specific findings may be useful for designing robots with tactile/haptic sensing capabilities.

8.1.4.2 Example of, and Framework for, Designing Working Multi-level, Integrated Systems

Also critical to our argument is the fact that biological systems are complete, multi-level integrated systems that work. Any of the overlapping problems selected (e.g., see section 8.1.4.1) from the domain of biological intelligence will demonstrate the complexity of considerations required by the roboticist. For our purposes, we include sensing, thinking, and motor control/planning in the definition of intelligence.

The study of biological systems also provides comprehensive multi-level conceptual frameworks within which to organise the overall task of designing sensor-based robotic perceptual/control systems. Both roboticists and biological scientists have found it easier to treat the different sensory modalities as independently functioning modules, or to disregard higher-level cognitive and motoric processing when studying sensing. However, the most recent work with various biological systems makes this fallacy, and its consequences, abundantly clear.

8.1.4.3 Application to the Design of Telerobotic Systems

As robotics has come to recognise the daunting complexity of developing intelligent autonomous systems, it has been suggested that, temporarily, the human operator could be retained in the control loop in an attempt to speed the development process by co-opting the considerable sensory, cognitive, and motor capacities of the human system. Those of us who study human systems are quick to point out that it is absolutely critical for the roboticist to appreciate fully the nature and complexity of these human subsystems with which the telerobotic system must be interfaced. No artificial system will work without a sufficient understanding of the human subsystems, regardless of the elegance of its design. While we devote most of our chapters to issues that pertain to autonomous robotic systems, the material should also be of value to those who design telerobotic systems with tactile or haptic sensing capabilities. Issues pertaining to how the sensory information from the remote robotic site can be presented most effectively to the human within the operator environment would presumably be most relevant. We do not, however, expect that the sensory information would necessarily be processed in identical ways in telerobotic versus autonomous robotic systems.

8.1.5 A Mini-tutorial on the Scientific Method in Biological Research

In engineering and robotics, it is traditional to model a system analytically, using differential equations with parameters set by the first principles of physics. However, since the systems are typically too complex to handle as such, approximations must be made. The reduced parameter selection is performed arbitrarily by the investigator.

The scientific method offers a considerably different approach to the study and design of sensor-based systems, both living and artificial. Regardless of the focus of study (e.g., cosmology, physics, biology), the scientific approach must involve careful, systematic, repeated observation. When in addition, we actually manipulate the parameters under study in a controlled fashion, we are said to be using the "experimental method". As we discuss below in greater detail, the scientific study of systems provides a systematic, coherent framework within which to perform one's investigations, regardless of the problem or system domain (living or artificial) involved. It provides formal methods (i.e., experimental and statistical) for systematically and rigorously testing the validity of one's hypotheses, based on empirical results. We see such issues as also being critical to the successful development of intelligent autonomous sensate robotic systems.

The rigorous principles and methodologies of the experimental method expose some of the inherent weaknesses and limitations of current robotic practices. We argue here that in designing sensor-based systems, the analytic methods used by roboticists can be powerfully complemented by including additional scientific tools, and by considering the implications of scientific knowledge of biological systems. We now consider in greater detail various aspects of this approach and implications for robotic design.

8.1.5.1 Constraining the Problem

The scientific method demands the precise specification of the general problem domain. Consider, for example, human (or robotic) texture extraction. The specification must clearly state both what the domain does and does not include (e.g., the texture extraction procedure may only apply to rigid surfaces, or to a specific texture dimension such as roughness, spatial density of the elements, or element regularity), so that other investigators can judge the appropriateness of the proposed solution for their own problem domains (e.g., compliant objects, or a different sub-dimension of texture). We argue that the judicious selection of clearly specified constraints can result in highly effective designs. This requirement should also advance the field more quickly given the greater likelihood that solutions will transfer successfully across laboratories.

8.1.5.2 Formulating Hypotheses

Having clearly specified the problem domain, the experimenter must next formulate specific hypotheses (e.g., which parameters affect the algorithm(s) by which humans or robots perceive surface texture, however that term may be defined). It is also incumbent upon the investigator at this stage to consider alternative hypotheses. Careful selection of the specific task and experimental design used to test these hypotheses, coupled with various statistical techniques for analysing the data, allow the investigator to evaluate the extent to which the proposed hypothesis, or some alternative, is supported by the results for this type of task (e.g., the hypothesis that the magnitude of perceived roughness of uni-dimensional gratings is some function of the spatial period of the surface elements, rather than, for example, its element width and inter-element spacing).

8.1.5.3 Designing an Experimental Study

In designing the experiment (or usually, a set of experiments), the investigators will manipulate the variable(s) critical to their hypothesis (e.g., if using uni-dimensional gratings to vary surface texture, perhaps varying element width, inter-element spacing, duty cycle, spatial period; see section 8.4), while simultaneously holding constant other parameters that might also affect the outcome of the experiment. We note that as there is a dearth of off-the-shelf technologies in the tactile field, this usually requires the investigator to custom-design the stimuli, the equipment for presenting them, and the apparatus used to measure the subject's responses. It is imperative that the subjects (e.g., human, monkey, robotic tactile sensor, or complete robotic system), the stimulus objects (e.g., gratings, 2-dimensional raised dot patterns, different materials), and both the task and measurement procedures (including the apparatus) all be carefully selected and fully described. As the experimental results must be repeatable, considerable effort is spent initially in piloting the experiment until a satisfactory set of specifications is obtained. Any published reports must contain sufficient information for any investigator to reproduce reliably the results in the same or another laboratory. Although tremendously time-consuming, all are necessary for the execution of high quality scientific research.

8.1.5.4 Experimental Bias in the Results

It is easy to bias the results experimentally if one is not careful in selecting the subjects used to evaluate a hypothesis. For example, arbitrarily using one person, especially the investigator, to test a prediction will produce highly biased results (this seems common in the field of robotics, as for example when the investigator alone judges the success of his or her analytic solution without the benefit of rigorous testing criteria). When a larger number of subjects is used to reduce potential bias, it becomes necessary to assign subjects randomly to different experimental conditions. Alternatively, when a subject performs all experimental conditions, the order in which the conditions are presented must be counterbalanced to avoid order effects, such as that due to fatigue over successive trials. Various order effects are also relevant to roboticists — sensor performance may decline with wear on the covering, or perhaps with changes in the thermal environment as a result of continued use. The same concerns must be expressed with respect to the number of, and procedures for, selecting the stimulus objects. Unwanted (and unrepresentative) effects can result from the anomalous characteristics of the particular objects selected. We note the tendency in robotics to draw broad conclusions from results obtained with just a few arbitrarily selected objects (e.g., many shape extraction algorithms have been tested on samples of only four or five objects). The experimental method not only forces the investigator to consider such potential sources of bias, but offers formal procedures for handling them.

8.1.5.5 Statistical Analysis of the Data

The experimental method also employs statistical tools to help describe and interpret the data collected. The statistical field can be viewed as the study of variation, because it offers formal tools for exploring the sources of variation in events and for drawing inferences about the causes that produce the variation. Although the investigator may wish to employ statistical techniques to explain performance of a given *population*, (e.g., all young adults aged 18 to 25 years, or all Lord Corporation Model LTS-210 array sensors), it is usually impractical or impossible to include all members of the population in the experiment. Thus, the investigator must adopt certain methods (e.g., random selection) to draw a representative sample from the specific population. Using statistical techniques, probabilistic statements may be made about the performance of populations (e.g., for young adults, the magnitude of perceived roughness of a specific set of textures varies as a function of the inter-element spacing), based on knowledge of the results obtained with just the one sample set used in the experiment (e.g., 15 subjects randomly selected from that same population).

This type of inductive reasoning is paramount in science, but we recognise that it is also subject to error. One important aspect of the scientific method is that the size of such error in drawing inferences can be measured with various quantitative statistical techniques. Hence, conclusions must be formally expressed in terms of probability statements about the statistical significance of the effect in question. Continuing with our texture experiment, let us suppose that the results of our statistical evaluation allow us to conclude that the effect of inter-element spacing on perceived roughness was significant at the 0.01 level of significance. This statement means that if we were to repeat the experiment not once, but 100 times, then *by chance alone* we would expect to find a perceived-roughness effect that is as large as that obtained in the actual experiment (i.e., the sample) in no more than 1 out of 100 such experiments. Another way of phrasing this is to say that we would predict that in 99 out of 100 such experiments, a change in the perceived roughness estimates of our designated population as large as that obtained in our actual experiment, must be due to the experimental manipulation of the parameter, inter-element spacing. We may be reasonably confident in concluding, therefore, that inter-element spacing influences perceived roughness, although we would be incorrect in our claims 1 out of 100 times. Taking another example from the robotic domain, we may be interested in the extent to which a specific tactile sensor (rather than human observer) can differentiate various textures within a set. Multiple measurements of each surface with the sensor can be similarly performed to evaluate statistically the effectiveness of that sensor in differentiating textures (for further suggestions on how to apply psychophysical methods to the evaluation of sensors and feature-extraction algorithms, see section 8.4).

Using various formal univariate and multivariate statistical tests, the experimenter can further draw quantitative conclusions about the relative contribution of various experimentally manipulated parameters to performance (in our example, the tactual perception of surface roughness). Where it is not feasible to control the other parameters (e.g., where varying the surface roughness by using different materials simultaneously alters thermal conductivity, compliance, and so forth), it is still possible through statistical techniques to separate their effects, provided their values can be measured. Finally, we wish to re-emphasise the fact that experimental (scientific) and statistical methods recognise *variability* in performance, both within (e.g., because of individual subject variation) and across experimental conditions (because of a specific experimental manipulation). Statistical tests determine the effectiveness of an experimental parameter by considering whether the variability across parametric conditions is greater than that within, and thus, whether variation in the values of a parameter "significantly" affects performance. These same concerns should be relevant for roboticists as well. For example, there will in fact be variation in the individual tactile sensors mounted on separate end effectors of a dextrous hand, or in the performance of a single sensor on multiple occasions. The critical consequences of knowing the variability in one's results, as well as the extent to which one's results may be generalised, are rarely formally addressed or documented within the field of robotics.

8.2 Skin Mechanics

As the mechanoreceptive transduction units (for a full description of the skin and its receptive units, see Chapter 7 by Moss-Salentijn in this book; we shall use the terminology she provides in both our chapters) receive signals through the mechanical medium of the skin, the influence of skin mechanics on tactile perception is, potentially, an important consideration. However, this effect is also quite complex, as human skin is a highly structured, non-linear, non-homogeneous, viscoelastic material. Many models of skin mechanics have been proposed; the interested reader is referred to a summary article by Lanir (1987), which also provides an overview of the properties of skin. Here, we will only mention work that directly considers the influence of the mechanical structure of skin on tactile perception (for additional details concerning perceptual consequences, see the remaining sections in this chapter).

The mechanics of the skin affect both spatial and temporal aspects of cutaneous information processing. Spatially, when a probe is statically applied to the skin, a primary indentation is produced beneath the face of the probe. However, in addition, the probe creates a much wider but shallower secondary depression that surrounds the primary one (Quilliam, 1978). For a stimulus with very high spatial detail, this mechanical spreading effect will produce a deformation pattern that is only a smeared facsimile of the stimulus. Such blurring has considerable impact on spatial information processing at the fingertip, where biological research has shown that peripheral (*cf.* central) neural factors limit resolution. In addition, stimuli which contain a temporal component can create travelling waves on or beneath the surface of the skin, further blurring the pattern of mechanical stimulation (Loomis, 1981b). Unwanted effects of mechanical blurring should also concern roboticists, since adjacent sensing elements within a tactile sensor may be passively triggered by the mechanical response of the sensor covering to deformation, including travelling waves.

Phillips and Johnson (1981b) have developed a mechanical model of fingertip skin to predict the specific local conditions (i.e., stress or strain components) monitored by each mechanoreceptor's endings, when static stimuli having a defined spatial profile (e.g., bars, edges and gratings) are applied. To make the model tractable, Phillips and Johnson assumed the skin's mechanical properties could be adequately approximated by those of a homogeneous, isotropic, elastic medium with infinite depth and a smooth flat surface, and that Hooke's law applies. They further assumed that each afferent unit terminates in a single ending, whose discharge rate is proportional to the magnitude of the strain at that point, no matter which component is considered. Despite these simplifications, there was a marked similarity between the model's predictions and the experimental results. Comparisons between the stresses and strains of the skin model and the experimental neural responses of monkeys (Phillips & Johnson, 1981a) suggested that SAI units are specifically sensitive to the local maximum compressive strain regardless of orientation; FAI units are specifically sensitive to horizontal tensile strain. More recently, van Doren (1989) has used similar assumptions to develop a model that predicts the tactile sensitivities of the psychophysical channels associated with the FAI and FAII units.

However, none of these skin models has addressed the contribution of the highly structured ridge organisation of the fingertips (see Chapter 7 by Moss-Salentijn, in this book) to cutaneous information processing. Cauna (1954) first proposed that the micro-lever action of the intermediate ridges, which move freely within the dermis layer, might enhance tactile spatial acuity by transmitting magnified signals from the surface of the skin to the mechanoreceptors below. Lederman (1978) subsequently suggested that this mechanism might also be involved in tactile texture perception.

Finally, Johansson and LaMotte (1983) have proposed that the ability to detect very small asperities (e.g., a 1 μm edge) with the fingertip during active exploration is due to local lateral deformation of the papillary ridges; the deformation that results from a raised dot moving over a ridge may be larger than the height of the asperity. In robotics, Fearing and Hollerbach (1985) have suggested moulding a textured pattern into the surface of the rubber skin of a robotic finger to increase detectability. Their proposed texture, parallel ridges, is similar to the biological papillary ridges of the primate hand.

8.3 Cutaneous Sensitivity

In the previous section, we indicated that skin mechanics can influence tactile perception. However, the subsequent stages of neural processing also play an important role. In principle, perception will be limited by the spatial and temporal filtering properties of the cutaneous nervous system. One biological experimental approach, which uses both psychophysical and neurophysiological techniques, involves determining characteristics of human (or monkey) cutaneous sensitivity. Stimuli typically include punctate and other relatively simple stimuli whose amplitude, spatial and/or temporal features can be precisely controlled.

We will first describe cutaneous sensitivity in terms of its amplitude, spatial and temporal features. As it has been difficult in the past to evaluate these factors conjointly, it has been traditional to study each separately. With the advent of more sophisticated technologies, interactive effects have been experimentally observed recently; where relevant to our current discussion, they are noted. These data address issues pertaining to the nature of the cutaneous spatiotemporal transfer function, which will then be briefly considered, as its filtering properties have profound effects on cutaneous processing.

For autonomous robotic systems, examination of the cutaneous system at this level of approach provides ideas for the construction of tactile sensors and sensing systems, which will be considered throughout this section. The results presented are also directly applicable to teleoperations as they specify what is required to provide tactile "telepresence". The information provided to the operator may not need to be this precise, as humans function well with minimal information; however, one should understand the consequences of depriving the human cutaneous system of this accuracy.

8.3.1 Amplitude Resolution

The amplitude resolution of the cutaneous system has been considered both in terms of pressure (i.e., force per unit area, or, mistakenly, just force, see below) and deformation of the skin; the designation of either as the "more fundamental" parameter remains in question. However, for either parameter, amplitude resolution is measured by determining the threshold of detection or its inverse, sensitivity, to a step function input. Two variables are typically considered in these measurements. The first relates to whether the stimulus is applied statically (i.e., as a constant value) or as a vibration (i.e., repeatedly applied at a given temporal frequency; for a detailed discussion of vibration, see section 8.3.3). The second variable relates to the type of resolution evaluated: absolute or differential. Absolute sensitivity is measured as the inverse of the minimum stimulus value that must be applied for the subject to just detect the stimulus some arbitrarily specified percentage of the time (e.g., 75%). The minimum stimulus value is referred to as the absolute threshold. In contrast, differential sensitivity is measured as the inverse of the minimum change in the stimulus value required to just notice a difference in the stimulus some specified percentage of the time (Lederman, 1991). The minimum change value is referred to as the difference threshold.

8.3.1.1 Psychophysics

Weinstein (1968) performed the definitive examination of pressure sensitivity, in which he found that absolute pressure sensitivity varied with body locus, sex and laterality. The absolute thresholds obtained ranged over 5 – 355 mg force across these factors (note that these measurements should have been reported in terms of force per unit area to reflect variation in pressure). The most sensitive position examined on the hand was found to have a threshold of about 16.0 mg force. Absolute sensitivity to vibration is also known to vary across body locus, although not in the same manner as static pressure. Differential pressure sensitivity has been described in terms of the Weber fraction, the ratio of the difference threshold to the initial stimulus value. Weber fractions of approximately 0.14 for static pressure and 0.20 for a vibratory frequency of 160 Hz have been found for the finger. In general, Weber fractions for pressure typically range between 0.20 and 0.25 over the frequency range of 20 – 300 Hz (Lederman, 1991).

For displacement, Johansson and Vallbo (1979) have determined the absolute static threshold of the finger to be about 10 μm (for a hemispherically tipped probe of diameter 0.45 mm). They have also noted that pre-indentations of 0.25 – 0.5 mm could decrease this threshold by a three- to five-fold factor. The size of the contactor (e.g., the probe tip) has also been found to affect the displacement threshold. Johansson and LaMotte (1983) have suggested that when lateral movement is allowed, a larger contactor size may produce larger local deformations of the papillary ridges, thus improving detectability (see section 8.2). In addition, contactor size is an important stimulus parameter in the determination of vibrotactile thresholds (see section 8.3.3). In fact, extensive studies using vibratory stimuli have shown that displacement thresholds vary with many stimulus parameters; under certain stimulus conditions, vibratory thresholds as low as 0.1 μm have been measured (Bolanowski *et al.*, 1988).

In contrast to the exquisite sensitivity to external stimuli, Hulliger and his colleagues (1979) have observed that the excitation of a large number of mechanoreceptive units due to voluntary finger movements does not result in similar strong conscious tactile sensations. They suggest that such mechanoreceptor responses are differentiated from those due to direct skin contact, and are redirected to assist in the control of motor reflexes. Although there remain many unknowns, this research area (see also Moberg (1983)) should provide valuable suggestions for the processing of tactile information during active exploration in the robotic domain. The contribution of cutaneous information provided about external objects to motor control will be discussed in the accompanying chapter.

8.3.1.2 Neural Coding

Johansson and Vallbo (1979), and Johansson *et al.* (1980) have compared psychophysical and physiological (in the neural periphery) absolute static thresholds for both force and displacement in the human hand. They found that only the Fast Adapting (FAI and FAII) units had thresholds low enough to account for the psychophysical detection of force and displacement, although circumstantial evidence indicated that solely the FAI units may provide the decisive inputs for at least displacement (force was not examined in these particular studies) (Johansson & Vallbo, 1979). Although capable of providing the requisite sensitivity, the FAI and FAII units do not necessarily set the detection capacity of the entire cutaneous system. Comparisons of the psychophysical and mechanoreceptor displacement thresholds in the human hand do suggest that the FAI and FAII units seem to set the detection capacity in the finger. However, in contrast, the detection capacity in the centre of the palm seems to be set by more central neural mechanisms (Vallbo & Johansson, 1984).

8.3.2 Spatial Sensitivity

The spatial sensitivity of the cutaneous system has been measured as the minimum discriminable spatial separation between elements. This threshold defines the spatial detection limits of the spatial transfer function (see section 8.3.4), which has been shown to have profound effects on the perception of spatial patterns (e.g., see section 8.5).

8.3.2.1 Psychophysics

In addition to pressure sensitivity, Weinstein (1968) also examined spatial sensitivity as a function of body location, laterality and sex. The measures used were the two-point touch test and point localisation; the two-point touch test determines the minimum distance at which subjects experience two separate stimuli, whereas point localisation measures the accuracy with which a point applied to the skin can be localised (Figure 8.1). Weinstein found that spatial sensitivity varied with body location (smaller parts tended to be more sensitive and there appeared to be an increasing proximal to distal gradient) and possibly with laterality. Some of the

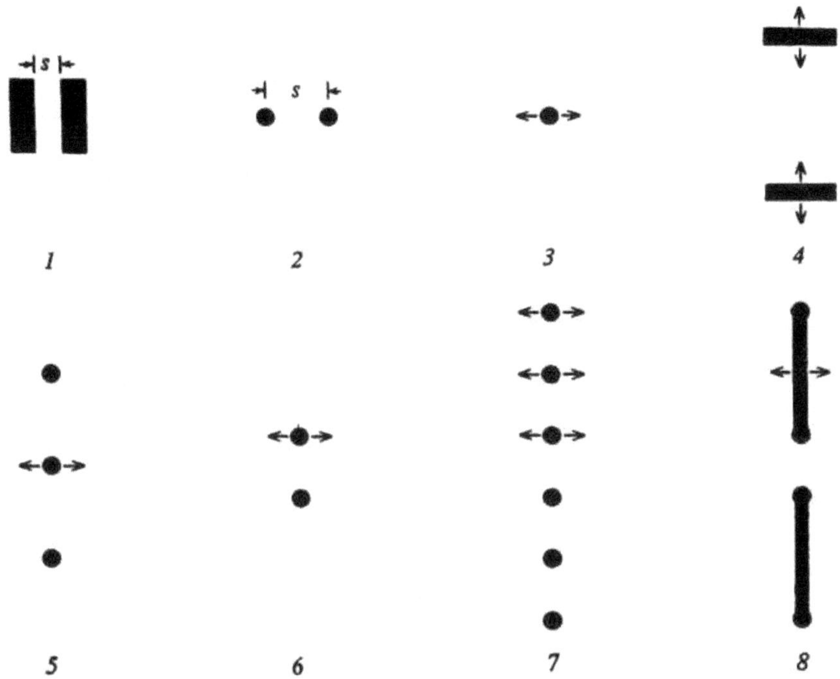

Figure 8.1: *Examples of spatial resolution and hyperacuity tasks. The first two tasks are measures of spatial resolution, the minimum distance at which subjects detect two separate stimuli;* **(1)** *is a König two bar target, and* **(2)** *is a two-point touch target. The remaining targets are examples of hyperacuity tasks.* **(3)** *is a point localisation target, which subjects judged to be either left or right of an internalised reference position.* **(4)** *is a spatial interval target, which subjects judged to be either wider or narrower than a standard interval.* **(5)** *is a misalignment target, where misalignment of the centre dot must be discriminated.* **(6–8)** *are vernier targets, where discrimination is of the displacement of the upper half of the targets. Reprinted with permission from Loomis (1981b).*

underlying factors of the variation with body location are: (1) static mechanical properties of the skin, (2) travelling waves of vibration resulting from dynamic stimulation, (3) innervation density and spatial sensitivity of the different classes of mechanoreceptor units, and (4) cortical magnification (i.e., the size of the cortical area per cutaneous surface area subserving each body part) (Loomis, 1981b).

Johnson and Phillips (1981) have used more refined techniques to determine the spatial resolution of the distal pad of the finger: (1) gap detection, (2) square wave grating orientation detection, and (3) raised letter recognition. All three tasks determined the spatial resolution to be 0.9 mm. A modified version of the third test was performed (Phillips *et al.*, 1983) to examine whether relative motion between the finger and stimuli affected spatial discrimination. In these studies, subjects were allowed to scan the stimuli freely. From their results, Phillips and his colleagues extrapolated that the resolution limit decreases to 0.7 mm when relative motion is allowed. However, this does not account for the major improvement in performance which normally occurs when scanning fine surface features (i.e., textures); the actual stimulus parameters involved will be considered in section 8.4.

Loomis (1974; 1979; 1981b), and Loomis and Collins (1978), found tasks in which cutaneous spatial discrimination capacities are well below the spatial resolution determined by Johnson and Phillips. These tasks are termed *hyperacuity* tasks (Figure 8.1), and the resulting behavioural effects are exhibited by both touch and vision. For touch, the threshold obtained for one such task, point localisation, was 0.17 mm, which is significantly below the spatial resolution (0.9 mm) and mechanoreceptor spacing (typically not less than 1 mm (Johansson & Vallbo, 1983)). Yet, although the discrimination accuracy for hyperacuity tasks is measured below the spatial resolution of the cutaneous system, the underlying mechanisms are still expected to be involved in the perception of suprathreshold stimuli in conjunction with the mechanisms being examined by Johnson and Phillips above (Loomis, 1979).

Loomis has proposed a model to explain the discrepancy between the spatial discrimination capabilities in the hyperacuity tasks and spatial resolution; the model is similar to the one proposed for visual hyperacuity (Loomis, 1979). In this model, spatial resolution is attributed to the low pass filtering of the spatial information. In contrast, hyperacuity is accomplished by using the phase information in the lower spatial frequencies; the discrimination capabilities are achieved by the cutaneous system's sensitivity to phase shifts (Loomis, 1974). Physiologically, Loomis proposed that localisation of a point stimulus (a hyperacuity task) is determined by the relative response magnitudes, possibly their ratio, of a small group of overlapping mechanoreceptor units around the stimulus. In the two-point touch task (a spatial resolution task), Loomis suggested that the discrimination is determined by equal responses in two sets of units and a smaller response in the units between these two sets, with the difference determined by some criterion value (Loomis & Collins, 1978; Loomis, 1979).

Loomis's model of hyperacuity has direct implications for the design of tactile sensors. It suggests that spatial discrimination can be increased by using the spatial phase information available from tactile arrays; thus, accurate discrimination is not restricted solely to a tactel spacing solution. In addition, the physiological mechanism proposed for hyperacuity offers some advantages: (1) the use of a ratio of adjacent units could avoid sensitivity to non-uniformities in the sensor "skin", and (2) the use of a small group of units could possibly reduce sensitivity to electrical noise, while not introducing a prohibitively expensive calculation from a computational perspective.

To determine the characteristics of the cutaneous spatial filter, ideally spatially varying sinusoids with control over both frequency and amplitude should be used.

Unfortunately, until recently, there was no such technology available (van Doren *et al.*, 1987). The data above all use aperiodic stimuli (e.g., points, bars, rectangular gratings), which introduce higher spatial frequency components. However, they probably serve reasonably well as a good first approximation of cutaneous spatial sensitivity.

8.3.2.2 Neural Coding

Examination of the peripheral neural coding of spatial patterns (Johnson & Lamb, 1981; Phillips *et al.*, 1990) has suggested that the SAI units, as well as possibly the FAI units, are responsible for the spatial resolution capabilities of the human hand (see section 8.5.2 for details concerning this work). This conclusion seems applicable to the SAI and FAI populations as a whole due to their small receptive fields and relative uniformity over a local area (Vallbo & Johansson, 1984). In addition, Vallbo and Johansson (1978) have found that the increasing spatial acuity gradient from the proximal to distal regions in the hand corresponds to the gradient of the sum of the FAI and SAI units. These results also suggest that the type I units are the essential populations involved in encoding detailed spatial features. Thus, as in foveal vision, fingertip spatial resolution appears to be determined by the spacing of the peripheral receptors. Foveal spatial resolution is primarily set by the spacing of the cones, and fingertip resolution is set by the FAI and SAI units (Johnson & Phillips, 1981).

8.3.3 Temporal Sensitivity

The large majority of studies have measured temporal sensitivity of the cutaneous system in terms of the absolute vibrotactile threshold as a function of temporal frequency. This varies with: (1) the contactor size used to vibrate the skin, (2) the presence or absence of a contactor surround (i.e., the restriction or allowance of surface waves), (3) the stimulus duration, (4) temperature, and (5) selective adaptation (i.e., the previous exposure to an intense vibration at a selected frequency) (Lederman, 1991).

8.3.3.1 Psychophysics

A model to account for the psychophysical variation of vibration thresholds with space and time was originally proposed by Verrillo (1968), and has subsequently been refined by him and his colleagues (Verrillo & Gescheider, 1979; Gescheider, 1979; Capraro *et al.*, 1979; Bolanowski *et al.*, 1988). The model has been developed by the examination of the vibrotactile threshold over the range of 0.4 – 500 Hz, primarily in the hairless skin of the human hand. This series of experiments included: selective adaptation, selective masking (i.e., the intensity of a "masking" stimulus of a given frequency necessary to just impair the perception of a test stimulus at

channel	**P**	**NPI**	**NPII**	**NPIII**
range	40 – 800 Hz	10 – 100 Hz	15 – 400 Hz	0.4 – 100 Hz
shape	U-shaped	partially sloping	gently U-shaped	relatively flat
sensitivity	max. ≈ 300 Hz	less to changes in freq. than P channel	much lower than P channel	comparable to the NPI channel
exhibits spatial summation	yes	no	not known	no
exhibits temporal summation	yes	no	yes	no
sensitive to changes in skin-surface temperature	yes, highly	possibly	yes, highly	yes
mechano-receptive unit involved	FAII	FAI	SAII	SAI

Table 8.1: *Summary of the properties of the four channels mediating the vibrotactile threshold characteristic, as described in Bolanowski* et al., *(1988).*

a selected frequency), modification of the skin-surface temperature, modification of the stimulus size, and modification of the stimulus burst duration.

Presently, it is believed that four distinct channels, P, NPI, NPII and NPIII, are involved in the determination of the vibration threshold. The properties of the four channels are summarised in Table 8.1. The overall psychophysical threshold at a particular frequency is normally determined by the channel having the lowest threshold at that frequency, with the threshold being a function of the stimulus size, duration and skin-surface temperature. The channels partially overlap in their absolute sensitivities, which makes it likely that with "everyday" stimuli more than one channel will be activated at the same time (Bolanowski *et al.*, 1988).

The properties of this four-channel model suggest a more realistic aim for the design of robotic sensors as, although the entire cutaneous system is sensitive to vibration over a very broad range of frequencies, each channel contributes to the threshold frequency response over only part of this range. This suggests that rather than having one "perfect" tactile sensor capable of being highly sensitive over a wide range of frequencies, more than one type of sensor could be used; each would be limited in accuracy to different frequency ranges, different stimulus dimensions and/or different functions. Thus, tradeoffs could be made in the design of each sensor, while not degrading the system's overall response. This separation into different channels would also provide for a certain degree of parallel processing of the data, alleviating computational bottlenecks which could occur (an important consideration for real-time systems). "Perception" would occur by the integration of the information from these different channels, as well as from other types of sensors.

In closing this section, we also note the use of other experimental methodologies for determining the temporal resolution of the cutaneous system. For example, the time between two 1 msec pulse stimuli that is necessary for an observer to experience them as successive is 5.5 msec (Gescheider, 1974).

8.3.3.2 Neural Coding

Freeman and Johnson (1982a; 1982b) have examined the response of the mechanoreceptor units in monkeys (SA, RA and PC units) to vibrations of varying amplitude and frequency. They found that the responses of each type of unit to vibratory stimuli could be described as temporal bandpass filters with a centre frequency of 20 Hz for the SA units, 40 Hz for the RA units and 250 Hz for the PC units.

Johansson and his colleagues (1982) have examined the responses of the mechanoreceptor units in humans (SAI, SAII, FAI and FAII units) to vibrations of varying amplitude and frequency. They found that at close to absolute threshold, the FAI units responded most easily within the range 8 - 64 Hz, the FAII units most easily above 64 Hz, the SAI units most easily between 2 - 32 Hz, and the SAII units most easily below 8 Hz; this concurs well with the above results obtained by Freeman and Johnson for monkeys. They also observed that the higher the stimulus amplitude used, the lower the frequency at which the response was maximal and there occurred a rapid increase in the frequency range covered by each unit.

It is important to note that, especially at higher intensities, there is considerable overlap of the range of frequencies at which the different types of units respond (Johansson & Vallbo, 1983). Thus, information is redundantly encoded by different processing channels, possibly to facilitate parallel processing of the data for differing purposes. This parallelism could allow for the exclusion of high density tactile information, such as provided by the FAI and SAI units, from feedback loops where high speed rather than spatial accuracy is necessary; this is true for the general detection of contact. In fact, certain properties of the FAII units suggest its use in detecting unanticipated contact (anticipated contact, in contrast, may be smooth and not always excite the FAII units, see section 9.3.1 on Cutaneous Information for Motor Control in our companion chapter in this book).

In particular, the ability of the FAII system to sum both in time and in space suggests that it could be sensitive to the beginning of contact (as well as to the breaking of contact); the initial displacement of the skin may not be great, but contact could be detected by summing over the area of contact and the initial time period. It should be noted that the specific contact location on a limb cannot be determined by the FAII units; however, to switch control modes (e.g., from position control to force control), the specific location of contact is not required.

8.3.4 Spatiotemporal Transfer Function

The stimulus and response of the cutaneous system can be described in terms of the dimensions of surface location and time; the functional relation among them is called the spatiotemporal transfer function. It has been traditional to assume that this function is linear (although more recently, some neurophysiological studies have considered the non-linear aspects of the temporal transfer function of tactile neurons in the femoral tactile spine of the cockroach to mechanical stimuli (French & Wong, 1977; Korenberg *et al.*, 1988; French & Korenberg, 1989)). The spatial and temporal dimensions of the transfer function have typically been treated separately, and we will therefore do so below.

The empirical data presented in the previous sections on cutaneous sensitivity offer some relevant information about the spatiotemporal transfer function. It is generally assumed that the transfer function derived from these relatively simple laboratory stimuli can also be used to reconstruct the response of the cutaneous system to more complex stimuli. To model the complete spatiotemporal transfer function, however, it would be necessary to specify its characteristics at each body locus and across the full amplitude range. Nevertheless, the existing data do include the most relevant body sites (e.g., the palm and fingers) and the most critical amplitude values for robotic and telerobotic design, respectively.

The amplitude sensitivity data help to define the boundaries of the amplitude range over which the spatiotemporal transfer function holds; they also specify the quantisation error, which varies over the amplitude range. These data would seem to be most relevant to telerobotic design.

The spatial resolution data specify the highest spatial frequency of the spatial transfer function. We also mention here additional approaches used by Loomis (1981a), and by Johnson and Phillips (Johnson & Lamb, 1981; Phillips *et al.* (1990)), which address the shape of the cutaneous transfer function. These two approaches are considered in greater detail in section 8.5. Finally, we note that the spatial resolution data and Loomis's model of cutaneous hyperacuity collectively provide some insight into the underlying mechanisms of the spatial transfer function; these suggest unique algorithms and implementations for autonomous robotic systems.

The temporal resolution data provide estimates of the upper temporal frequency boundary of the temporal transfer function, which is most relevant to telerobotics. In contrast, the models of vibrotactile sensitivity (both psychophysical and neurophysiological) may be more relevant to robotics, as they emphasise that the temporal transfer function in fact derives from the parallel activity of four independent subchannels, each with its own transfer function characteristics. The multichannel approach permits an efficient solution to the computationally intensive problem of tactile processing by the fingertip.

The reader is referred to Loomis and Lederman (1986) for further consideration of the cutaneous spatiotemporal transfer function and constraints on interpretation. Its consequences for tactile perception will be considered in sections 8.4, 8.5, and 8.7 on texture perception, spatial pattern perception, and the processing of spatially distributed inputs, respectively.

8.4 Texture Perception

The term "texture" has been used in a variety of ways. The domain has been broadly defined to include all physical properties of objects, liquids, and materials, with the exception of macroshape and thermal properties; some examples are roughness, hardness, stickiness, elasticity, slipperiness and oiliness (Loomis & Lederman, 1986). It has also been more narrowly defined to include only very fine surface micropatterns, whose element positions cannot be behaviourally or neurally resolved spatially; we shall use this narrower definition. Roboticists might well group these "texture patterns" with larger fingertip- (i.e., sensor surface-) sized patterns, and apply spatial algorithms for feature extraction. However, to date, biological research has found that intensity-based, not spatial, algorithms are likely used for processing texture.

The sense of touch is particularly effective at judging the texture of objects, and in many cases is better than vision. It can be a diagnostic feature of some object classes when identified by touch (see our companion chapter in this book). It may also be used to speed object recognition, because the textural homogeneity of many objects permits the observer to sample an area of the object quickly and precisely by a quick rub, compared with the relatively slow and imprecise contour extraction methods used to derive geometric models of objects. Texture may also serve to define the parts of an object. Finally, it may prove important for grasping and manipulation.

8.4.1 Psychophysics

The most extensive and systematic behavioural research on texture has investigated the perception of surface roughness, one of its most prominent dimensions. Owing to technological limitations, early experimental studies used fabrics and sandpapers (e.g., Katz (1989), and Stevens and Harris (1962)); more recently, however, investigators have used precisely controlled uni-dimensional gratings, and 2-dimensional (x, y), regularly varying or jittered (in which the position of each element randomly varies within a predefined circular area), raised-dot arrays. We will therefore restrict our discussion to the more quantitative results obtained with the gratings and raised-dot patterns.

8.4.1.1 Estimating Roughness Magnitudes

With respect to the geometry of 1-dimensional rectangular-wave gratings, Lederman and her colleagues (Lederman & Taylor, 1972; Lederman, 1974; Lederman, 1983; Taylor & Lederman, 1975; see also work concerning monkeys by Sathian *et al.* (1989)) found that perceived roughness strongly increased with increasing groove width (i.e., inner edge to inner edge), while it decreased only slightly with an increase in ridge width. Neither the spatial period nor the duty cycle had any significant effect on roughness judgements, when they were independently manipulated experimentally. The fingertip force applied to the stimulus grating played an important secondary role to groove width, with apparent roughness increasing as a function of increasing fingertip force (range 28 – 700 g). Speed of motion between the skin and stimulus grating had little effect on roughness estimates, the latter tending to increase only slightly with as much as a 25-fold change in speed (although motion was necessary; see section 8.4.3). The quantitative relation between perceived roughness (R) and the primary parameters, groove width (G) and fingertip force (F), was approximated by a power function FG^4. This model predicted and experimentally confirmed that large variations in the static coefficient of friction between skin and surface would have almost no effect on the perception of roughness. Taylor and Lederman (1975) therefore found it necessary to distinguish between (micro)friction caused by slipping materials, and the considerable effects of (macro)friction caused by specific geometric features (i.e., groove width) of the slipping objects. In contrast, roboticists do not appear to differentiate the two parameters in terms of their potential effects.

Given the relatively negligible role of speed of motion, Taylor and Lederman (1975) argued against the need for a dynamic model. They produced a quasi-static skin mechanics model, which showed that total skin deformation summed across the area of contact, an intensive parameter, best predicted the perceived roughness judgements obtained by Lederman (1974), rather than a temporal or explicitly spatial code. The experimental findings of Lederman and her colleagues further argued against Katz's earlier claim (Katz, 1989; first published in 1925) that vibration is the prerequisite for human roughness perception. Supporting evidence was obtained using several experimental paradigms: surface grating geometry held constant while speed of relative motion was varied, speed of relative motion held constant while surface grating geometry was varied, and selective adaptation of separate cutaneous channels that code vibration (Lederman *et al.*, 1982). In all cases, changes in the dynamic aspects of vibratory impulses produced in the fingertips failed to affect human roughness estimation, thus supporting the quasi-static model.

This of course does not preclude the idea of "dragging" a robotic "nail" across a surface to generate variations in temporal vibratory patterns that could be used for fine texture discrimination (algorithmically based on a temporal rather than an intensive code). While humans do not naturally choose to investigate textures in this way, Katz (1989) has shown that people can perform quite efficiently by dragging a hand-held probe across surfaces.

2-dimensional dot patterns with the same spacings in the x-y plane tended to yield results similar to the 1-dimensional gratings above (Lederman *et al.*, 1986). However, recently, Connor *et al.* (1990) showed that with increases in spatial period beyond about 3.5 mm, perceived roughness actually declined. Consequently, they argued for a quadratic relation between perceived roughness and spatial period. But, this interpretation remains open to question, as there appears to be some ambiguity about the meaning of the percept associated with the downward portion of the quadratic function. Similar experiments in our laboratory (Lederman, unpublished) indicated that for these larger inter-element spacings, subjects increasingly denied any experience of "roughness". Instead, they preferred to focus on the spatial percepts of the pattern elements, which were perceived to lie on a "smooth" background.

8.4.1.2 Texture Discrimination

Morley *et al.* (1983) have shown that humans are capable of discriminating gratings with variations in spatial period as small as 5%, while Lamb (1983a; 1983b) found that subjects could differentiate variations as small as 2% in the spatial period of 2-dimensional dot patterns. Although subjects may have been judging roughness, no specific texture dimension was specified in the instructions.

The standard psychophysical techniques employed here to obtain input/output functions relating various physical dimensions of texture to the magnitude of perceived roughness (as well as those for obtaining input/output functions for cutaneous sensitivity, as described in section 8.3) can also be used by roboticists (see, e.g., Fuma and Bajcsy (1988)). They may be employed, for example, to determine the response characteristics of any new tactile sensor, which in this case would serve as the "subject". They may be used to compare the relative performance of different sensors, given a common texture extraction algorithm, by contrasting the slopes of the obtained "psychophysical" functions; the steepness of the obtained slope(s) is critical in determining the response range, and the sensitivity of response within that range of each sensor/algorithm. Finally, they may be used in conjunction with a single sensor to compare the performance obtained with different algorithms.

8.4.2 Neural Basis

The psychophysical techniques mentioned above focused on behavioural measures of the perceptual experiences of the entire organism to various stimulus parameters describing the set of textured surfaces. In contrast, the following neurophysiological work focuses on which stimulus parameters are actually neurally encoded, and on the types of codes available. Potential neural codes include those based on intensity, time, or space. The type of code possibly employed can be independent of the dimensions of the physical parameters; any of the available neural codes may underlie the perceptual experience of a given physical parameter.

Most of the work to date has dealt with coding at the first level of the nervous system, in peripheral mechanoreceptive units arising in the glabrous (hairless) skin of anaesthetised monkey hands (this assumes that neural processing in anaesthetised animals is the same as that in fully awake animals). Various textured surfaces have been moved across the immobilised fingers of anaesthetised monkeys under controlled force and speed conditions.

8.4.2.1 Peripheral Coding

Working with anaesthetised monkeys, Sathian *et al.* (1989; for summary, see Sathian (1989)) moved gratings in a sinusoidal fashion to emulate the movements used by humans when judging texture (Lederman & Taylor, 1972). The purpose of these experiments was to examine the possible neural codes for the stimulus parameters experimentally varied in the human roughness perception work. First, they considered possible intensive codes (e.g., mean firing rate of a mechanoreceptive population, a ratio of the mean rates of different populations, etc.). They found that the mean discharge rate in all three mechanoreceptive populations of monkeys (RA, SA, and PC) increased as a function of *groove width* (with ridge width held constant). The mean rate of discharge by each of the three populations was also affected by *contact force*. *Spatial period* could be coded by the mean rate of SA firing, or by the ratio of the mean discharge rates of RA and PC units. Increasing *ridge width* only resulted in a minor decrease in PC response rate; there was no consistent response in either SA or RA units, although Sathian and his colleagues discussed the possibility of a code based on the number of impulses per spatial period of the grating in all three mechanoreceptive populations.

In addition to intensive codes, Sathian and his colleagues found that a temporal code was available in the temporal frequency of the afferent signal bursts, which, in all three classes, was equivalent to the fundamental temporal frequency of the grating (defined as velocity/grating period). The third alternative, some kind of spatial code, did not seem likely from these data.

Regarding 2-dimensional dot patterns, Johnson (1983) suggested that, for spatial periods below about 1 mm, texture may be coded by the relative discharge rates in the three peripheral mechanoreceptive populations. Above that value, Lamb (1983b) found that the mean discharge rate of the different classes of primary afferents code the important stimulus parameters for 2-dimensional patterns in a similar manner to those for 1-dimensional gratings (see above). However, more recently, Connor *et al.* (1990) argued that a code based on mean firing rate could not fully account for the human psychophysical results described above. They proposed that the previously mentioned results, for both 1-dimensional gratings and 2-dimensional dot patterns, might be accounted for either by temporal or spatial variations in firing rate (i.e., by a single spatial/intensive or temporal/intensive code for perceived roughness).

8.4.2.2 Cortical Coding

Finally, we mention the only study known to us that considers, from cortical recordings, available cortical codes for the parameters affecting psychophysical roughness judgements in humans (see above). Sinclair and Burton (1988) measured the single unit activity response of neural units in area 3b (a subarea of cortical area SI — the primary sensory area, concerned mainly with mechanoreceptor inputs) in an alert monkey, who was taught to scan gratings sinusoidally with his fingertips. Amongst other findings, they found cells that showed a graded increasing response to increases in groove width, suggesting that a simple intensive code, namely mean discharge rate, might serve as a code for the effects of groove width on perceived roughness in the cortex. The authors also noted that these graded-response cells increased, rather than decreased, in firing rate in response to increasing spatial period. In keeping with Lederman (1983), they argued that a vibration-based model of roughness perception seems unlikely, since increasing the spatial period should in turn produce a decreasing fundamental temporal frequency signal. For such a model, this should have resulted in a decrease in the neural firing rate, yet the rate actually increased under such conditions.

8.4.3 Static Touch vs. Scanning

It is important to note that without relative motion between skin and surface, human texture estimation and discrimination are limited to simple binary distinctions (Katz, 1989). The full range of perceptual differentiation is only possible when there is relative tangential motion between skin and textured surface. When textured surfaces (as well as embossed letters and step edges; see section 8.5) are applied normal to the skin of monkeys without lateral scanning, neither of the transient detectors (RA or PC) respond; moreover, SA activity is only about one-tenth that during scanning, although the spatial response profiles produced by the two modes remain quite similar (Johnson & Lamb, 1981). This may explain the poor human texture discrimination (and pattern recognition) capabilities observed under a static mode of contact (see also Drucker (1988), for a recent computational approach).

In robotic tactile sensing, the traditional use of static tactile sensors (usually without temporal resolution capabilities) has limited reliable discrimination to a binary judgement. The results of the biological experimental and computational work reported here, however, suggest the value of developing dynamic sensors for tasks in which finer discrimination is required. Howe (1990) has used biological arguments to justify convincingly the promise of sensors that use tangential motion to give information about contact conditions. He argues that because such sensors respond to time derivatives of physical quantities, they can measure and respond to considerably finer variation in both the space and time domains than can static tactile sensors. He describes a stress rate sensor, which appears particularly appropriate for differentiating textures, and also a skin acceleration sensor for detecting incipient slip. Howe further proposes that both sensors might be used together to provide a measurement of texture variation on an even smaller scale, reminding us of the manner in which the parallel operation of selectively sensitive neural channels serves to increase discriminative capacity (e.g., peripheral coding of visual colour and of tactile texture). Given relative motion, however, it makes little perceptual difference whether the surface moves across a passive finger, or the finger moves actively across the stationary surface, provided finger force is controlled (Lederman, 1981; Lamb, 1983b). Hence, it has been argued that roughness perception is based solely on the processing of cutaneous, as opposed to kinesthetic, inputs. However, texture discrimination is typically better when the change in spatial period of gratings is along rather than across the direction of the hand movement (Lamb, 1983b). Presumably, this is because of the strong changes in tangential stress produced as the skin catches on the leading edges of raised surface features, noted and used to guide the design of Howe's (1990) stress rate sensor (see also Johansson and LaMotte (1983) for biological work). The orientation-specific differences in texture discrimination are also relevant to the planning of efficient scanning trajectories across uni-dimensionally varying surfaces for the robot end effector/sensor.

In summary, the biological work on texture offers ideas for sensor design and its evaluation, and for texture extraction algorithms and their implementation.

8.5 Processing 2-dimensional Patterns

The following section concentrates primarily on the perception and neural coding of small (i.e., fingertip-sized) patterns contacting the skin; for the perceptual tasks used, local element position is important (*cf.* texture processing). The stimuli used in these tasks have been various raised character sets: Braille or Braille-like patterns, Roman letters, Japanese characters, and graphics symbols.

Loomis (1981b) has considered evidence for a number of factors that constrain tactile pattern recognition performance. He has argued for effects of spatial resolution (as affected by skin mechanics, travelling waves on the skin, neural organisation), various spatial interactions among stimuli separated by more than the spatial resolution threshold, temporal resolution, perceptual integration, and limited processing capacity. For summary details and references, the reader is referred to Loomis (1981b), and Loomis and Lederman (1986). In the following section, we address behavioural and neural issues that we believe are relevant to the design of computational tactile perception systems for sensate robots.

8.5.1 Behavioural

8.5.1.1 Pattern Masking Effects and Spatial/Temporal Integration

Research with human subjects has indicated that tactile pattern detection and recognition may be impaired ("masked") by the presence of other patterns that overlap or are adjacent to the target pattern in space and/or time.

For example, Loomis (1990) has demonstrated lateral (spatial) masking effects on the fingertip. He found that spatially contiguous, temporally overlapping features (such as a border around a letter) impaired letter recognition compared with performance with no mask. Loomis attributed such a deficit to the limited bandwidth of the cutaneous spatial filter on the fingertip. More generally, he has argued that at any selected locus on the skin, whenever the tactile spatial resolution (and/or temporal resolution for patterns separated in time) is about the size of the features in the stimulus patterns presented, considerable masking may occur due to the inability of the skin to resolve the target and masking patterns.

Craig (e.g., 1978; 1980; 1981) has studied vibrotactile spatial pattern recognition in the presence of stimuli that temporally precede or follow the pattern target. Impairment in the former condition is described as temporal forward masking; in the latter condition, it is described as temporal backward masking. Typically, greater backward masking has been found for short stimulus onset intervals (i.e., less than 200 - 300 msec), while stronger forward (*cf.* backward) masking occurs when the stimulus onset intervals are larger than that. Evans and Craig (1986) have attributed the backward masking effects to the skin's inability to resolve two physical patterns that are separated by less than about 100 msec. They attributed forward masking effects to the integration of an internal sensory representation that is formed of the first pattern presented (an accurate copy that lasts for about 1200 msec in memory) with the target that follows. In both cases, the integration of the target and mask into a composite form obscures the target pattern that must be identified.

Finally, "metacontrast" effects have also been demonstrated (Weisenberger & Craig, 1982; Loomis, 1990), a form of masking in which the target and its mask do not overlap, but are spatially contiguous in both space and time.

Masking studies provide one experimental paradigm for studying the processes that underly the consequences for performance of temporal integration by the tactile system. In this case, integration causes a degradation in performance.

Another paradigm adopted by Craig and his colleagues showed conditions under which temporal integration will actually improve performance. The task involved the presentation of two halves of a vibrotactile pattern whose onsets are separated by varying temporal intervals. Subjects had to combine the halves so as to identify the complete configuration. Craig (1982) determined the course of temporal integration as a function of stimulus onset asynchrony (SOA), the temporal separation between onset times of the two halves. The results indicated that subjects could fully integrate the two halves when the SOA was less than 10 msec; recognition performance declined until it asymptoted at 50 - 100 msec. Clearly, compression of tactile inputs begins at a very early stage, if the inputs are delivered within the temporal integration window set by the cutaneous system. This, of course, can be an advantage or a hindrance, depending upon whether the inputs must be kept separate for subsequent processing. The implications of studies that require integration of parts delivered to separate finger sites are considered in section 8.7.

Effects of masking and spatiotemporal integration on tactile robotic sensing should also be anticipated. Sensing elements contiguous to those triggered by the target feature may be inadvertently activated. For example, they may be activated by feature(s) that lie adjacent to the target and fall within the sensor's "receptive field". The current results also suggest the critical impact of time delays in telerobotic performance. For tasks requiring the integration of sensory feedback to the human operator, the information must be presented within his/her temporal integration window. Failing this, experiences with the systems could prove disjoint and difficult to use effectively.

8.5.1.2 Tactile Presentation Modes

Static vs. scanned touch: Phillips *et al.* (1983) found that scanning the surface over the skin typically produced more accurate performance than impressing static patterns into the skin, although there were very strong correlations in the error patterns produced by the two different modes of touch. Clearly, dynamic sensing (relative motion between skin and stimulus pattern) improves character recognition performance as well as texture estimation and discrimination (for explanations, see section 8.4).

Active vs. passive scanning: Vega-Bermudez *et al.* (1991) more recently showed that, provided there is relative motion between the skin and raised fingertip-size patterns, pattern recognition performance (as previously documented with texture) is identical, whether produced by the subject actively scanning a stationary surface or by the movement of raised fingertip-sized patterns across the subject's stationary hand. The authors take this to imply similar cutaneous representations of the target pattern in the nervous system.

Other ways of presenting tactile information: the relative effectiveness of several different ways of displaying 2-dimensional information to the skin (e.g., simultaneous static, sequential full scan, sequential scan through a narrow slit) have been investigated in the interests of substituting a functioning tactile system for a deficient visual system. For details, the interested reader should consult papers by Loomis (Apkarian-Stielau & Loomis, 1975; Loomis, 1974; Loomis & Lederman, 1986), and by Craig (1981). Their research directly provides a scientific approach for creating a more principled basis for selecting efficient combinations of display mode, skin site and type of pattern in a telerobotic task, where patterns extracted remotely must be delivered in a usable form to the human operator's skin. For those designing tactile sensors for autonomous robots, these results may provide some idea of the consequences of one's design decisions, such as choosing to scan a robotic tactile sensor array serially, without having to build the actual sensor.

8.5.1.3 Similarities Between Vision and Touch

A number of experiments have emphasised the strong similarities between the visual and tactile human recognition of fingertip-size spatial patterns, suggesting the involvement of similar representations and processing mechanisms (see below). Such findings are relevant to those who wish to design machine pattern recognition systems based on correlated inputs derived from visual and tactile sensors.

The additional tactile array information might be used, at various levels of processing, to facilitate vision, as for example, in separating figure from ground, and the illumination source from surface albedo. We also remind the reader that strong correlations are not always obtained (e.g., see discussion of the recognition of extended spatial patterns in our next chapter).

Craig (1979), Loomis (1981a; 1982), and Johnson and his colleagues (Johnson & Phillips, 1981; Phillips *et al.*, 1983) have all obtained extremely high correlations between visual and tactile recognition accuracy for a variety of Roman and Braille character sets, and between tactile and visual confusion matrices. These results collectively suggest that character recognition is primarily limited by the spatial bandwidth of the cutaneous and visual systems.

In the Loomis studies, tactile and visual spatial resolution were first equated by applying a spatial low-pass filter to the optical stimuli before viewing. Given the strong correlations obtained, Loomis (1981a) further suggested that visual low-pass spatial filtering could be used, to a first approximation, as a model of tactile recognition of raised fingertip-sized patterns.

More recently, Loomis (1990) has developed a two-stage model of tactile and visual character recognition. The stimulus is first transformed by linear, low-pass spatial filtering, then by non-linear compression of the stimulus intensity. Second, the transformed stimulus is compared via template matching with every stored character representation, and response selection is performed according to Luce's (1963) unbiased choice model.

An additional intriguing thought is suggested by Loomis's proposal that fingertip touch can be viewed as blurred vision. It actually suggests certain processing advantages during object manipulation of biological touch over vision in providing perceptual information to achieve a stable grasp and to perform various manipulatory tasks. Being a low-pass filtered version of vision means that there will be less information to process and transmit with tactile inputs.

To the extent that tactile sensing is used for real-time control during grasping and manipulation, one can achieve a fast control loop (i.e., sense → process → output control commands) by trading-off resolution against computation time (*cf.* vision, which has made trade-offs in the other direction for the extraction of precise contour information). This advantage adds to such benefits as tactile sensors being positioned closer than vision sensors to the site of the action, thereby reducing sensor error.

8.5.2 Neural Coding

The reader is referred to a review article by Sathian (1989) for an outline of neural (peripheral and cortical) tactile sensing of surface features, including complex and periodic patterns discussed here and in section 8.4. In this same article, Sathian also considers viable peripheral neural codes for local surface features, such as curvature and single asperities on an otherwise smooth surface. The paper could prove valuable for those designing computational tactile perception systems for sensate robots. It may suggest reasonable tactile primitives for computational models of artificial tactile pattern recognition, and possible representations based on neural pattern recognition.

8.5.2.1 Peripheral

There is considerable evidence to suggest that SAI (in humans; SA in monkeys, where no further differentiation among SA units has been possible) and FAI units are both involved in processing patterns, although the type of stimulus (static indentation vs. stationary, but vibrating) and mode (static vs. scanned) are both relevant. For example, Johnson and Phillips (1981) showed that monkey SA afferents resolved the fine spatial details of gaps and bars pressed into the skin, while FAI units responded only to the transient portions of the skin deformation, and were not as acute as the SA units. Thus, only SA units seem to code the stationary psychophysical threshold responses described by Johnson and Phillips (1981) and by Loomis (1981a).

However, corresponding psychophysical and neurophysiological studies using the Optacon (a reading machine for the blind that converts visually sensed print patterns into corresponding vibratory patterns on the fingertip) produce results that implicate the FAI units in stationary tactile pattern recognition. So also do the known differences in innervation density of SAI and FAI units in the distal portion of the human fingertip, since the FAI units are almost twice as dense as the SAI units. It appears possible that different neural processing channels, with different spatial resolution capacities, are responsible for the tactile pattern recognition of static versus vibratory patterns. Johnson and Lamb (1981) have demonstrated that peripheral SA units encode Braille-like patterns scanned over the distal portion of a monkey's stationary finger most precisely; FAI responses were less spatially distinct, although spatial patterning was still evident. As in other single-unit recording studies, FAII units failed to resolve the spatial details of the patterns; this is not surprising given their very large receptive fields. Similar results have been found by Phillips *et al.* (1990), recorded from the four peripheral mechanoreceptor afferent populations that can be distinguished in human subjects. SAII units, which can be distinguished from SAI units in humans, also failed to resolve the Braille characters (and have relatively large receptive fields). Only SAI units can spatially code (statically or scanned) patterns with spacings as fine as 1.5 mm, although the innervation density of FAI units is higher. Below this value, an intensity code must be used, since none of the mechanoreceptor population spatially modulate its responses (see section 8.4).

Static vs. scanning modes of touch: possible reasons for the behavioural advantage of tactile scanning over stationary contact have been considered in section 8.4.

Similarities between touch and vision: as we have previously explained in section 8.3.2, both foveal and fingertip spatial resolution and pattern recognition (Johnson & Phillips, 1981; Phillips *et al.*, 1983) appear to be limited primarily by the spatial filtering properties of the optical and cutaneous peripheral systems, which, in turn, are affected largely by the spacing of cones in the fovea and of FAI and SAI units in the human fingertip.

8.5.2.2 Cortical

Mechanical inputs are transmitted from the periphery through several further neural levels, which may further transform the inputs, up to and within the cortex. The relatively low number of connections (synapses) used to transmit information from the mechanoreceptors provides a fast, direct transmission route to the primary sensory and motor cortical areas (*cf.* the slower, multisynaptic routes used to transmit thermal and all high intensity, i.e., pain, inputs). Unfortunately, relatively little work has been performed as yet on the cortical processing of micropatterns.

However, Phillips *et al.* (1988) have found that SI units in area 3b (a subarea of the primary sensory cortical area concerned mainly with mechanoreceptor inputs) of unanaesthetised monkeys show a sustained response to stimuli impressed into the skin. These cortical units, therefore, must receive inputs from peripheral SA units, the only ones that respond to sustained stimulation. These slowly adapting (SA) cortical units also show the most precise spatial and temporal acuity to scanned letters and dot patterns, followed by the less precise rapidly adapting (FAI) cortical units. About one third of the SA units show isomorphic neural images of embossed letters scanned across the finger, like the peripheral SAs. Other cortical SA units in this area produce considerably less isomorphic responses, although they are still highly structured. FAI units in area 3b, as well as both SA and FAI units in area 1 (behind area 3b), show much less structure in their response profiles. Therefore, it is argued that SA units in area 3b probably code spatial patterns. Bankman *et al.* (in press) have begun to use neural network simulations to study the manner in which a representation that originally mapped in a 1:1 fashion to the stimulus is transformed into a completely distributed representation. They noted that those network responses that most closely resemble neurophysiological responses are derived from receptive fields containing oriented bands of excitation and inhibition, much like simple cells in the visual cortex. This led them to emphasise, once again, the possibilities of similar tactile and visual mechanisms for processing small patterns.

8.6 Thermal Sensing

Thermal sensing serves several different functions in biological systems. Infrared sensing may be used by a predator to detect accurately the spatial position of its prey (e.g., vipers and pit snakes (Newman & Hartline, 1982)). Thermal sensing can be used to identify materials through their different thermal properties (e.g., thermal conductivity, capacity, diffusivity, etc.). Finally, the capacity to sense temperature extremes serves a protective function, allowing an organism to withdraw quickly from a potential source of damage. For a summary of the psychophysical and neurophysiological findings on biological thermal sensing, the reader is referred to articles by Sumino and Dubner (1981), and Kenshalo (1984). Here, we briefly present the highlights of that material.

8.6.1 Psychophysics

Human thermal responses may be described as either static or dynamic, depending on whether they are affected by steady or changing skin temperature, respectively. The range of approximately 30 – 36°C is known as "physiological zero", as no thermal sensations are experienced when the skin temperature is changed very slowly within this region (i.e., thermal adaptation is complete). When changed rapidly within this zone, however, sensations of warmth and cool are experienced as skin temperature is increased and decreased, respectively. When skin temperatures are raised above the zone of physiological zero, up to about 45°C, warm sensations are always experienced; within this region, sensations become more, or less, warm with increasing and decreasing skin temperature, respectively, but never cool. Below the physiological zero zone, down to about 18°C, sensations are always of coldness; they become more or less cold, as the skin temperature is decreased and increased, respectively, but never warm.

Human thermal sensitivity to changes in skin temperature is affected by three stimulus parameters: the static or adapting skin temperature, the rate of temperature change, and the surface area to which the thermal stimulus is applied. The influence of the last two parameters has been considered for both threshold and suprathreshold sensations.

Once again the human system has effectively subdivided the task of sensing, this time with regard to thermal sensing. There are two separate thermal channels that function independently, each selectively sensitive to its own temperature range (rather than a single "super-sensor" that performs all functions). However, the system achieves good thermal sensitivity at a cost. We further note that humans localise thermal spatial details far less precisely than they do cutaneous details derived from skin deformation (Taus *et al.*, 1975). The system has thus traded localisation of thermal events for good thermal sensitivity through the process of spatial summation (as also occurs for FAII mechanoreceptors, with respect to spatial and temporal resolution, respectively; see section 8.3.3).

8.6.2 Neural Coding

Two separate classes of thermally-sensitive peripheral afferent units in monkeys have been shown to respond to graded thermal intensity. These are described as "warm" and "cold" fibres (see Chapter 7 by Moss-Salentijn in this book). There are no known non-neural terminal structures that aid thermal transduction (*cf.* FAI, FAII, SAI and SAII mechanoreceptor units, whose encapsulated endings are involved in mechanical transduction). The response profiles of both populations to steady state and changing skin temperature have been determined for many of the same parameters considered for human thermal sensations (for details, check thermal references above). Sumino and Dubner (1981) claim that the parallels observed between the response characteristics of the warm and cold peripheral afferent populations and human sensitivity to thermal stimulation are "... directly related to the characteristics of thermal primary receptors" (pg. 120).

8.6.3 Robotic Applications

The hybrid tactile/thermal sensor designed by Siegel *et al.* (1986) offers an interesting approach that incorporates the advantage of the separate functions of the densely spaced mechanoreceptor populations used for functions requiring the extraction of fine spatial details (FAI and SAI) versus the two peripheral thermal units (warm and cold fibres). (However, no attempt was made to differentiate further by design the different kinds of mechanoreceptor or thermal populations described earlier.) Siegel and his colleagues selected the spacing of the tactile elements in the array sensor such that they were twice as close as those for the thermal elements. The design of this hybrid sensor optimises the limited space available on dextrous robot hands, while permitting relatively fine spatial discriminations through force sensing and material distinctions (e.g., for object identification; see our companion chapter in this book) through thermal sensing. Should finer distinctions in material be required (e.g., for the stable grasping and manipulation of unknown objects), it might be possible to increase the sensor's sensitivity through spatial summation over a denser array of thermal elements. The sensitivity obtained through spatial summation also offers a simple method for developing a protective function for the robot system that may be activated well before any actual damage occurs.

8.7 Processing Spatially Distributed Inputs

Handling objects for purposes of perceptual exploration, grasping and manipulation typically involves stimulation of more than one site on the hand. How do people process information presented to multiple skin sites? Do they do this preattentively and in parallel across processing channels? Perhaps they focus their attention on a single channel, and switch from one to another. Or maybe they divide their attention across the channels (although perhaps with some cost)? Does the processing style further depend on the type of tactile features in the inputs and on other task demands?

8.7.1 Fingers as Separate Processing Sites

Craig (1983; 1985), and Craig and Xu (1990) have argued that the fingers may serve as separate processing channels, although whether they behave as separate channels is likely to prove task-specific. Craig's argument is based on the results of several experimental tasks; in all of them, two patterns were presented to adjacent sites on the same finger or to adjacent fingers of the same hand. Subjects were required to identify one of two patterns presented together (a spatial masking task), to integrate two separately presented halves, or to make temporal order judgements. One criterion that might be used to define the single finger as a "channel" is evidence of greater interactions within, as opposed to across, channels. With each experiment, Craig found greater interactions between vibrotactile patterns presented to a single-finger site than to adjacent-finger sites in all tasks, although performance varied with the task. The single-finger condition was poorer than adjacent-finger condition in the masking task, and better in the pattern integration and temporal order tasks. Craig attributed the performance differences to the differential effects of masking and to the cost of switching attention in a limited-capacity system. Such issues may also prove relevant when considering the processing requirements of a specific robot hand system for different task demands.

8.7.2 Are Two Hands Better Than One?

Lappin and Foulke (1973) initially showed a bilateral advantage in a Braille reading task when using one finger from each hand as opposed to two adjacent fingers on the same hand. More recently, Craig (1985), and Craig and Xu (1990), also documented this two-hand advantage using some of the tasks mentioned above. Patterns were presented to the index and middle fingers of different hands or to the same hand, respectively. Craig (personal communication) attributes the two-hand superiority to greater processing flexibility provided by additional resources for the second hand, possibly a separate limited-capacity processor.

8.7.3 Limited Cutaneous Field of View

In conjunction with developing methods for improving Braille reading rates, researchers have suggested that speed might be improved if the cutaneous field of view were expanded, by allowing characters to be read across two adjacent fingers. However, Craig (in the single-finger vs. adjacent-finger comparisons above), Lappin and Foulke (1973), and Hill (1974) all failed to find any improvement in performance. It would appear that the human tactile processing of characters is basically limited to one character at a time, unlike visual reading where a number of characters can be processed in a single glance. We shall also show in our accompanying chapter that the limited cutaneous field of view further restricts the haptic recognition of extended raised 2-dimensional spatial displays of common objects. However, we wonder whether it might, in contrast, be possible to process low-level cutaneous inputs, such as friction, more quickly during grasping and manipulation tasks with an extended field of view. Such savings could prove valuable in the robotic domain as well.

8.7.4 Selective Attention

Is there some benefit to knowing the site at which tactile information will subsequently appear? That is, can subjects selectively attend to information presented to multiple skin sites? In another experimental paradigm used by Craig (1985), before the stimuli were simultaneously presented to two fingers, subjects were cued as to which finger would receive the pattern they would have to report. The results indicated that directing the observer to focus his or her attention initially on a specified finger improved pattern recognition performance, compared with a condition in which pre-cueing was not used.

8.7.5 Preattentive vs. Attentive Tactile Processing

Lederman and Klatzky (in prep.) recently investigated automated (parallel, preattentive) vs. focused attentive (serial) processing of multiple texture inputs (jittered raised dot arrays). On any trial, from one to six textured surfaces were presented simultaneously to any single or combination of the middle three fingers of both hands. Subjects were required to detect whether a "rough" ("smooth") target was present amongst varying numbers of "smooth" ("rough") distractors, that is, non-target surfaces. The results indicated that with a coarse discrimination task, subjects were able to detect the presence or absence of a rough target in the same amount of time, regardless of the number of smooth background items in the display, i.e., the search time functions were flat. Perhaps surprisingly, this indicates that the tactile/haptic system is indeed capable of performing the roughness detection task in parallel, that is, at a fairly early stage that does not demand serial, focused attention.

In contrast, when the smooth surface was the target, search times increased with each additional item in the display, suggesting that serial attentive processing was necessary. Such results have important ramifications for both speed and accuracy of performance, and address issues that are relevant for any limited-capacity processing system, whether biological or machine.

8.7.6 Features for Computational Models of Early Tactile Processing

The set of studies by Lederman and Klatzky immediately above also constitutes part of a research programme aimed at determining a reasonable set of tactile "features" that may be used to model computationally the early stages of tactile processing, following initial fingertip contact. Potential dimensions (e.g., texture, size, shape) are initially selected on the basis of supporting tactile psychophysical and neurophysiological evidence. Empirical evidence of perceptual (a)symmetries in detecting the presence/absence of a target are subsequently used to guide the designation of features (rough vs. smooth; small vs. large, etc.), as suggested by Treisman and Souther (1985) in modelling early visual processing. Targets that produce relatively flat search functions in coarse discrimination tasks are considered reasonable features (e.g., roughness). It is argued that when such a feature is made the target, the signal to noise ratio will be very high; the "feature" creates a strong signal that is little influenced by the addition of low-noise background items, since the latter represent the "absence of the feature". In contrast, when the absence of that feature (e.g., smooth) is made the target, there will be a relatively weak signal against an increasingly noisy background produced as background items are added. The increasingly precise detection task presumably requires attentive processing, and should therefore produce a search function with positive slope. When perceptual symmetry is obtained and the slopes are both flat, both targets are considered to be reasonable features. In the experiment mentioned above, it appeared that "roughness", in contrast to "smoothness", should be considered the feature of the given dimension.

While it is only recently that researchers have begun to investigate issues pertaining to the tactile processing of spatially distributed inputs in humans or machines, the topic should critically concern those who design telerobotic and autonomous tactile perception systems. The empirical findings with human subjects make it clear that there are sensory (e.g., masking, limited field of view) and attentional constraints on processing tactile (visual, or auditory) patterns that must be considered when selecting the type of information and display modes for telerobotics in order to optimise the speed and accuracy with which the human operator can process incoming signals. Considering similar constraints (e.g., restricted field of view, attentional constraints) may also be relevant for robotic systems, in which limited transmission speeds and processing power impose restrictions on tactile processing. The methodologies and results of biological investigations also offer a principaled, scientific basis for selecting "features" for computational models of early tactile processing. Such features are likely to prove critical in designing telerobotic systems, since a human operator is involved. In contrast, for autonomous robotics we emphasise the experimental methodology used to determine a reasonable feature set. Such an experimental approach can be used to determine the limits of a tactile sensor and to select processing techniques accordingly.

8.8 Conclusion

We have argued that the scientific study of biological touch can further the development of current robotic tactile systems in many different ways. We argue that the scientific method can be used to complement current analytic techniques. Such work addresses many of the same problem domains, provides an example of, and framework for, designing working multilevel integrated systems, and provides valuable suggestions for presenting robotically extracted sensory information effectively to a human operator for purposes of teleoperation.

The mini-tutorial on the scientific method highlights the value of properly constraining the problem, formulating testable hypotheses, designing unbiased experimental tests of the hypotheses, and using statistical techniques for assessing the validity, reliability, and generality of the experimental findings. The scientific approach used to study biological systems highlights the fact that robotic thermal and tactile sensing areas would benefit from the use of rigorous methodologies for systematically evaluating and delimiting the response characteristics of any thermal or tactile sensor. The latter part of the chapter deals with substantive scientific findings that pertain to problems of mutual interest to those who study living and robotic tactile systems. We conclude that roboticists have much to gain from studying biological touch systems without being constrained by, or limited to, anthropomorphic design.

Acknowledgements

This chapter was written with the financial assistance of the Institute for Robotics and Intelligent Systems Centre of Excellence of Canada, the Information Technology Research Centre of Excellence of Ontario, the Manufacturing Research Corporation of Ontario Centre of Excellence, the Natural Sciences & Engineering Research Council of Canada, and the Alberta Heritage Foundation. The authors wish to express their gratitude to Cheryl Wilson for her assistance with the manuscript and figures.

References

Apkarian-Stielau, P. & Loomis, J.M. (1975). A comparison of tactile and blurred visual form perception. *Perception & Psychophysics, 8(5)*, 362-368.

Bankman, I.N., Hsiao, S.S. & Johnson, K.O. (In Press). Transformation of neural images in the somatosensory system of the monkey: Comparison of neurophysiological observations with the operations of a neural network model. *Cold Spring Symp. on Quant. Biol.*

Bolanowski Jr., S.J., Gescheider, G.A., Verrillo, R.T. & Checkosky, C.M. (1988). Four channels mediate the mechanical aspects of touch. *J. Acoustical Society of America, 84(5)*, 1680-1694.

Capraro, A.J., Verrillo, R.T. & Zwislocki, J.J. (1979). Psychophysical evidence for a triplex system of cutaneous mechanoreception. *Sensory Processes, 3*, 334-352.

Cauna, N. (1954). Nature and functions of the papillary ridges of the digital skin. *Anatomical Record, 119(559)*, 449-468.

Clark, F.J. and Horch, K.W. (1986). Kinesthesia. In K.R. Boff, L. Kaufman & J.P. Thomas. (Eds.), *Handbook of Perception & Human Performance, Vol. 1 Sensory Processes and Perception* (pp. 13-1-13-62). New York: John Wiley.

Connor, C.E., Hsiao, S.S., Phillips, J.R. & Johnson, K.O. (1990). Tactile roughness: Neural codes that account for psychophysical magnitude estimates. *J. Neuroscience, 10*, 3823-3836.

Craig, J.C. (1978). Vibrotactile pattern recognition and masking. In G. Gordon. (Ed.), *Active touch: The mechanisms of recognition of objects by manipulation. A multidisciplinary approach* (pp. 229-242). Oxford: Pergamon Press.

Craig, J.C. (1979). A confusion matrix for tactually presented letters. *Perception & Psychophysics, 26(5)*, 409-411.

Craig, J.C. (1980). Modes of vibrotactile pattern generation. *J. Experimental Psychology: Human Perception and Performance, 6(1)*, 151-166.

Craig, J.C. (1981). Tactile letter recognition: Pattern duration and modes of pattern generation. *Perception & Psychophysics, 30(6)*, 540-546.

Craig, J.C. (1982). Vibrotactile masking: A comparison of energy and pattern maskers. *Perception & Psychophysics, 31(6)*, 523-529.

Craig, J.C. (1983). The role of onset in perception of sequentially presented vibrotactile patterns. *Perception & Psychophysics, 34(5)*, 421-432.

Craig, J.C. (1985). Attending to two fingers: Two hands are better than one. *Perception & Psychophysics, 38(6)*, 496-511.

Craig, J.C. & Xu, B. (1990). Temporal order and tactile patterns. *Perception & Psychophysics, 47(1)*, 22-34.

Drucker, S.M. (1988). Texture from touch. In W. Richards. (Ed.), *Natural Computation* (pp. 422-429). Cambridge,MA: MIT Press.

Evans, P.M. & Craig, J.C. (1986). Temporal integration and vibrotactile backward masking. *J. Experimental Psychology: Human Perception and Performance, 12*, 160-168.

Fearing, R.S. & Hollerbach, J.M. (1985). Basic solid mechanics for tactile sensing. *Int. J. Robotics Research, 4(3)*, 40-54.

Freeman, A.W. & Johnson, K.O. (1982a). Cutaneous mechanoreceptors in macaque monkey: Temporal discharge patterns evoked by vibration, and a receptor model. *J. Physiology, 323*, 21-41.

Freeman, A.W. & Johnson, K.O. (1982b). A model accounting for effects of vibratory amplitude on responses of cutaneous mechanoreceptors in macaque monkey. *J. Physiology, 323*, 43-64.

French, A.S. & Korenberg, M.J. (1989). A nonlinear cascade model for action potential encoding in an insect sensory neuron. *Biophysics Journal, 55*, 655-661.

French, A.S. & Wong, R.K.S. (1977). Nonlinear analysis of sensory transduction in an insect mechanoreceptor. *Biological Cybernetics, 26*, 231-240.

Fuma, F. & Bajcsy, R. (1988). *Comparative studies between mechanical and biological tactile sensors.* Tech. Rept. 1-18, Dept. of Computer and Information Science, Univ. Pennsylvania, Philadelphia, PA.

Gescheider, G.A. (1974). Temporal relations in cutaneous stimulation. In F. Geldard (Ed.), *Cutaneous communication systems and devices.* Austin, Tex.: The Psychonomic Society.

Gescheider, G.A. and Verrillo, R.T. (1979). Vibrotactile frequency characteristics as determined by adaptation and masking procedures. In D.R. Kenshalo (Ed.), *Sensory Functions of the Skin of Humans.* New York: Plenum Press.

Hill, J.W. (1974). Limited field of view in reading lettershapes with the fingers. In F.A. Geldard (Ed.), *Cutaneous communication systems and devices.* Austin, Tex.: The Psychonomic Society.

Howe, R.D. (1990). *Dynamic tactile sensing.* Ph.D. thesis, Stanford University.

Hulliger, M., Nordh, E., Thelin, A.E. & Vallbo, A.B. (1979). The responses of afferent fibres from the glabrous skin of the hand during voluntary finger movements in man. *J. Physiology, 291*, 233-249.

Johansson, R.S. & LaMotte, R.H. (1983). Tactile detection thresholds for a single asperity on an otherwise smooth surface. *Somatosensory Research, 1*, 21-31.

Johansson, R.S., Landstrom, U. & Lundstrom, R. (1982). Responses of mechanoreceptive afferent units in the glabrous skin of the human hand to sinusoidal skin displacements. *Brain Research, 244*, 17-25.

Johansson, R.S. & Vallbo, A.B. (1979). Detection of tactile stimuli. Threshold of afferent units related to psychophysical thresholds in the human hand. *J. Physiology, 297*, 405-422.

Johansson, R.S. & Vallbo, A.B. (1983). Tactile sensory coding in the glabrous skin of the human hand. *Trends in Neuroscience, 6*, 27-32.

Johansson, R.S., Vallbo, A.B. & Westling, G. (1980). Thresholds of mechanosensitive afferents in the human hand as measured with Von Frey hairs. *Brain Research, 184*, 343-351.

Johnson, K.O. (1983). Neural mechanisms of tactual form and texture discrimination. *Federation Proc., 42(9)*, 2542-2547.

Johnson, K.O. & Lamb, G.D. (1981). Neural mechanisms of spatial tactile discrimination: Neural patterns evoked by Braille-like dot patterns in the monkey. *J. Physiology, 310*, 117-144.

Johnson, K.O. & Phillips, J.R. (1981). Tactile spatial resolution. I. Two-point discrimination, gap detection, grating resolution, and letter recognition. *J. Neurophysiology, 46(6)*, 1177-1191.

Johnson, K. & Phillips, J. (1984). Spatial and nonspatial neural mechanisms underlying tactile spatial discrimination. In C. von Euler, O. Franzen, U. Lindblom, & D. Ottoson (Eds.), *Somatosensory Mechanisms. Vol. 41* (pp. 237-248) in Wenner-Gren Centre Int. Symposium Series, London: Macmillan.

Katz, D. (1989). *The World of Touch* (L.E. Krueger, Trans). Hillsdale, N.J.: Lawrence Erlbaum Assoc. (Original work published 1925).

Kenshalo, D.R. (1984). Cutaneous temperature sensitivity. In W.W. Dawson & J.M. Enoch. (Eds.), *Foundations of Sensory Science* (pp. 419-464). Berlin: Springer-Verlag.

Korenberg, M.J., French, A.S. & Voo, S.K.L. (1988). White-noise analysis of nonlinear behavior in an insect sensory neuron: Kernel and cascade approaches. *Biological Cybernetics, 58*, 313-320.

Lamb, G.D. (1983a). Tactile discrimination of textured surfaces: Psychophysical performance measurements in humans. *J. Physiology, 338*, 551-565.

Lamb, G.D. (1983b). Tactile discrimination of textured surfaces: Peripheral neural coding in the monkey. *J. Physiology, 338*, 567-587.

Lanir, Y. (1987). Skin mechanics. In R. Skalak & S. Chien. (Eds.), *Handbook of Bioengineering* (pp. 1-25). New York: McGraw-Hill.

Lappin, J.S. & Foulke, E. (1973). Expanding the tactual field of view. *Perception & Psychophysics, 14(2)*, 237-241.

Lederman, S.J. (1974). Tactile roughness of grooved surfaces: The touching process and effects of macro- and microsurface structure. *Perception & Psychophysics, 16(2)*, 385-395.

Lederman, S.J. (1978). Heightening tactile impressions of surface texture. In G. Gordon (Ed.), *Active Touch: The mechanism of recognition of objects by manipulation. A multi-disciplinary approach.* Oxford: Pergamon Press.

Lederman, S.J. (1981). The perception of surface roughness by active and passive touch. *Bulletin of the Psychonomic Society, 18(5)*, 253-255.

Lederman, S.J. (1983). Tactual roughness perception: Spatial and temporal determinants. *Canadian J. Psychology, 37(4)*, 498-511.

Lederman, S.J. (1991). Skin and touch. In R. Delbucco. (Ed.), *Encyclopedia of Human Biology, Vol. 7* (pp. 51-63). San Diego: Academic Press.

Lederman, S.J. & Klatzky, R.L. (1987). Hand movements: A window into haptic object recognition. *Cognitive Psychology, 19(3)*, 342-368.

Lederman, S.J. & Klatzky,R.L. (1991). Processing texture vs. orientation in a haptic search task. Manuscript in preparation.

Lederman, S.J., Loomis, J.M. & Williams, D.A. (1982). The role of vibration in the tactual perception of roughness. *Perception & Psychophysics, 32(2)*, 109-116.

Lederman, S.J. & Taylor, M.M. (1972). Fingertip force, surface geometry and the perception of roughness by active touch. *Perception & Psychophysics, 12(5)*, 401-408.

Lederman, S.J., Thorne, G. & Jones, B. (1986). Perception of Texture by Vision and Touch:Multidimensionality and Intersensory Integration. *J. Experimental Psychology: Human Perception and Performance, 12(2)*, 169-180.

Loomis, J.M. (1974). Tactile letter recognition under different modes of stimulus presentation. *Perception & Psychophysics, 16(2)*, 401-408.

Loomis, J.M. (1979). An investigation of tactile hyperacuity. *Sensory Processes, 3*, 289-302.

Loomis, J.M. (1981a). On the tangibility of letters and braille. *Perception & Psychophysics, 29(1)*, 37-46.

Loomis, J.M. (1981b). Tactile pattern perception. *Perception, 10*, 5-27.

Loomis,J.M. (1982). Analysis of tactile and visual confusion matrices. *Perception & Psychophysics, 31(1)*, 41-52.

Loomis, J.M. (1990). A model of character recognition and legibility. *J. Experimental Psychology: Human Perception and Performance, 16(1)*, 106-120.

Loomis, J.M. & Collins, C.C. (1978). Sensitivity to shifts of a point stimulus: An instance of tactile hyperacuity. *Perception & Psychophysics, 24(6)*, 487-492.

Loomis, J.M. and Lederman, S.J. (1986). Tactual perception. In K. Boff, L. Kaufman & J. Thomas (Eds.), *Handbook of Perception and Human Performance* (pp. 1-41). New York: John Wiley.

Luce, R.D. (1963). Detection and recognition. In R.D. Luce, B.R. Bush & E. Galanter (Eds.), *Handbook of Mathematical Psychology (Vol. 1)*. New York: John Wiley.

Moberg, E. (1983). The role of cutaneous afferents in position sense, kinaesthesia, and motor function of the hand. *Brain, 106*, 1-19.

Morley, J.W., Goodwin, A.W. & Darian-Smith, I. (1983). Tactile discrimination of gratings. *Experimental Brain Research, 49*, 291-299.

Newman, E.A. & Hartline, P.H. (1982). The infrared "vision" of snakes. *Scientific American, 246(3)*, 116-127.

Phillips, J.R., Johansson, R.S. & Johnson, K.O. (1990). Representation of Braille characters in human nerve fibres. *Experimental Brain Research, 81*, 589-592.

Phillips, J.R. & Johnson, K.O. (1981a). Tactile spatial resolution. II. Neural representation of bars, edges, and gratings in monkey afferents. *J. Neurophysiology, 46(6)*, 1192-1203.

Phillips, J.R. & Johnson, K.O. (1981b). Tactile spatial resolution. III. A continuum mechanics model of skin predicting mechanoreceptor responses to bars, edges, and gratings. *J. Neurophysiology, 46(6)*, 1204-1225.

Phillips, J.R., Johnson, K.O. & Browne, H. (1983). The equivalence of visual and tactile letter recognition. *Perception & Psychophysics, 34*, 243-249.

Phillips, J.R., Johnson, K.O. & Hsiao, S.S. (1988). Spatial pattern representation and transformation in monkey somatosensory cortex. *Proc. Nat. Acad. Sci. USA, 85*, 1317-1321.

Quilliam, T.A. (1978). The structure of finger print skin. In G. Gordon (Ed.), *Active Touch: The mechanism of recognition of objects by manipulation*. Oxford: Pergamon Press.

Sathian, K. (1989). Tactile sensing of surface features. *Trends in Neuroscience, 12*, 513-519.

Sathian, K., Goodwin, A.W., John, K.T. & Darian-Smith, I. (1989). Perceived roughness of a grating: correlation with responses of mechanoreceptive afferents innervating the monkey's fingerpad. *J. Neuroscience, 9*, 1273-1279.

Siegel, D., Garabieta, I. & Hollerbach, J.M. (1986). An integrated tactile and thermal sensor. *Proc. IEEE Int. Conf. on Robotics & Automation (San Francisco)*, 1-16. Washington, DC: IEEE Computer Soc. Press.

Sinclair, R. & Burton, H. (1988). Responses from Area 3b of somatosensory cortex to textured surfaces during active touch in primate. *Somatosensory Research, 5*, 283-310.

Stevens, S.S. & Harris, J.R. (1962). The scaling of subjective roughness and smoothness. *J. Experimental Psychology, 64*, 489-494.

Sumino, R. & Dubner, R. (1981). Response characteristics of specific thermoreceptive afferents innervating monkey facial skin and their relationship to human thermal sensitivity. *Brain Research Reviews, 3(2)*, 105-122.

Taus, R.H., Stevens, J.C. & Marks, L.E. (1975). Spatial localization of warmth. *Perception & Psychophysics, 17(2)*, 194-196.

Taylor, M.M. & Lederman, S.J. (1975). Tactile roughness of grooved surfaces: A model and the effect of friction. *Perception & Psychophysics, 17(1)*, 23-36.

Treisman, A. & Souther, J. (1985). Search asymmetry: A diagnostic for preattentive processing of separable features. *J. Experimental Psychology: General, 114(3)*, 285-310.

Vallbo, A.B. and Johansson, R.S. (1978). The tactile sensory innervation of the glabrous skin of the human hand. In G. Gordon (Ed.), *Active Touch: The mechanisms of recognition of objects by manipulation. A multi-disciplinary approach* (pp. 29-54). Oxford: Pergamon Press.

Vallbo, A.B. & Johansson, R.S. (1984). Properties of cutaneous mechanoreceptors in the human hand related to touch sensation. *Human Neurobiology, 3*, 3-14.

van Doren, C.L. (1989). A model of spatiotemporal tactile sensitivity linking psychophysics to tissue mechanics. *J. Acoustical Society of America, 85(5)*, 2065-2080.

van Doren, C.L., Pelli, D.G. & Verrillo, R.T. (1987). A device for measuring tactile spatiotemporal sensitivity. *J. the Acoustical Society of America, 81(6)*, 1906-1916.

Vega-Bermudez, F., Johnson, K.O. & Hsiao, S.S. (1991). Human tactile pattern recognition: Active versus passive touch, velocity effects and patterns of confusion. *J. Neurophysiology, 65(3)*, 531-546.

Verrillo, R.T. (1968). A duplex mechanism of mechanoreception. In D.R. Kenshalo (Ed.), *The Skin Senses*. Springfield, IL: Charles C. Thomas.

Verrillo, R.T. and Gescheider, G.A. (1979). Psychophysical measurements of enhancement, suppression and surface gradient effects in vibrotaction. In D.R. Kenshalo. (Ed.), *Sensory Functions of the Skin of Humans* New York: Plenum Press.

Weinstein, S. (1968). Intensive and extensive aspects of tactile sensitivity as a function of body part, sex, and laterality. In D.R. Kenshalo (Ed.), *The Skin Senses*. Springfield, IL: Charles C. Thomas.

Weisenberger, J.M. & Craig, J.C. (1982). A tactile metacontrast effect. *Perception & Psychophysics, 31*, 530-536.

Whitney, D.E. (1986). Real robots don't need jigs. *Proc. IEEE Int. Conf. on Robotics & Automation (San Francisco)*, 746-752. Washington, DC: IEEE Computer Soc. Press.

Chapter 9

Lessons From The Study Of Biological Touch For Robotic Haptic Sensing

Susan J. Lederman, Roberta L. Klatzky and Dianne T. Pawluk

9.1 Introduction, Focus and Chapter Outline

In our companion chapter in this book, we consider tactile and thermal sensing in biological systems. Here, we examine the nature of biological haptic sensing, and implications for the design of efficient haptic robotic systems for use in highly unstructured environments.

To appreciate fully the potential contributions of the biological study of haptic, as well as tactile sensing, it is critical that the reader first read the complete introduction to our previous chapter. There we provide a mini-tutorial on the scientific approach to studying biological sensing and its potential contributions to the field of robotics. There are many empirical results that may prove valuable to the roboticist, and we consider them in both chapters. However, we do not advocate the blind acceptance of an anthropomorphic approach, in which an attempt is made to copy biological systems. Rather, we stress the importance of the scientific method in forcing the investigator to conceptualise the most important issues with respect to any given problem in a systematic fashion and to formulate proper research questions. It also provides rigorous methodologies for determining critical parameters, and for testing the validity, reliability, and generality of any empirical finding. As such, the scientific approach can be used to complement the analytic methods currently employed by roboticists.

In section 9.2, we consider recent work on human haptic perception and recognition of 2- and 3-dimensional objects; both the limitations and the capabilities of the haptic processing system are considered (sections 9.2.1 and 9.2.2). In section 9.2.3, we introduce the notion of systematic patterns of manual exploration and their specific roles in the extraction of object properties. Section 9.2.4 discusses critical characteristics of these manual patterns and their selection consequences; it also considers the nature of the underlying representations of objects in memory, as derived from haptic exploration with and without vision. In section 9.2.5, we describe sequences of human exploration performed during the course of object recognition. Section 9.2.6 introduces our recent computational approach to constraint-driven exploration and haptic object identification.

The remaining sections on haptic perception (9.2.7 and 9.2.8) briefly introduce the reader to complementary approaches (i.e., the development of haptic exploration and perception in infants, and haptic perception through dynamically wielding a rod). To conclude, section 9.3 discusses several contributions of cutaneous/haptic inputs to motor control during the course of grasping, lifting and manipulation.

9.2 Haptic Perception

Traditional robotic practice has been to use one or more static "touch images", usually obtained from a single sensor, for tactile sensing. Any "movement" is typically created through a series of discrete, stationary contacts (*cf.* dynamic sensing) along a surface or an edge. From a biological perspective, this form of sensing can provide only impoverished tactile inputs, as well as impoverished thermal information from thermal sensors, if available.

In contrast, those of us who study living organisms are impressed by the breadth and consummate skill with which some biological systems can perceive, recognise, stably grasp and manipulate objects within highly unstructured environments. Such performance engages the haptic system (Loomis & Lederman, 1986), which relies to a significant extent not only on cutaneous and thermal inputs, but also on kinesthetic information about the positioning of the limbs of the arm/hand or leg/foot systems. We shall use the term haptic sensing, when considering implications of the biological work for robotic applications.

In addition, the biological literature (e.g., Gibson (1966)) has convincingly argued that all perceptual systems (and most concretely demonstrated by haptics), require more than sensory inputs. Purposive (active) exploration is invaluable for perception within unstructured environments. Paralleling this argument in the robotic field, Bajcsy (1989) has coined the term "exploratory robotics" to describe such an approach. She has argued that active robotic exploration is mandatory for extracting the perceptual information needed to perform complex perceptual and manipulatory tasks in highly unstructured environments (see also Allen, this book; Stansfield (1988)).

9.2.1 Processing Raised 2-dimensional Spatial Patterns & Homogeneous 3-dimensional Objects

When roboticists consider the relative merits of computational touch and vision systems, they often prejudge the outcome by selecting geometry-based tasks. Perhaps it is understandable that the tactile field has chosen such tasks, because of their significant role in models of visual processing. However, such biased selection prejudges the outcome in favour of vision. We will begin our discussion of biological haptics by demonstrating a similar finding with humans. Next, we will contrast it with results obtained with tasks involving multi-dimensional objects. The highly competent performance obtained in the latter tasks serves to highlight the significant contributions that the human tactile/haptic system can and does offer.

How well do people haptically (vs. visually) extract information about the features and identity of 2-dimensional planar objects and large raised-outline depictions of real objects? Note that for these displays the primary source of information is provided by raised contours, and only occasionally, by textured surfaces. The visual system is highly skilled at using *geometric* information derived from the 2-dimensional layout of edges, vertices, parts, and surfaces. However, the haptic system does not fare as well with such impoverished displays.

Lederman *et al.* (1985) found considerable haptic distortion in subjects' estimates of the inferred linear distance separating the ends of a raised line pathway, with increasing overestimation as the length of the "detour" pathway actually explored increased. Subsequent work by Lederman *et al.* (1987) indicated that the distortion was the result of subjects including in their computations the length of the pathway actually explored and the duration of the hand movements, although the latter was weighted considerably less. Note that such a task is necessary whenever the object in question is too large to be grasped in its entirety, that is, when sequential exploration is necessary to extract the required spatial information. (A similar visual length distortion effect has been demonstrated with temporally extended displays, although its magnitude was considerably less (Balakrishnan *et al.*, 1989).)

Other work by Lederman *et al.* (1990) (see also Magee and Kennedy (1980)) has shown that people have considerable difficulty in haptically identifying prototypical raised-outline drawings of common objects that are easily accessible to vision. In the Lederman *et al.* study, subjects were not allowed to examine the particular drawings prior to the experiment, and were only told the domain used to select the set of sample objects (i.e., "... all common objects varying in size in everyday life from a pair of pants to a ring"). Under such conditions, the mean accuracy in identifying the given objects was 37%, and the mean response time was about 90 seconds for haptics without vision (a maximum of 120 seconds was permitted).

Klatzky *et al.* (1991) and Lederman *et al.* (1991a) further confirmed the difficulty with which people can derive geometric information about haptically explored planar objects made of a single material. In this study, subjects were required to perform a number of tasks with one or both hands, as they wished. They were to identify the entire spatial configuration, to identify a specific part, or to recall the position of a specific part relative to other parts also present. Based on the initial exploration, parts or their positions were correctly identified only about 75% of the time, while the overall configuration could not be correctly identified above chance levels.

Finally, we mention earlier work (e.g., Bryant and Raz (1975); Cashdan (1968); for a summary of the major findings, the reader should consult Jones (1981)) that documents additional conditions under which haptic performance is considerably inferior to that of vision; such tasks required subjects to identify or match unfamiliar objects that varied only in 3-dimensional shape. We emphasise the fact that as above, subjects were forced to rely on geometric information, since the objects were identical in their *material* properties (e.g., texture, thermal characteristics, and compliance).

The studies in this section have emphasised the negative consequences of constraining human haptic processing to contour extraction through sequential exploration of edges. But what factors limited performance, and how? We note that while subjects were free to explore the 2-dimensional displays as they wished, the large display size denied them the opportunity to grasp, a method of exploration that provides simultaneous information, albeit coarse, about 3-dimensional structure. Rather, the observer was forced to explore the contours sequentially, a procedure that produces a highly impoverished spatial representation. Recall from the companion chapter (in section 8.7 on processing spatially distributed inputs) that successful pattern (contour) extraction is necessarily limited by the reduced cutaneous field of view, compared with vision. Thus, to obtain the entire envelope, the observer was forced to integrate multiple segments, each of which was coded with considerably less precision than by vision owing to the skin's relatively poor spatial resolution. Moreover, these segments were usually extracted over an extended period of time. The considerable memory load must have further reduced the accuracy with which the pieces were recalled for purposes of spatial integration.

The preceding limitations certainly constrained performance in many of the tasks with raised displays. Yet even when it was possible to grasp an object, as with the 3-dimensional objects made of a homogeneous material, fine geometric discriminations could not be performed competently, whether a fingertip or palmar grasp was employed. The entire envelope was unavailable to a fingertip, while the spatial resolution of the palm was insufficient for discriminating the object's spatial details, respectively. Under these circumstances too, then, the subject was typically forced to explore contours sequentially (e.g., Lederman and Klatzky (1987)), with consequent limitations on haptic performance as described above.

We also emphasise the fact that the stimulus displays in these experiments were themselves all highly impoverished. As the stimuli were made of the same material, there was no variation in surface texture, thermal conductivity, compliance, not much in weight, and little if any variation in the third dimension. Yet results with such displays are highly instructive in that we learn the consequences of depriving the haptic system of much of the information that it is capable of extracting and processing in a highly efficient manner, as we shall see below (section 9.2.2). For all the reasons above, Klatzky and Lederman (1987) and Lederman *et al.* (1990) have emphasised that even if the haptically derived inputs were translated into a visual image for additional processing by visual processors, success would not be guaranteed.

In summary, the results of the studies on haptic processing of 2-dimensional spatial configurations and 3-dimensional objects that vary only in geometric properties clearly demonstrate conditions under which people often process haptically derived information rather poorly. While there is no doubt that people could improve with practice, the poor performance obtained in these haptic tasks is highly instructive, because it highlights specific constraints on haptic processing (see also Lederman *et al.* (1990); Klatzky and Lederman (In Press)). Many of the same factors (although we can exclude poor sensor resolution) may constrain robotic performance as well, not only because of physical constraints that limit sensor placement, but because of the traditional emphasis on contour-based object models (but see Stansfield (1988)). We consider the importance of non-geometric, material properties in the next section on haptic processing of common objects.

9.2.2 Processing Common 3-dimensional Objects

In striking contrast to the relatively poor haptic performance documented in the tasks described in section 9.2.1, people are remarkably skilled at haptically identifying common real objects. Klatzky *et al.* (1985) showed that adult observers recognised 100 common objects (e.g., fork, hammer) with almost perfect accuracy and, most often, within just a few seconds. What constraints are imposed on haptic processing of real objects that permit people to perform so capably?

9.2.2.1 Haptic Recognition and Age

Another informative aspect of human haptics is its ability to identify accurately 3-dimensional common objects despite a considerable decrement in cutaneous sensitivity and motor dexterity. Lederman and Jessel (1992) found that elderly subjects (above 65) were considerably less sensitive than young adults in measures of cutaneous sensitivity (spatial, pressure) and motor dexterity. However, the elderly subjects were not meaningfully different from the young adults in their haptic recognition of common hand-held objects without vision. Elderly subjects were correct about 88% of the time, compared with 93% for young adults. Although robotic tactile sensors in an industrial environment degrade more rapidly than human cutaneous sensitivity over a lifetime, ideas could still be gleaned from how the human haptic system manages when tactile sensing and motor dexterity are impaired.

The success with which elderly subjects could identify common objects questions the common assumption that robots must possess good tactile acuity and considerable dexterity to perform tasks currently performed by humans. This is an important consideration for current robotic tactile sensing systems, where acuity must be traded-off against transmission speed. However, additional findings suggest that the assumption above cannot automatically be rejected either. For example, Dellon and Kallman (1983) found very strong positive correlations between the accuracy of object identification and spatial sensitivity when the stimulus objects were very small and usually made out of a uniform material. Thus, at least for some tasks, robotic systems may require dexterity and acuity equal to or even significantly beyond human capabilities.

9.2.3 Patterns of Manual Exploration: "Exploratory Procedures"

Given the initial demonstration (Klatzky *et al.*, 1985) of the haptic system's competence, Lederman, Klatzky and their colleagues have pursued a systematic and comprehensive scientific examination of haptic processing. They began by examining the nature of manual exploration, since these observable behaviours provide an ideal method for uncovering the nature of haptic representations of objects in memory, as well as the processes by which these representations are created. Lederman and Klatzky (1987: Expt. 1) first demonstrated that humans use their hands in a systematic and highly intelligent fashion to extract various properties of multidimensional objects. Subjects were required to examine manually (i.e., without vision) a "standard" object for a particular property, for example, texture. Then they selected, from three "comparison" objects, the one that best matched the standard in terms of the named dimension (texture). Analysis of the videotaped hand movements indicated that people performed highly stereotypical movement patterns, which were called "Exploratory Procedures" or "EPs". Further, people selected different procedures, depending upon the particular type of property match required (texture, hardness, thermal, etc.). Figure 9.1 shows typical versions of each procedure, as well as the type of property information with which it is most commonly associated. A complete description of the EPs is available in the original paper, while further parameterisation is now available (Klatzky & Lederman, 1991).

It should be noted that the level of analysis at which the EP/property associations are documented does not specify a specific end-effector configuration or area of contact. There are a number of ways in which each of these EPs may be performed, although some versions will be more precise and/or easier to execute than others. Clearly, this must vary with the specific hardware and object domain involved.

The important point of the experiment above is that people actively and systematically explore their environment, selecting EPs and corresponding perceptual information in what appears to be a highly efficient manner. As previously mentioned, research and theory on living perceptual systems clearly underscore the need for active, purposive exploration for effective performance in unstructured

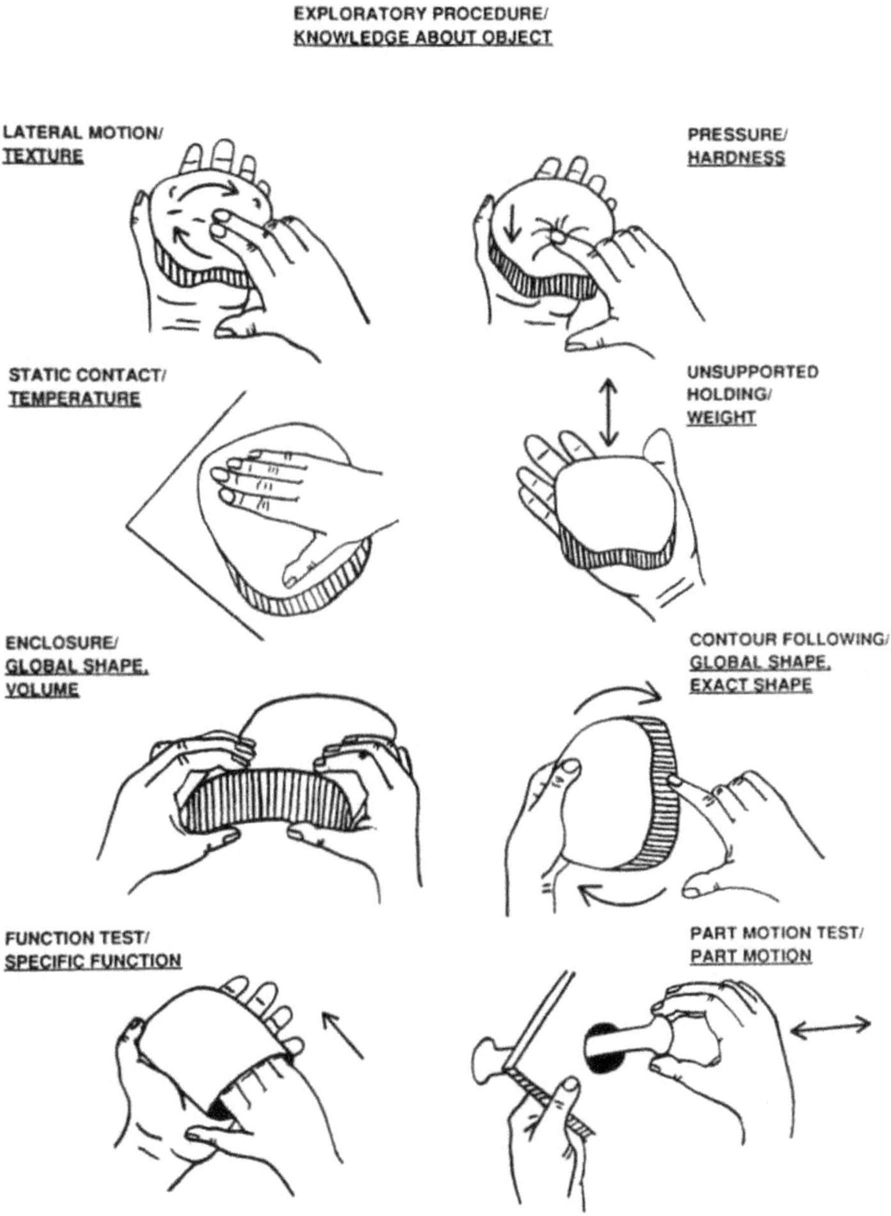

Figure 9.1: *Exploratory procedures and associated object properties. (Reprinted with permission from Lederman (1991), which was revised from Lederman and Klatzky (1987).)*

environments (Gibson, 1966). The same is almost certain to be true for the complex tasks we expect robots to perform within highly demanding environments: underwater, outer space, unstructured industrial workspaces, disposal of radioactive waste, and agriculture (e.g., fruit picking). Efficient action depends upon the continual extraction and updating of perceptual information; the converse is also true, in as much as action (exploratory procedures, grasping, and manipulation) is necessary for perception.

In sections 9.2.4 - 9.2.6, we show how other work in Lederman and Klatzky's research programme has been used to develop a conceptual model of human haptic object recognition; the architecture, which is derived from these empirical findings, is treated as a system of constraints that guides manual exploration (EP selection) during object identification. The network approach to constraint satisfaction accounts for several other empirical results. The research described in these sections is valuable in several respects. First, it raises critical issues that must be addressed by any field regarding haptic object exploration and recognition (i.e., with vision versus without vision). Second, in keeping with the ideas presented in the introduction to our companion chapter, the experimental approach offers the field of robotics a principled and rigorous methodology for addressing these same issues. Third, it provides substantive findings using a multilevel, fully functioning system.

9.2.4 Constraints on the Selection of Exploratory Procedures

9.2.4.1 Necessity, Sufficiency, and Optimality

In a highly informative variant of the experiment mentioned above (Lederman & Klatzky, 1987: Expt. 2), subjects were constrained in each trial to execute one designated EP. By combining each EP with each property instruction across the overall experiment, it was possible to determine critical characteristics of each EP based on relative performance (accuracy and in the case of ties, duration) within each task. The results are shown in Table 9.1 in the form of an EP/property weight matrix as follows. "0" was assigned when the designated EP could not perform the designated task with above-chance accuracy. "1" was assigned if that EP was *sufficient*, but not optimal, for performing the task above chance level. "2" indicates the EP that achieved the best performance of all sufficient EPs for that task (i.e., it was *optimal*). Finally, that EP was assigned "3" if it was not only optimal, but *necessary* for the task. These reflect the relative strengths of the interconnections between each EP and the property knowledge extracted, and have been used to provide corresponding weights in a network that uses a constraint satisfaction algorithm to select EPs (Klatzky & Lederman, 1991; Lederman & Klatzky, 1990a); see also section 9.2.6.

EP	Property*							Generality	EP Duration†
	tex.	hard.	temp.	wt.	vol.	global shape	exact shape		
Lateral Motion	2	1	1	0	0	0	0	3	3
Pressure	1	2	1	0	0	0	0	3	2
Static Contact	1	0	2	0	1	1	0	4	2
Unsupp. Hold.	0	1	1	2	1	1	0	5	2
Enclosure	1	1	1	1	2	2	0	6	3
Contour Foll.	1	1	1	1	1	1	3	7	11

* EP-Property weights based on relative EP performance in Expt. 2 of Lederman & Klatzky (1987):

- **0** – chance-level;
- **1** – sufficient, but not optimal;
- **2** – optimal, but not necessary;
- **3** – necessary.

† Durations (in seconds) from Expt. 1 of Lederman & Klatzky (1987).

Table 9.1: *EP-Property weightings, EP generality, and EP durations (adapted in part from Lederman & Klatzky (1990a) and Klatzky & Lederman (1991)).*

9.2.4.2 Relative EP Generality

Another EP constraint that the experimental methodology enabled us to determine is the degree to which each EP is broadly sufficient for extracting few versus many different properties. The quantitative measure of EP *generality* reflects the number of non-zero cells in each property row of Table 9.1. Values are reported in the second to last column, with higher numbers indicating greater generality. Note that whereas Lateral Motion and Pressure were the least general EPs, Enclosure (or grasp) and Contour Following were the most general, i.e., they were capable of extracting information about almost all property classes in the experiment. The performance implications will be discussed in the remainder of this section, as well as in sections 9.2.5 and 9.2.6.

9.2.4.3 EP Duration of Execution

Another important constraint on EP implementation is the relative time for execution. Mean *durations* derived from tasks performed in Expt. 1 of Lederman and Klatzky (1987) (shown in Table 9.1) permit an ordinal measure of relative EP efficiency in terms of execution time. These show primarily that Contour Following (an edge-following procedure) took considerably longer than any of the other EPs. However, the Enclosure (or grasp) was not only capable of extracting much useful coarse information, it could also do so within a relatively short period of time. The implication of this finding for producing and predicting human EP sequences will be discussed in sections 9.2.5 and 9.2.6.

9.2.4.4 Multiple Sufficiency & Compatibility of EPs Influence Property Integration

A recent series of experiments (Klatzky *et al.*, 1989; Reed *et al.*, 1990; Lederman *et al.*, 1992) has clearly shown that humans are capable of integrating information from different object properties that are used to define a set of object classes redundantly, thereby reducing the time required to learn an object-classification task.

Several experiments employed a set of custom-designed planar, wafer-like objects that varied in selected texture, hardness, shape, and size values. Different pairs of property values redundantly defined object classes within a given subset (e.g., all A's were both rough and hard; all B's were of intermediate roughness and hardness; all C's were smooth and soft). In one task, after subjects had learned to classify the objects by two dimensions that were redundantly associated as described, one of these dimensions was withdrawn (i.e., the value along the withdrawn dimension now remained constant). Dimension integration is experimentally demonstrated if the response times increase following withdrawal of one dimension. The results indicated reciprocal integration of texture and hardness, and of shape and size (i.e., a bidirectional withdrawal effect). The extent to which size was integrated with either texture or hardness was smaller and unidirectional, that is, it only occurred when size was withdrawn (other results have further confirmed the importance of size in haptic processing). In contrast, there was little or no integration between shape and either texture or hardness for planar objects.

These results highlight the importance of several factors that permit efficient haptic extraction of stimulus redundancies. We merely note them here, and refer the reader to the experimental papers for detailed explanations. Firstly, *a single EP that is multiply sufficient* for extracting the redundant properties may be used, particularly when coarse discriminations are required. An Enclosure (grasp) is the most obvious example, and has the added motoric advantage of being required both to stabilise and to manipulate objects. Secondly, when more precise information about these properties is required, the observer may be forced to execute the associated optimal EPs. In this case, *motoric compatibilities* may speed execution, since they can be performed together as a hybrid (as was found with Lateral Motion and Pressure). Thirdly, these studies clearly show the importance of *object region compatibilities* (i.e., where precise information about the properties in question is available over the same region of the object) in taking advantage of stimulus redundancies. Strong integration was usually demonstrated when the two dimensions varied within the same regions of the object, e.g., texture and hardness within a large surface area, and size and shape at the edges. This was confirmed in additional experiments performed with custom-designed, 3-dimensional shapes (ellipsoids of revolution) that varied factorially in texture, hardness, and shape. The most interesting departure from the previous planar object experiments was the bidirectional integration now shown between texture and shape. Note that there were no edges on these objects; rather, local shape information was available, along with texture, over the entire object surface. Results from this set of experiments concerning the speed of object classification and incidental learning about object properties confirm predictions based on the network approach to EP selection, as will be described in section 9.2.6.

We further note that object perception and identification may not be the only tasks that benefit from the initial rapid extraction of multiple dimensions. It may also assist stable grasping and manipulation, such as when the highly general Enclosure is used to extract multiple sources of information about the geometric and/or material properties of an object, before the object is lifted away from a supporting surface (see section 9.3.2).

9.2.4.5 Input Modality, EPs, and the Nature of Object Representations

Experiments by Klatzky *et al.* (1987), and more recently by Summers (1991) and Summers *et al.* (1992), have clearly demonstrated that motoric/processing costs can also influence EP selection, which in turn influences the nature of the object representations.

Subjects were required to sort multi-dimensional objects into "similar" piles, where objects perceptually varied uniformly along both material (e.g., texture, hardness) and geometric (e.g., shape, size) dimensions. With these objects, there was no *a priori* class membership. The different object dimensions were thus perceptually pitted against each other, with subjects free to make their judgements of similarity based on any of the four dimensions. Each of four groups of subjects was given a different set of instructions, to determine the nature of the object representations when haptics was either used alone or with some form of visual information (real or visual imagery).

The "unbiased haptics" and "haptically-biased haptics" instruction groups were directed to sort objects into similar piles (i.e., with no verbal bias toward processing the available haptic inputs), or to sort objects into piles that *felt* similar (i.e., with specific verbal emphasis on haptic processing), respectively. Although the other two conditions also involved haptic sorting, different forms of visual information were additionally available. Subjects in the "touch + visual imagery" condition sorted according to the similarity of the *visual images* of the handled objects (i.e., with deliberate bias toward visual processing). Subjects in the "touch + real vision" condition were simply instructed to sort objects into similar piles (i.e., with no strong processing bias toward either the haptic or visual inputs).

Input modality and EP selection: with respect to EP selection, the results indicated that subjects tended to select EPs that are highly efficient (re: accuracy and speed) for the task. The unbiased and haptically-biased haptics groups both preferred to execute EPs that are optimal for extracting the material properties of objects (see also section 9.2.3). These material-extracting EPs are both precise *and* fast, compared with those that optimally extract geometric information. In general, the haptic system is particularly well suited and efficient at extracting properties derived from the material from which an object is constructed. In contrast, when subjects were permitted to use both haptics and vision (real or visually imagined), they performed little manual exploration or chose to execute contour extraction EPs, respectively; as we shall describe shortly, they weighted shape heavily in their sorting judgements. These results also seem appropriate because when vision is available, it is typically faster than any haptic EP, and is particularly accurate at extracting geometric, compared with material, information about objects.

Haptics, vision and the nature of object representations: the experimental results indicated that EP selection can also have associated processing consequences. Properties associated with EPs that were frequently executed in the study were weighted more strongly in the representations of the stimulus objects, as indicated by the sorting judgements. When haptics was used alone, the sorting judgements of both groups (biased and unbiased) were based more on differences in material than on geometry. In contrast, when imagined visual similarity was used to sort the objects haptically, subjects weighted the geometric properties (particularly shape) more strongly than they did material properties. When both haptics and vision were used, shape was primary, presumably because of the efficacy of visual processing (although texture was secondarily important). The unbiased haptics results were considerably more similar to those of the biased haptics group (emphasis on material) than to those of the haptics + visual imagery group (emphasis on shape). Clearly, then, haptic processing does not always involve visual mediation (i.e., processing the visual image produced from the haptically-derived inputs (Klatzky & Lederman, 1987; Lederman *et al.*, 1990)).

The results of the Klatzky *et al.* study have since been confirmed with fully 3-dimensional objects that involved putting material variation (aluminium vs. pine) into conflict with variations in shape and size (Summers, 1991; Summers *et al.*, 1992). Of additional interest is the fact that during this sorting task, people usually selected only an Enclosure, which is capable of extracting coarse information about many dimensions relatively quickly. Presumably the Enclosure was the most efficient EP to use, because the values along each dimension were perceptually very different, compared with those in (Klatzky *et al.*, 1987). Nevertheless, people persisted in differentially weighting dimensions much as they did in the previous study, suggesting that the patterns of dimensional preference have become very well established.

These results suggest the importance of using vision and haptic robotic systems to extract and process complementary information, with each modality extracting what it can do most efficiently. In developing robotic systems, the division of labour will clearly depend upon the relative performance characteristics of the specific sensing systems. But while the solutions may prove different for robotic and human systems, both domains need to assess systematically the relative efficiency with which each modality can perform a variety of tasks (for further discussion, see for example, Bajcsy *et al.* (1989)).

In concluding this section, we would also like to mention additional empirical results from the Lederman and Klatzky study (Lederman and Klatzky, 1990b: Expt. 1; see also section 9.2.5), which provide additional information concerning the nature of haptically derived object representations. The results consist of a database of knowledge-based diagnosticity weights for a set of 114 common objects at the generic "basic", and 114 objects at the less inclusive "subordinate", levels of classification (Rosch, 1978); e.g., hammer and ball-pane hammer, respectively. Subjects were required to select properties needed to identify each object, in order of importance, from a closed set derived from Lederman and Klatzky (1987). The experimental data were also considered across all objects within each of the the two classification levels to determine whether certain properties are more diagnostic than others, depending on the level of class inclusiveness. At the basic level, shape was found to be highly diagnostic, while size and texture were next highest and equal in strength. (No comparable comparison at the subordinate level was possible, given the current experimental design.) Across levels of inclusion, material properties (e.g., thermal) proved to be more diagnostic at the subordinate (*cf.* basic) level whilst shape proved to be less diagnostic at this level. The implication for computational haptic object models is considered at the end of the next section.

9.2.5 Two-stage Sequence of Manual Exploration During Haptic Object Recognition

Lederman and Klatzky (1990b) have also investigated whether there is some principled sequence in which humans select EPs during an object classification task. Subjects were required to decide whether a common object placed in their hands was a member of the object class named in the experimenter's question (e.g., "Is this abrasive surface further a piece of sandpaper? (Y/N)"). Hand movements during the trial were videotaped and subsequently analysed as a sequence of EPs. Subjects consistently performed a two-stage sequence, regardless of the specific question. Stage 1 consisted of some variant of a grasp-and-lift (Enclosure-and-Unsupported Holding) routine, thus involving relatively general EPs. Stage 2 consisted of more precise (optimal) EPs, the specific selection predicted by the attribute previously determined in a separate experiment (Expt. 1) to be most diagnostic of the class of object named in the question (e.g., for sandpaper, where subjects had judged texture to be most diagnostic, they tended to execute Lateral Motion).

We also note an interesting parallel in the robotic domain. For intuitive reasons, Allen and Michelman (1990) recently incorporated this two-stage (coarse-to-fine) search sequence into the robotic haptic exploration of shape. They implemented a version of the human Enclosure EP to obtain gross shape, followed by two types of the more specialised Contour Following procedures, used to test surface planarity and to extract detailed edge information, respectively. We advocate using the human experimental paradigm from Expt. 2 of Lederman and Klatzky (1990a) to develop a more general robotic search solution for EP selection when multidimensionally varying objects are involved. This experimental approach would allow the determination of the relative performance characteristics of the set of robotic EPs selected, which could then be used to select viable EP sequences.

Consider the following scenario. We begin with the view of an EP as a motoric routine that is optimal for extracting one property, although it may be sufficient for extracting several others. The properties and associated EPs will depend upon the particular robotic end effector and sensing system selected, and could be quite different from that which humans use (unless an anthropomorphic design has specifically been adopted). Having selected a set of properties and EPs, one can now experimentally test the relative constraints on EP performance (as defined by the roboticist). The constraints on human EPs listed in this section may be applied to any exploring system, along with any others that are specifically relevant to the robotic domain. The experiment described in Expt. 2 of Lederman and Klatzky (1987) can be used as a methodological guide for systematically evaluating the relative performance of each EP in extracting each property; tasks should involve different levels of property discrimination, if EP performance is to be assessed properly. The results can then be used to rank robotic EPs for use in associated computational models of EP selection.

9.2.6 Towards a Computational Model of Constraint-driven Exploration and Haptic Object Identification

Klatzky and Lederman have developed a conceptual model of the human haptic system with respect to haptic exploration and recognition of multi-dimensional objects (Klatzky & Lederman, 1991). The model describes major architectural components, including representations of haptically accessible object properties and exploratory procedures, as derived from empirical data reported in sections 9.2.3 and 9.2.4. The components relate to one another in processing-specific ways. Each property component is connected with EPs to the extent to which the EP provides information about that property. The EPs are associated with each other according to their compatibility, based on parameterisation by motor execution and interactions with the object and the workspace. The resulting architecture is treated as a system of constraints that guide exploration during the course of object identification. The selection of the next step in the exploratory sequence requires that constraints be optimally satisfied. A network approach to constraint satisfaction has been implemented, in which constraints that are derived from empirical data determine the weights on the EP-Property and EP-EP connections (Lederman and Klatzky, 1990a; sections 9.2.3 and 9.2.4 of this chapter).

The model accounts for a number of our previous empirical phenomena concerning the two-stage course of exploration (Lederman and Klatzky, 1990b; section 9.2.5 of this chapter), speed of object classification, and incidental learning about object properties (Klatzky *et al.*, 1989; section 9.2.4.4 of this chapter). Vision may be treated as an EP with strong links with geometric properties (see section 9.2.4, paragraph on input modality, EPs and the nature of object representations). Finally, we suggest that the database of object properties described in the aforementioned section might be used to extend the breadth of our computational model of EP selection during object recognition by providing the relative weights on the links between properties and a large number of specific objects.

9.2.7 The Development of Manual Exploration and Haptic Object Perception

The work reported up to this point has concerned haptic exploration and object processing in highly experienced human adults. Of course, by this age haptic perception and recognition have become highly consolidated and finely tuned. To address issues pertaining to the development of haptic perception, it can be instructive to study relatively inexperienced infants and very young children. By what age can infants extract information about specific object properties? What factors constrain the time course of haptic sensitivity to these properties? Such questions should be of considerable interest to the roboticist, who works with systems that functionally resemble the young infant.

The reader is therefore referred to a recent review by Bushnell and Boudreau (1991). Briefly, using extant empirical findings, they determined the temporal course for the appearance of infant haptic sensitivities to individual object properties. (To date, most of the experimental tasks have been performed haptically in the presence of vision.) They found that property differentiation could begin from as early as birth - 3 months to as late as 12 - 15 months. The associated developmental progression was as follows: temperature and size, texture and hardness, weight, and finally shape. Guided by our conceptualisation of exploratory procedures as prerequisites for haptic perception, Bushnell and Boudreau successfully predicted the above temporal course of haptic property differentiation by considering the developmental sequence of manual movements, which are differentially required for the execution of specific EPs. For further details concerning infant haptic object recognition and manipulation skills, the reader should also consult work by Ruff (e.g., 1982; 1989).

9.2.8 Haptic Perception Through Wielding a Rod

We broadly address haptics as an information processing subsystem, which, by definition, involves sensory, perceptual, cognitive, and motor components. Turvey and his colleagues (Solomon, Turvey & Burton, 1989) have adopted a systematic and complementary approach that specifically targets the sensory and motor levels. Their tasks require human subjects to judge the properties of a wielded rod, as well as the spatial layout of surfaces within the explored space. They analyse the hand/rod system as a dynamical system, and perform psychophysical experiments to test the validity of the proposed dynamical parameters which, it is argued, serve as the basis for perceptual judgements during wielding, the critical parameter being dependent on the given haptic task (e.g., moments of inertia for extent of a rod or some segment, or ratios of the moments of inertia for the rod's shape). For further details, the reader is referred to such work as that of Solomon *et al.* (1989). The research programme may be of interest to those designing telerobotic systems, in as much as the rod may be thought of as a tool whose dimensions and orientations must be continuously monitored, both with respect to the hand manipulating the tool and to the environment upon which the tool is acting.

9.3 Sensory-guided Motor Control

Up to this point, we have considered issues related primarily to haptic perception and recognition. We now address the (potential) role of tactile and haptic sensory cues in human motor control tasks, many of which are of considerable interest to roboticists as well (e.g., for grasping, lifting, and manipulation).

9.3.1 Cutaneous Information for Motor Control

The reader who is unfamiliar with the terminology, distribution and function of the mechanoreceptor populations distributed within fingertip skin may wish to consult additional sources (e.g., Moss-Salentijn, this book; Lederman and Browse (1988); Lederman and Pawluk, this book).

Accumulating evidence in neurophysiology has suggested that signals from the mechanoreceptive units in the human hand may play important roles in triggering motor responses and providing information about relevant physical properties of an object that is manipulated for motor control. Johansson and Westling (1990) have suggested that this is because the mechanoreceptive units are the most efficient in providing signals about various mechanical events at the object/hand interface, due to their high innervation density, sensitivity to external events, and location close to the points of contact with the manipulated object.

Johansson and Westling have contributed significantly to this understanding through their studies of tactile influences on motor control during precision grip (Johansson & Westling, 1984; 1987; 1988a; 1988b; Westling & Johansson, 1984; 1987). The interested reader is referred to the excellent review of their work (Johansson and Westling, 1990) for a more detailed explanation of the motoric aspects.

The following discussion primarily emphasises the contribution of cutaneous (tactile) information to the basic grasp-and-lift task studied by Johansson and Westling. Human subjects were required to grasp and lift an object with a precision grip, which involves pinching by the thumb and index finger. The friction and the weight of the object were systematically varied. Visual discrimination was minimised. In some experiments, the skin of the fingertips was anaesthetised or electrically stimulated.

Johansson and Westling (1984) have described the task as a sequence of coordinated events: (a) the preload phase (consisting of an increase in the grip force as the subject securely grasps the object), (b) the loading phase (in which the load force, i.e., the vertical lifting force, and grip force increase in parallel during isometric conditions), (c) the transitional phase (i.e., the lifting movement), (d) the static phase (in which the object is held stationary in space; during this phase, the two forces and the position of the object are nearly constant), (e) the replacement phase, and finally, (f) the unloading phase (consisting of the two forces isometrically declining in parallel until the object is released). During the actual lift, subjects maintained a safety margin in the ratio of the grip to load force above the minimum required to prevent slips; however, slips sometimes occurred, triggering an upgrading of the force ratio to prevent further slips. The sequence of events and the associated tactile responses are schematically shown in Figure 9.2.

Johansson and Westling's results suggested that the coordination of the forces during the grasp-and-lift task is defined by preprogrammed muscle commands, internally represented in sensorimotor memories. The suggested contribution of the tactile afferent signals consists of intermittent updating of the sensorimotor memory when inappropriate force coordination is encountered (e.g., slips), and triggering the sequential release of the motor commands during the lifting tasks (Johansson & Westling, 1984; Westling, 1986).

9.3.1.1 Response to Initial Contact and Friction

At the very moment the fingers contact the object, all FAI units and many SAI were found to show a distinct response; many FAII units responded, but in general the response was weaker and more varied than those of FAI and SAI units (Westling & Johansson, 1987). Further evidence, from studies using topical anaesthesia (Johansson & Westling, 1984), suggested that the FAI and/or the SAI units provide the decisive information in triggering the loading phase.

In general, during the grasp-and-lift task it is important that the appropriate adaptation to frictional conditions be made. If the grip is too weak, the object will slip; if it is too strong, however, the object or hand may be damaged, and/or the hand may be fatigued more than necessary (Westling, 1986). Johansson and Westling showed that adaptation to frictional conditions did occur and that a fairly small safety margin was used to prevent slips (Johansson & Westling, 1984; Westling & Johansson, 1984). Further studies (Johansson & Westling, 1984; 1987; Westling & Johansson, 1984; 1987) provided evidence that the mechanoreceptor units play a decisive role in this adaptation.

Johansson and Westling (1987) found that an initial adjustment to new frictional conditions can occur soon after the object is touched. Evidence suggested that this information was signalled by the FAI units, as some of these units were strongly influenced by the surface material. This initial response could possibly be an estimate of the unknown friction using available haptic attributes, such as texture

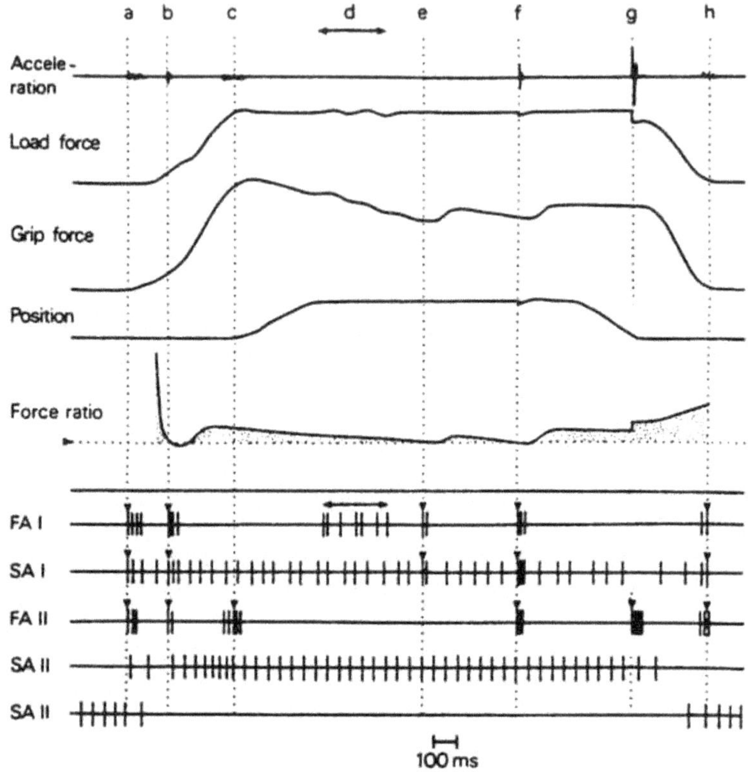

Figure 9.2: *Schematic of the sequence of events during grasping and the associated tactile responses.*
Obligatory events (occurring in all trials):
a – *initial contact, which may cause responses in all mechanoreceptors.* c – *the start of the vertical lifting movement, eliciting a response in the FAII units.* g – *contact with the table, causing a response in the FAII units.* h – *loss of contact with the object, triggering responses in the FAI, SAI and FAII units.*
Possible additional events (occurring only intermittently throughout the grasp and lift trials):
b – *slip during the loading phase, to which the FAI, SAI and FAII units respond; subsequent to the slip response there is an upgrading of the force ratio.* d – *FAI responses during periods of pronounced physiological tremor.* e – *localised slips (i.e., small slips localised to only part of the skin area in contact with the object), eliciting activity in the FAI and SAI units; subsequently there is an upgrading of the force ratio.* f – *slip, triggering responses in the FAI, SAI and FAII units; again, followed by an upgrading in the force ratio. (Reprinted with permission from Westling (1986).)*

(see section 8.4 on texture in our accompanying chapter, in this book). However, it is important to emphasise that friction and texture are not equivalent properties. For example, when people wash their hands, friction is altered as the sweat is removed, but the texture of a touched object does not change.

9.3.1.2 Slip Response

Despite the initial adjustment to frictional conditions described above, Johansson and Westling (1984) found that sometimes small slips of the object between the fingers occurred because the safety margin in the force ratio was insufficient (Figure 9.2). They also found that small slips localised to only part of the skin area in contact with the object occurred during periods of declining force ratio or at force ratios close to the minimum needed to prevent slip. Following the slips, there was an upgrading in the ratio of the grip to load force to a higher and more stable value. Johansson and Westling (1987) further showed that the FAI, FAII and SAI units responded to the overall slips, whereas only the FAI and SAI units responded to localised slips.

9.3.1.3 Response at Phase Transitions

Johansson and Westling also found that one or more of the mechanoreceptor populations responded quite distinctly to the following phase shifts: (a) the initial response (as considered previously), (b) the start of the lifting movement, (c) the end of the lifting movement, and (d) the final release of the object (Westling & Johansson, 1987).

The results of Westling and Johansson's experiments suggested that the FAII units are responsible for the triggering of the subsequent motor responses at the start and end of the lifting movement (Westling & Johansson, 1987). In additional studies, Johansson and Westling (1988a) showed that such trigger signals were particularly important in adapting to unexpected weight changes between trials. The first of their studies showed that the force rate profiles during the grasp-and-lift were preprogrammed on the basis of stored weight acquired during the previous trial (as well as friction). The remaining studies showed that the somatosensory signals triggering the start of the lift automatically altered erroneous motor commands caused by unexpected changes in weight. The weight information was then updated during the lift, as no erroneous results occurred in the unloading phase.

The loss of contact with the object in the final release phase was found (Westling & Johansson, 1987) to be signalled well by the response of the FAI units and half of the SAI units, and less reliably by the FAII units. Johansson and Westling suggested that this signalling indicates when parts of the fingers are free for further tasks; it may also forewarn loss of contact, in addition to the slip sensing described above.

Although no certain role has been found for the SAII units, and Westling and Johansson (1987) have suggested that SAII units may play a role in providing information on the magnitude and direction of the load force (e.g., for weight estimation) and on the current balance between the grip and other manipulative forces (e.g., for regulating force coordination).

9.3.1.4 Applications to Robotics and Teleoperations

In addition to the variants on the basic grasp-and-lift task described above, Johansson and Westling (1984; 1988b) have found that during further manipulative actions superimposed on the basic grasp-and-lift task, similar compensatory actions derived from tactile afferent signals occur. These results suggest that tactile sensing can provide valuable information for indicating changing conditions (e.g., contact, slip), and for updating object parameters needed for accurate manipulation. This is particularly significant for robotic systems, as dealing with these changing conditions has proved to be one of the most difficult problems in the field.

The skin acceleration sensor developed by Howe (1990) shows many similarities to the tactile afferent responses (discussed above) in the basic grasp-and-lift task. The sensor, which consists of an accelerometer attached to the outer rubber skin of a robot finger and mechanically isolated from the robot manipulator, can be used to detect vibrations of the outer skin. Howe has shown that it is similar to the human FAII units in its abilities to detect slip and to announce the beginning and end of the lifting movement. However, unlike the FAII units, the acceleration sensor was also reliable in signalling both the initial contact and the final loss of contact; as mentioned above, in humans this appears to be signalled by the FAI and/or the SAI units, rather than the FAII units. Consequently, the use of this sensor shows considerable promise in the signalling of changing system conditions for mechanical manipulators.

The greater detection capabilities of the skin acceleration sensor, compared with the FAII units, may be due to the sensor's mechanical isolation from the remainder of the finger (unlike the FAII units); this would allow it to be more sensitive to external conditions than the FAII units. Howe's approach, in this respect, nicely illustrates the advantages of learning from biological systems without being constrained by the limits of an anthropomorphic design.

In addition to the implications for autonomous robots, Johansson and Westling's work with local anaesthesia of the skin (Johansson & Westling, 1984; Westling & Johansson, 1984) elucidates the importance of providing tactile information in telemanipulation tasks. Toward this end, Edin and his colleagues (1991) have recently developed a mechanical method of producing a non-habituating response in an operator, similar to that for natural slips, in order to maintain a slave robot's grasp of an object; these slip responses were elicited by controlled, small (0.5 – 2.0 mm) sliding movements of plates contacting the operator's fingers.

9.3.2 Haptic Volumetric Attributes for Lifting an Unknown Object

A complete analysis of the contribution of sensory inputs to grasp-and-lift sequences (as well as implications for robotic/telerobotic systems) may further need to consider the effect of concurrent visual and haptic cues on the pre-programmed estimates; these cues were absent in all of Johansson and Westling's experiments.

The following psychophysical results suggest a possible role for haptic (and visual) volumetric cues when humans prepare to lift an object of unknown mass; algorithms describing the processes that underlie the human estimation of object weight using haptics (with and without visual feedback) were determined from empirical findings by Ellis (1990; 1993). In the absence of *a priori* knowledge, the human Enclosure EP (or grasp) provided haptic (probably, mostly kinesthetic) information about volume, which was used to compute initial estimates of an unknown mass before the lift; other experimental conditions indicated that visual volumetric information was only used when corresponding haptic cues were denied (i.e., when subjects viewed an object, and lifted it via a string). This suggests a general, data-driven robotic strategy for lifting such objects. Volumetric computations based on the initial grasp could be used to provide the first estimate of what is needed to optimise the subsequent lift action. The cutaneous inputs from the mechnoreceptors (see previous section) and the dynamics of the subsequent arm movements could then be used to fine tune the lift sequence. Presumably, the tactile/haptic cues would optimise the initial lift action, in as much as the same effector configuration is employed during the entire perception-grasp-lift cycle.

9.3.3 Contour Exploration Procedures (CEPs) for Grasping/Manipulation and the Role of the Second Hand

Contour extraction is of considerable importance to the field of robotics because of its relevance to grasping and manipulation, as well as to object description and recognition. In order to grasp, it is frequently necessary to isolate parts; these parts may represent projections to be grasped, or regions within which forces can be applied for manipulation, as in disassembly. However, part isolation in turn requires knowledge of object contours. Further, when manipulation involves large and/or heavy objects or complex activities (e.g., pick-and-place), additional considerations that pertain to the coordination of activity between two effector systems arise. Klatzky *et al.* (1991) and Lederman *et al.* (1991) have recently investigated the exploration of contour in the absence of vision. Task goals and intrinsic system capacities were assumed to constrain the breadth of processing and the precision with which a contour was encoded, thus determining parameters of exploration and ultimately producing movement synergies or "Contour Exploration Procedures" (CEPs). Details are provided concerning CEP descriptions and movement characteristics (i.e., number of effector(s) typically exploring simultaneously, whether there was coordinated or independent activity between multiple effectors, number of object parts contacted by one CEP, and extent of persistence on an explored segment). There was a striking preference for bilateral exploration that was symmetric, when possible. These results suggested that the use of multiple effectors, coordinated in space and/or time, may be a prerequisite for the extraction of relatively complex information about an object's structure. Such speculation regarding the optimal parameterisation of hand movement procedures used to explore and process contour information, and to grasp and manipulate, may also be of value to roboticists.

9.4 Conclusion

We have argued that the study of biological touch can further the development of current robotic tactile systems in a variety of ways. As many of the issues discussed in our companion chapter are equally relevant to our discussion of haptic sensing, we note them here as well; the reader is particularly directed to the introductory section in the chapter by Lederman and Pawluk.

As specifically considered there, biological work on tactile and haptic sensing addresses many of the same problem domains (e.g., sensor performance, sensor fusion, selection of primitives for scene segmentation and object recognition, feature extraction algorithms, active vs. passive perception, selective perception, control and planning, manual exploration, grasping and manipulation, sensory-guided motor control). It provides an example of, and framework for, designing working multilevel integrated systems. It also offers valuable suggestions for presenting robotically extracted sensory information effectively to a human operator for purposes of teleoperation. The mini-tutorial on the scientific method in that chapter highlights the value of properly constraining the problem, formulating testable hypotheses, designing unbiased experimental tests of the hypotheses, and using statistical techniques for assessing the validity, reliability, and generality of the experimental findings. We argue that the scientific method can be used to complement current analytic techniques.

The current chapter on haptic sensing deals with substantive issues and scientific findings which we believe will become important in future robotic design. We particularly address the nature and need for active (purposive) haptic exploration during object perception, recognition, and manipulatory tasks. We suggest that because of differences in the relative efficiency with which most vision and haptic systems extract specific object properties, where possible the haptic system should be used to extract material properties while vision extracts geometric properties. We also explain how to adapt the scientific method to the robotic domain for purposes of selecting effective sequences of differentially constrained exploratory procedures during manual robotic exploration within highly unstructured environments.

Finally, we consider the role of cutaneous and haptic inputs for motor control. From our studies, we conclude that roboticists have much to learn from examining biological touch systems without being constrained by, or limited to, anthropomorphic design.

Acknowledgements

This chapter was written with the financial assistance of the Institute for Robotics and Intelligent Systems Centre of Excellence of Canada, the Information Technology Research Centre of Excellence of Ontario, the Manufacturing Research Corporation of Ontario Centre of Excellence, the Natural Sciences & Engineering Research Council of Canada, and the Alberta Heritage Foundation. The authors wish to express their gratitude to Cheryl Wilson for her assistance in preparing the manuscript and figures.

References

Allen, P.K. & Michelman, P. (1990). Acquisition and interpretation of 3-D sensor data from touch. *IEEE J. Robotics & Automation, 6(4)*, 397-404.

Bajcsy, R. (1989). Active perception and exploratory robotics. In I. Plander (Ed.). *Artificial intelligence and information-control systems of robots.* New York: North Holland.

Bajcsy, R., Lederman, S.J. & Klatzky, R.L. (1989). Machine Systems for exploration and manipulation a conceptual framework and method of evaluation. *Tech. Rept. MS-CIS-89-03, Dept. of Computer and Information Science, Univ. Pennsylvania, Philadelphia, PA.*

Balakrishnan, J.D., Klatzky, R.L., Loomis, J.M. & Lederman, S.J. (1989). Length distortion of temporally extended visual displays: Similarity to haptic spatial perception. *Perception & Psychophysics, 46(4)*, 387-394.

Bryant, P. & Raz, I. (1975). Visual and tactual perception of shape by young children. *Developmental Psychology, 11(4)*, 525-526.

Bushnell, E.W. and Boudreau, P.R. (1991). The development of haptic perception during infancy. In M.A. Heller & W. Schiff (Eds.), *The Psychology of Touch.* New Jersey: Lawrence Erlbaum Associates.

Cashdan, S. (1968). Visual and haptic form discrimination under conditions of successive stimulation. *J. Experimental Psychology: General, 76(2)*, 215-218.

Dellon, A.L. & Kallman, C.H. (1983). Evaluation of functional sensation in the hand. *J. Hand Surgery, 8*, 865-870.

Edin, B.B., Howe, R., Westling, G. & Cutkosky, M. (1991). Relaying frictional information to a human teleoperator. *Tech. Rept., Dept. Physiology, University of Umea, Sweden.*

Ellis, R. (1990). *The haptic size-weight illusion: With and without vision.* Master's thesis, Queen's University, Canada.

Ellis, R. & Lederman, S.J. (1993). The haptic size-weight illusion with versus without vision. *Perception & Psychophysics.*

Gibson, J.J. (1966). In *The senses considered as perceptual systems.* Boston: Houghton Mifflin.

Howe, R.D. (1990). *Dynamic tactile sensing.* Ph.D. thesis, Stanford University.

Johansson, R.S. & Westling, G. (1984). Roles of glabrous skin receptors and sensorimotor memory in automatic control of precision grip when lifting rougher or more slippery objects. *Experimental Brain Research, 56*, 550-564.

Johansson, R.S. & Westling, G. (1987). Signals in tactile afferent from the fingers eliciting adaptive motor responses during precision grip. *Experimental Brain Research, 66*, 141-154.

Johansson, R.S. & Westling, G. (1988a). Coordinated isometric muscle commands adequately and erroneously programmed for the weight during lifting task with precision grip. *Experimental Brain Research, 71*, 59-71.

Johansson, R.S. & Westling, G. (1988b). Programmed and triggered actions to rapid load changes during precision grip. *Experimental Brain Research, 71*, 72-86.

Johansson, R.S. and Westling, G. (1990). Tactile afferent signals in control of precision grip. In M. Jeannerod (Ed.), *Attention and performance XIII.* Hillsdale, N.J.: Lawrence Erlbaum Assoc.

Jones, B. (1981). The developmental significance of cross-model matching. In R.D. Walk & H.L. Pick, Jr. (Eds.), *Intersensory Perception and Sensory Integration* (pp. 109-142). New York: Plenum Press.

Klatzky, R.L. and Lederman, S.J. (1987). The intelligent hand. In G. Bower. (Ed.), *Psychology of learning and motivation: Advances in research and theory* (pp. 121-151). New York: Academic Press.

Klatzky, R.L. and Lederman, S.J. (In Press). Spatial and nonspatial avenues to object recognition in human haptics. In N. Elian, R. McCarthy, & W. Brewer (Eds.), *The nature of spatial representation.* Cambridge, UK: Basil Blackwell.

Klatzky, R.L. & Lederman, S.J. (1991). Toward a computational model of constraint-driven exploration and haptic object identification. *Tech. Rept. #RPL-TR-9104, Queen's University, Canada.*

Klatzky, R.L., Lederman, S.J. & Balakrishnan, J.D. (1991). Task-driven extraction of object contour by human haptics: Part 1. *Robotica, 9*, 43-51.

Klatzky, R.L., Lederman, S.J. & Metzger, V. (1985). Identifying objects by touch: An "expert system". *Perception & Psychophysics, 37(4)*, 299-302.

Klatzky, R.L., Lederman, S.J. & Reed, C. (1987). There's more to touch than meets the eye: relative salience of object dimensions for touch with and without vision. *J. Experimental Psychology: General, 116(4)*, 356-369.

Klatzky, R.L., Lederman, S.J. & Reed, C.L. (1989). Haptic integration of object properties: texture, hardness and planar contour. *J. Experimental Psychology: Human Perception & Performance, 15(10)*, 45-57.

Lederman, S.J. (1991). Skin and touch. In R. Delbucco. (Ed.), *Encyclopedia of human biology, Vol. 7* (pp. 51-63). San Diego: Academic Press.

Lederman, S.J. and Browse, R.A. (1988). The physiology and psychophysics of touch. In P. Dario (Ed.), *Sensors and sensory systems for advanced robots.* (pp. 71-91). Springer-Verlag: Berlin: NATO ASI Series F: Computer and Systems Science.

Lederman, S.J. and Jessel, A. (1992). Cutaneous sensitivity, motor dexterity, and haptic recognition of common objects in young and elderly adults. Manuscript in preparation.

Lederman, S.J. & Klatzky, R.L. (1987). Hand movements: A window into haptic object recognition. *Cognitive Psychology, 19(3)*, 342-368.

Lederman, S.J. & Klatzky, R.L. (1990a). Flexible exploration by human and robotic haptic systems. *Proc. Ann. Int. Conf. IEEE/Engineering in Medicine & Biology Society, 12(5)*, 1915-1916.

Lederman, S.J. & Klatzky, R.L. (1990b). Haptic classification of common objects: Knowledge-driven exploration. *Cognitive Psychology, 22*, 421-459.

Lederman, S.J., Klatzky, R.L. & Balakrishnan, J.D. (1991). Task-driven extraction of object contour by human haptics: Part 2. *Robotica, 9*, 179-188.

Lederman, S.J., Klatzky, R.L. & Barber, P.O. (1985). Spatial and movement-base heuristics for encoding pattern information through touch. *J. Experimental Psychology: General, 114(1)*, 33-49.

Lederman, S.J., Klatzky, R.L., Chataway, C. & Summers, C.D. (1990). Visual mediation and the haptic recognition of two-dimensional pictures of common objects. *Perception & Psychophysics, 47(1)*, 54-64.

Lederman, S.J., Klatzky, R.L., Collins, A. & Wardell, J. (1987). Exploring environments by hand or foot: Time-based heuristics for encoding distance in movement space. *J. Experimental Psychology: Learning,Memory and Cognition, 13(4)*, 606-614.

Lederman, S.J., Klatzky, R.L. & Reed, C. (1992). Orthogonal vs. redundant information about 3D objects: Constraints on dimensional integration. Submitted.

Loomis, J.M. and Lederman, S.J. (1986). Tactual Perception. In K. Boff, L. Kaufman & J. Thomas. (Eds.), *Handbook of perception and human performance* (pp. 1-41). New York: John Wiley.

Magee, L. & Kennedy, J. (1980). Exploring pictures tactually. *Nature, 283*, 287-288.

Reed, C.L., Lederman, S.J. & Klatzky, R.L. (1990). Haptic integration of planar size with hardness, texture and plan contour. *Canadian J. Psychology, 44(4)*, 522-545.

Rosch, E. (1978). Principles of categorization. In E. Rosch & B. Lloyd. (Eds.), *Cognition and categorization* (pp. 27-48). New York: Lawrence Erlbaum Associates.

Ruff, H.A. (1982). Role of manipulation in infants' responses to invariant properties of objects. *Developmental Psychology, 18(5)*, 682-691.

Ruff, H.A. (1989). The infant's use of visual and haptic information in the perception and recognition of objects. *Canadian J. Psychology, 43(2)*, 302-319.

Solomon, H.Y., Turvey, M.T. & Burton, G. (1989). Perceiving extents of rods by wielding: Haptic diagonalization and decomposition of the inertia tensor. *J. Experimental Psychology: Human Perception and Performance, 15(1)*, 58-68.

Stansfield, S. (1988). A robotic perceptual system utilizing passive vision and active touch. *Int. J. Robotics Research, 7(6)*, 138-161.

Summers, C. (1991). *Haptic exploration with and without vision: Property encoding and object representation.* Ph.D. thesis, Queen's University, Canada.

Summers, C., Lederman, S.J. & Klatzky, R.L. (1992). Haptic object similarity with and without vision: Geometric vs. material properties. Submitted.

Westling, G. (1986). *Sensori-motor mechanisms during precision grip in man.* Ph.D. thesis, University of Umea, Sweden.

Westling, G. & Johansson, R.S. (1984). Factors influencing the force control during precision grip. *Experimental Brain Research, 53*, 277-284.

Westling, G. & Johansson, R.S. (1987). Responses in glabrous skin mechanoreceptors during precision grip in humans. *Experimental Brain Research, 66*, 128-140.

Chapter 10

Object Recognition Using Active Tactile Sensing

Peter K. Allen

10.1 Introduction

When we think of using our hands to explore an object's structure, we most often assume an active mode, in which our fingers are moved over an object's surfaces using a variety of different strategies (see Chapter 8). While we sometimes obtain a static, passive "fingerprint" or "handprint" of an object (usually a flat, planar object, smaller than a finger in the former case, smaller than a hand in the latter) we note that this is the exception rather than the norm in human haptic object exploration. More often, we use an active strategy that moves fingers and hands over the objects to be recognised.

This distinction is quite important, since it defines two major problems in active tactile sensing: strategies and control. In active sensing, one needs to have a strategy to explore and understand an object's structure. Once a strategy is decided upon, it is necessary to create a control algorithm that will effectively carry out that strategy. Once control is mentioned in a robotics environment, the problems of real-time response and error detection/recovery must be addressed. This control problem is an important reason why active tactile sensing has proven so difficult to achieve in a robust and general manner. Active tactile sensing also increases the demands placed upon tactile transducers, which need to be much more robust when actively pushed, poked, stroked, and rubbed across hard surfaces.

The strategies that have been used in active tactile sensing are quite varied. To date, no single strategy appears to have been isolated. Rather, it appears that the strategies chosen are related to the nature of the task, the particular object modelling system chosen, the nature of data derivable from the sensor and the recognition engine chosen. This has led to many *ad hoc* (but useful) methods.

The focus of this chapter is higher-level strategies and methods for recognising and discriminating objects using active tactile sensing. Most of these methods either assume or implement a series of much lower-level tactile primitives that form the basis for active methods of tactile exploration. Progress in object recognition using touch is tightly coupled to developing robust low-level tactile primitives. Some of the systems described below have been tested on simulated rather than real data, and in controlled environments. A key research issue that remains is the development of robust low-level tactile sensing systems. Given such a system, many of the methods below show promise in building an integrated system for recognition.

This chapter is organised as follows: section 10.2 is a brief discussion on some early pattern recognition approaches to active tactile object recognition. Section 10.3 is a review of the interpretation tree methods of model matching which are well suited for sparse tactile data. Section 10.4 describes intrinsic tactile sensing using force/torque sensing on fingertips. Section 10.5 discusses methods of shape interpretation using active touch. Section 10.6 describes tactile sensing methods that are used in conjunction with other sensors (vision, range). Section 10.7 is an overview of the current state of the field.

10.2 Early Work in Active Tactile Object Recognition

An early effort at pattern recognition with tactile sensors was the work of Kinoshita *et al.* (1975). They utilised a five-fingered hand containing 22 binary sensors to discriminate among objects. Each object was grasped from a number of different vantage points and the resulting binary pattern recorded. A discriminating plane was calculated in the sensor space from these learning samples. Then, the object was grasped a number of times and its membership in the discrimination space computed. This work was able to distinguish a square pillar from a cylinder with 90% reliability. A similar approach was taken by Okada and Tsuchiya (1977) who used an 11 degree-of-freedom three-fingered hand to grasp objects and form binary patterns with the hand's contact sensors.

Another example of tactile recognition was the work of Ozaki *et al.* (1982). In this work objects were treated as containing parallel slices which were sensed by a special gripper. The gripper consisted of seven contact surfaces with tactile sensors (one palmar segment and two three-segment fingers) which were wrapped around an object's contour and reported the unit normal distribution along the contour. This distribution was then matched with a set of model distributions to try to discriminate shapes. The system would not work well with objects that could not easily be described as a series of slices.

Overton (1984) described a tactile sensor organised in a rectangular array capable of yielding grey value information proportional to force exerted on each sensor in the array. Simple vision array operators were used to distinguish patterns of tools from static sensing. A similar effort was reported by Hillis (1982) who used a very high spatial resolution tactile sensor to distinguish small objects (screws, clips, bolts, etc.). His approach also was to use traditional grey level processing techniques on the array values to find bumps and holes on the surface. Because the sensor was larger than the object, static sensing was used with one notable exception: he also implemented a measure of the ease with which an object could be rolled, a truly active approach.

10.3 Interpretation Tree Methods

The sparse nature of tactile data, and the accompanying cost of hand/arm movement to obtain such data, dictates a strategy that tries to minimise the number of sensing probes necessary to be able to disambiguate objects. Most systems that have been built are model-based, and use the models to supply information that can guide further sensing. This forms the basis of many strategies currently used in active tactile sensing. A goal of such model-based recognition systems is stated by Hutchinson and Kak (1989): "The system should be able to dynamically assess its hypotheses about object identities and poses, and select a sensing operation which will yield the greatest reduction in the ambiguity in that information".

A series of papers (Gaston & Lozano-Pérez, 1984; Grimson & Lozano-Pérez, 1984; Grimson, 1986; Grimson & Lozano-Pérez, 1987) has dealt with the use of model-based techniques to recognise databases of polygonal and polyhedral objects from sparse data (not limited to tactile, but the methods appear well-suited to tactile sensing). The primary active sensing strategy is to rely only on position and surface normal data taken in a sequence at a finite number of separate points on the surface of the object. No use is made of any other features. The goal is to match the data to one object out of a given collection of known objects (discrimination), and to determine its position and orientation (localisation). This work assumes sensors can provide two pieces of information: 3-dimensional contact points and surface normal direction.

This method explicitly eschews feature recognition (i.e., finding prominent, identifiable features such as edges, vertices, holes and cavities) in place of finding *constraints* inherent in the data that can be used to prune possible interpretations. It is most effective when there is a small, finite set of objects to be recognised. It appears to be less well suited for exploring, discovering and recovering the 3-dimensional shape of arbitrary objects.

10.3.1 Eliminating Interpretations by Constraints

This method can be characterised as a search for a consistent matching between contacts on a surface of a target object and a model object. An interpretation is a pairing of each data point with one of the model object edges (two dimensions) or faces (three dimensions). The interpretations for each object in the given set can be arranged in a data structure called an interpretation tree, which consists of all possible pairings of data points to possible model components. The tree is constructed in a depth first fashion. The first contact point is assigned to all possible edge/face pairings. Each succeeding contact is assigned at the next level of the tree to all of its possible pairings. Given N contact points and M possible edges/faces in the database, there are N^M possible interpretations. The size of this tree can quickly become quite large unless whole subtrees of the search tree are eliminated quickly through constraints. Some of the constraints which are applied to eliminate interpretations are:

Distance constraint: the distance between the absolute positions of the data points must be possible for some pair of points on the respective pair of edges/faces.

Angle constraint: the known angle between the surface normals on the given object model must lie in the possible range of the angles between the normals of the corresponding data points.

Direction constraint: this is a constraint on the angle between the surface normal and the direction vector between one data position and another.

Triple product constraint: the sign of the triple product of the measured normals at three points must be consistent with the model normal's triple product.

Approach constraint: the approach path for a data point must not pass through some other edge of the object (i.e., the portion of the path before contact must be unoccupied).

Friction constraint: the approach path must lie in the friction cone for the surface.

Model constraint: there must exist some position and orientation of the given object model for which the position given by each data point can be accounted for by its corresponding edge in the model (i.e., there is a set of known stable poses which exist for an object).

This approach explicitly restricts the information and strategies used to acquire the information. No use is made of adjacency relations (topology) or of features such as edges (which separate two faces or surface patches), and no explicit use is made of the length or width of an edge or face. In addition, segmentation methods to group data points together are not used. The sensing paths for the tactile probes are through free space, instead of tracing along the object's surface. After testing it in simulated sensing, it was applied to real range images. Gaston and Lozano-Pérez (1984) handle polyhedral objects, but restrict their treatment to a planar cross-section, so their domain is essentially 2-dimensional. Grimson and Lozano-Pérez (1984) deal with polyhedra in three dimensions, taking 7 – 9 data points, and include positional error. Ellis (1987a; 1987b) has implemented a similar approach, using real actuators and sensors for haptic exploration in two dimensions. Siegel

(1991) has implemented a 2-dimensional variant of this approach using constraints on the internal joint angle and torque sensors of the Utah-MIT hand rather than Cartesian space constraints.

10.3.2 Strategies for Active Search

Given the general interpretation tree method, a logical extension is to develop more sophisticated strategies for determining where to search the object. The crux of this method is to generate new sensing points that are used to disambiguate between consistent interpretations of one or more objects and their poses. For example, a random strategy, in which multiple interpretations of objects are disambiguated by randomly generating a new tactile sensor probe direction, is only effective when there is a sizable unsensed volume of one or the other interpretation that can be detected by new rounds of random sensing. Grimson (1986) has described a method for strategically choosing a new sensing move. In this work, the implicit assumptions that are used are:

1. There is an initial, finite set of interpretations. This assumption is necessary due to the inherent combinatorial explosion of arbitrary interpretations on shape and pose. (Methods for developing initial hypotheses from sparse tactile data are discussed in Chapter 11.)

2. There is a restricted set of sensing directions possible. In general, choosing a sensing ray is a four degree-of freedom problem. This work restricts the choice of ray to two degrees of freedom.

3. Polyhedral object models are assumed. This assumption is quite common in model-based recognition. It is often argued that curved surface objects can be approximated by these models, although using polyhedral models to approximate curved surfaces may lead to a large increase in the number of possible contact faces and, correspondingly, an increase in the breadth of the interpretation trees.

If it is known that the number of objects in the environment is exactly one out of a known set, then the configuration space has seven parameters: one discrete choice among the given finite set of possible objects, and six real position/orientation parameters. It is assumed that the constraints from the previous sensing moves have reduced the set of possible configurations down to a finite set of points (or at least, tightly bounded regions) in the configuration space.

A sensing move is specified by a sensing ray, which could represent a linear path of a tactile probe. Each sensing move provides a position and surface normal direction, together with error bounds on each (in the form of $\pm\epsilon$ of angle or distance). The strategy is to minimise the sum of the cost of all sensor moves that can discriminate an object. Each sensor move's cost is calculated as the positional distance of the sensing ray from a possible error of false positive or negative, while eliminating some interpretations from the possible set.

A sensing ray in 3-space has four degrees of freedom. These may be separated into two for direction (the *sensing direction*) and two for position (in the *sensing plane*, a plane in physical space which is perpendicular to the sensing direction and does not intersect the object). This is analogous to the 4-parameter representation for a line used by Roberts (1988). For each sensing direction (which may be reduced to a finite number by a tesselation of the unit sphere), find the optimal position as follows: from each point on the sensing plane, project a ray parallel to the sense direction. Note the position and surface normal at each point where the ray intersects one of the possible object interpretations (or note the lack of intersection). If either the positions or the surface normals from two interpretations are different by greater than some error bound, then the interpretations can be distinguished by this sensing ray. The points with such distinguishing rays can be collected into distinguishing areas on the plane. To minimise the chance of error, choose the ray which, over all sensing directions and over all distinguishing areas, has the largest distance on its sensing plane from its point to the nearest point which is not in its distinguishing area.

Grimson (1986) then does a careful analysis of the error bounds. This is needed because a possible interpretation will not normally be a single point in the configuration space, but a small region, due to errors in the previous measurements. This must be taken into account, along with error in the current sensing move and in the error bounds used for testing the distinguishability of two interpretations. Also, if the size of the distinguishing area for the chosen sensing ray is too small with respect to the existing error, the ray may miss an intended surface altogether, or hit an unintended surface, in which case the algorithm would fail.

The algorithm has been implemented with simulated sensing, including a variety of simulated sensing errors, with reasonable success. The algorithm appears to have a high computational cost. One could imagine reducing this by adding some heuristics. Also, simply counting the number of sensing moves may not be an accurate measure of cost. Moves that require a large change in direction or position could be very costly. This could be taken into account with heuristics, while retaining the fundamental approach and careful error analysis.

Another difficulty with the method is that surface normals are difficult to sense accurately with current tactile sensor technology (see Fearing's work (Fearing & Binford, 1988; Fearing, 1990) for a sensor and method that tries to compute both curvature and normals). A commonly used technique to estimate a surface normal direction is to sample a surface patch of an object with point contacts, and then fit a plane to this data and approximate the surface normal direction in this manner. While computationally efficient, this method can be prone to error given current tactile sensor capabilities.

When the approach was applied to scenes with multiple objects, the cost of the algorithm became very large. To alleviate this difficulty, a number of methods were developed (Grimson & Lozano-Pérez, 1987): (1) heuristics for cutting off search; (2) coupled constraints that made explicit use of the length of edge boundaries and the shape and size of polygonal surface patches; (3) Hough clustering to identify similar interpretations; (4) grouping data points together into extended linear features.

10.3.2.1 Other Approaches

Schneiter and Sheridan (1990) developed a variant on the position/normal approach. Instead of data points, each sensory instance provides position and surface normal for a data patch which is of finite, non-zero extent. The Multi-Interpretation Image (MII) is an overlay in physical space of the images of object boundaries from the set of possible interpretations (again assumed to be finite). The idea is to examine the topology of the MII, and look for sections where the boundaries for the different interpretations do not coincide. Then a sensing path (possibly curved) which passes through such a section will distinguish between interpretations. No formal measure of sensing cost is given. But obviously bad sensing paths can be avoided, and obviously good moves can be taken. The approach has been implemented in simulation with 2-dimensional polygons, and no measurement error. Handling errors and extending to three dimensions will severely increase the computational cost.

Cameron (1989) has proposed a method that uses probabilistic models and statistical decision theory to evaluate the utility of sensing operations and can then be used to guide where to sense the environment next. The main idea here is that once a sensor has made contact with an object, some knowledge of the location of the object is available. If the contact is assumed to be on an edge, this constrains the object's location in the direction along the edge, and it also allows for the computation of a probabilistic membership function along that edge. Using Bayes' theorem, we can compute the probability of membership given prior information about the object's location. In addition, a sensor uncertainty function and a transformation uncertainty function (from sensed to canonical model coordinates) can be convolved with the point membership function to express the probability of a sensed point lying within the object's area (2-dimensional polygonal models are assumed).

Now, consider the task of identifying a sensed object from a set of candidate models. For each possible sensing location that might be chosen, we can sum the utility of that particular sensing operation by measuring the amount by which it increases the probability membership function. Sensing operations which do not increase the utility function are less useful than those that increase the utility function. Summing these utilities over all candidate objects, we can pick the sensing location that provides the most useful information.

Cameron argues that this method is potentially more useful than an interpretation tree method. The sensing is guided (by the utility function) and avoids the combinatorial explosion of assigning each new data point to all possible interpretations. On the other hand, iterating over all possible sensing locations

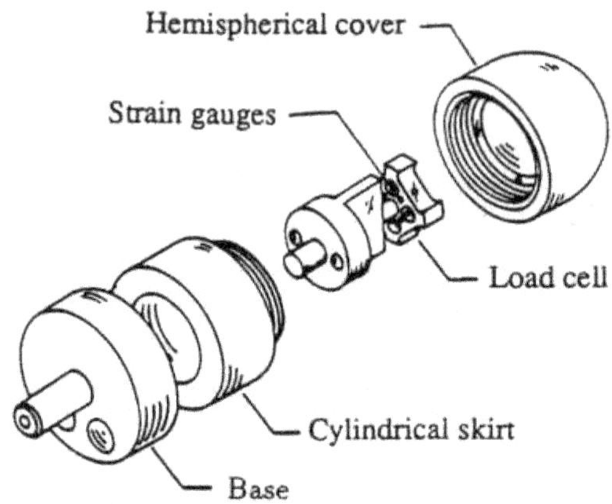

Figure 10.1: *Intrinsic tactile sensor developed by Brock and Chiu (1985).*

can be equally expensive. Two other points to consider are that this method provides a mechanism to handle "no contact" points (not easily done within the interpretation tree framework) and also provides a basis of discriminating among feasible interpretations.

Roberts (1990; 1991) has extended the interpretation tree to include some additional constraints based upon sensing larger and more stable primitives such as edges. The approach is to acquire tactile data in a serial, sequential manner, similar to human strategies of moving a hand over an object. In this work, (also discussed in Chapter 11) point contact data is collected from a Utah-MIT dextrous hand and is segmented into planar contacts that can be intersected to hypothesise edges and vertices. Roberts's approach is inherently 3-dimensional and has been tested on real objects using the Utah-MIT hand.

10.4 Intrinsic Tactile Sensing

Intrinsic tactile sensing is the measurement of forces and torques at the hand's fingertips. In this method, sensors are used to read the force and torque resultant vectors generated by the distributed contact pressure acting between two touching bodies (fingertip and object surface). Figure 10.1 is an example of an IT sensor developed by Brock and Chiu (1985). It is a six axis force-torque sensor which is placed at the base of the fingertip of a robotic hand. This design allows the fingertip, which does not contain any sensors, to be manufactured using a variety of different shapes and materials.

Intrinsic tactile sensors cannot provide direct information about the actual shape, size or type of the contact area. However, closed form methods are available to solve for the contact position giving reasonable assumptions about the finger and contacted object's local geometries. Intrinsic tactile sensing can be thought of as providing information similar to a human wearing a thick glove: gross level forces and torques are available, but accurate local positional information may not be. An apparent advantage of this method over traditional tactile sensor arrays is that it avoids many of the problems associated with these devices: too many wires, non-linearities, hysteresis, and less than robust performance.

Another advantage in this method of sensing is that the forces and moments measured by intrinsic tactile sensors are exactly what is needed in maintaining stable grasps and performing active force control with a robotic hand. Active force control can therefore be used as a means both to grasp stably *and* to recognise an object's shape. This type of sensing specifically addresses the control problem which is one of the major problems in active sensing.

Salisbury has performed some of the earliest work in this area. His approach was to use a simple and accurate strain gauge sensor that would report forces and moments of a contact. These forces and moments can then be used to solve for a contact position (Mason & Salisbury, 1985). A simple example from Salisbury's work shows how this is done. Consider an intrinsic tactile sensor which is spherical in nature and whose coordinate system's origin is at the sphere's centre. If we measure a set of forces and moments $(a_1, a_2, a_3, a_4, a_5, a_6)$ with this sensor, we can determine the line of action along which these forces must be acting. If the direction cosines of this line are l, m, n, then the intersection of this line with the sphere's surface (where the contact point we are looking for must lie) can be easily found. Figure 10.2 shows the geometry of the situation. The equation of the line of action is:

$$x = x_0 + \lambda l; y = y_0 + \lambda m; z = z_0 + \lambda n \tag{10.1}$$

where (x_0, y_0, z_0) is the point of closest approach of the line of action to the sphere's centre. Substituting this into the equation for a sphere:

$$x^2 + y^2 + z^2 = r^2 \tag{10.2}$$

yields a value for λ that is:

$$\lambda = \pm\sqrt{r^2 - x_0^2 - y_0^2 - z_0^2}. \tag{10.3}$$

The two values of the solution correspond to the points on the spherical sensor intersected by the line of action of the forces. These two solutions may be disambiguated by realising one solution represents a force pushing the objects into contact and the other represents a force pulling them apart. A similar analysis can be used for other sensor geometries such as a cylinder (Brock & Chiu, 1985).

10.4.1 Surface Recovery Using Intrinsic Tactile Sensing

Based on the analysis above, intrinsic tactile sensing can be used for exploratory searching of objects. A basic force control task is to perform surface following:

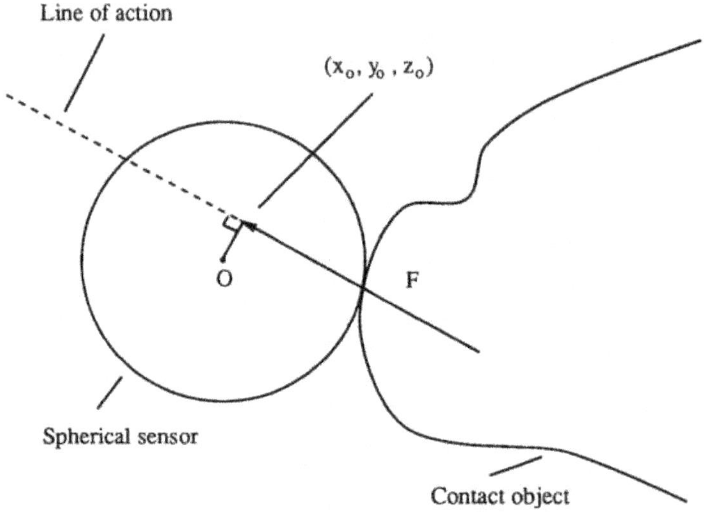

Figure 10.2: *Locating contact point on a spherical sensor.*

moving the fingers along an arbitrary surface while maintaining a force contact upon the surface. If positional contacts (or centroids of positional contact from a fingertip force contact) can be determined, then the 3-dimensional structure of the object can be recovered. This procedure can perform geometric feature extraction, a precursor to higher-level recognition methods. Using IT, geometric parameters of a surface can be recovered. The parameters can be local, such as surface normals and curvature, or global depending upon the length of the surface path followed.

Balestrino and Bicchi (1989) have proposed a method for recovering the shape of an object with an intrinsic tactile sensor. Their method calculates the contact point centroid and the surface normal direction. If the surface is modelled as a quadratic form, a point on the surface $\mathbf{r}$ must satisfy the following constraint:

$$S(\mathbf{r}) = \mathbf{r}^T\mathbf{Q}\mathbf{r} + 2\mathbf{p}^T\mathbf{r} - 1 = 0 \tag{10.4}$$

where $\mathbf{p}$ is a vector establishing the base frame and $\mathbf{Q}$ is a matrix of quadratic coefficients. An intrinsic tactile sensor can provide an estimate of the centroid of the contact area $\mathbf{c}$ and the unit surface normal vector $\mathbf{n}$. Substituting these values in the surface equation creates the following constraints:

$$\mathbf{c}^T\mathbf{Q}\mathbf{c} + 2\mathbf{p}^T\mathbf{c} = 1 \tag{10.5}$$

$$2\mathbf{Q}\mathbf{c} + 2\mathbf{p} + \kappa\mathbf{n} = 0. \tag{10.6}$$

These equations state that the contact centroid must be on the surface, and that the normal must be parallel to the gradient of **S** at **c**. These constraint equations are a system of four equations in ten unknowns (six elements q_{ij} of **Q**, three elements of **p**, and a scale factor κ). Sampling with N contact points creates an overdetermined system of $4 \times N$ equations in ten unknowns that can be solved in a least square sense. However, the resulting matrix representing these equations may be ill-conditioned and yield errorful approximations.

Balestrino and Bicchi have suggested an alternative method to overcome this problem. This method uses a dynamic approach, in which the shape parameters can be continuously estimated with an adaptive control algorithm during active tracing. In this approach, a parameter error vector (using the same set of parameters estimated above) can be computed with a positive definite gain matrix, and this can lead to a convergence of the parameter set for a steady-state solution. This approach may have special appeal due to its inherently tight coupling with the actual movement control algorithm along the surface.

Another example of this method is the work of Tsujimura and Yabuta (1989). They employed a similar method of analysis to Brock, using a six-axis force-torque sensor mounted on the end of a robotic manipulator. In this work, they used two different kinds of probes (i.e., different contact surface geometries): a linear bar probe and a hemispherical probe, each attached to the force-torque sensor. The experiments consisted of moving around known objects in a predefined sensing path, and contacting the probe with the object's rigid surfaces. Using the kinematic transforms of the robotic arm and the contact analysis described above, they were able to recover very accurately the positional contacts on the object. This information can be used by a higher-level system to recover the 3-dimensional shape of an object.

The two objects used were a telephone receiver whose contour was probed by the linear bar and a cylindrical cable of 10 cm diameter which was probed by a hemispherical probe. In both cases, the recovery was very good, with average measurement errors of 0.4 mm with the linear probe and 2.4 mm with the hemispherical probe. It should be noted that the system moved quite slowly between contact points in order to obtain a stable signal (on the order of 30 s). Reproducing normal human tactile sensing rates remains a difficult and challenging problem in active tactile sensing, given the problems of sensor delay, hysteresis, and coupled movement of the sensor and the hand/arm containing it.

Tsujimura and Yabuta also performed experiments that changed the external force applied to the sensor to see how the system's response would change. In these experiments, the point of application of the force, the force direction, and the force magnitude were all varied and the sensor response measured. In all cases, the sensor was able to report the correct values with only small errors, confirming the usefulness of the contact model and the approach.

10.5 Shape Interpretation Using Active Touch

Recovering the shape of an object from very sparse tactile data is the goal of Fearing's work (Fearing & Binford, 1988; Fearing, 1990; Berkemeier & Fearing, 1990). This work has touched upon many of the important aspects of active tactile sensing and shape recovery. As an example, consider a three-fingered dextrous robotic hand. Is it possible to recover the shape, or some parameters of a certain class of modelled shapes, from these three local tactile surface contacts?

A central idea in Fearing's work is to compute local surface geometry measures accurately from the tactile sensor responses (see Chapter 4 of this volume for a discussion on planar elasticity for tactile sensing). At the sensor-level, a capacitive array sensor is used which is mounted on the end of a finger, covered by a compliant rubber surface. This sensor is formed from the junctions of perpendicular copper strips, spaced 3.3 mm apart and at 18° around the circumference of the finger. In the experiments described below, an 8×8 subwindow of the 8×20 array was used.

A stress-strain model of contact is used, in which sensor strains are predicted by convolving the spatial impulse response of the compliant rubber skin with the assumed surface pressure distribution derived from a Hertzian contact model. Using gradient search, the parameters of the second-order shape that best fits the sensor data are found. This method is capable of determining radius estimation (curvature) within 10%, orientation (surface normal) within 10° and sub-tactel localisation within 3% of the element spacing (Fearing, 1990). Once these local surface geometry measures can be computed accurately, shape recovery becomes a much simpler problem since shape models can be used to constrain the possible surfaces being explored. Building a dextrous manipulation system for *a priori* unknown objects motivates the choice of so-called weak object models, which can give an initial estimate of the object's shape quickly and can guide further sensing.

The class of modelled objects considered is right linear straight homogeneous generalised cylinders (RLSHGC). This class of objects consists of a convex constant cross-section function which is translated along an orthogonal straight axis and scaled by a linear sweeping rule. Parameterising these objects by z and θ we can express them in canonical form as:

$$\mathbf{X}(z,\theta) = (Az + B)\rho(\theta)(cos\theta\mathbf{i} + sin\theta\mathbf{j}) + z\mathbf{k} \tag{10.7}$$

where A and B are constant scaling parameters ($A = 0$ for a regular cylinder and $B = 0$ for a cone). Given this model, we can compute the differential geometry measures of tangent directions on the surface, surface normal direction, principal directions and principal curvatures. Because the tactile sensor provides estimates of these measures at each contact point, the process of shape recovery then becomes one of solving sets of constraint equations to recover the parameters of the modelled shape from the measurements. In addition, there are the global unknowns of rotation and translation parameters that need to be recovered since the models are in a canonical coordinate system.

With this class of models, the finger contacts can be broken down into types by curvature: vertex, edge, face or parabolic point. Each type of contact provides a number of different constraint equations, and depending on where the fingers contact the object, different subsets of these contact types are sufficient for recovering the object's shape and location in space.

For example, two meridian edges and a surface normal from an end face are sufficient to recover the orientation and origin of a cone. There are other combinations which are also sufficient, as well as many subsets of contacts which will not be sufficient. A related problem is determining if edges are either meridian or end edges and if faces are end faces or side faces on the object. Knowledge of these positional contacts will affect determination of the origin and orientation. Using the geometric constraints of RLSHGC, it is possible to test these contacts and match them to the desired face or edge type.

10.5.1 Related Methods

Allen and Michelman (1990) have developed two methods of active shape recovery using a Utah-MIT hand (discussed in detail in Chapter 11). These methods have been patterned after human object recognition strategies (see Chapter 8). In the first method, grasping by containment, the hand grasps an object reporting finger contacts from the hand's enclosing volume. These sparse tactile contacts are used to recover the shape of objects as superquadrics (Pentland, 1986). The number of parameters needed to recover a superquadric is small (11 or 13 depending upon whether deformations of the superquadric are considered). A non-linear least squares recovery technique developed by Solina (1987) is used to recover the shape of a number of different objects using this method. In each case, the hand grasps the object in an enclosing grasp a number of times and fits the data to a superquadric model. The method is quite useful as a means of generating a first hypothesis of the object. Besides recovering shape, matching techniques for performing object recognition have been developed using the recovered superquadric parameters.

The second method also tries to recover a form of generalised cylinders by recovering the 3-dimensional contours of elongated objects and fitting these to a generalised cylinder model. The method traces the contours of objects using the thumb and index finger of the Utah-MIT hand, and fits them to a generalised cylinder model. By obtaining two such contour curves that are on either side of the object, the axis of the surface of revolution of the generalised cylinder can be estimated. This axis along with the contour curves are sufficient for shape recovery. An extension to this method (Allen, 1990) uses stereo information to determine the object's axis, and guide the hand in the contour sensing.

10.6 Integrated Methods

Tactile sensing can also be used in conjunction with other sensing modalities. Humans are particularly adept at integrating touch with other sensing modalities such as visual and auditory sensing. A number of robotic systems have attempted to integrate active touch with other sensors for object recognition.

10.6.1 Recognition Using Vision and Touch

Allen (1987; 1988) describes a robotic system for object recognition that uses passive stereo vision and active exploratory tactile sensing. The complementary nature of these sensing modalities allows the system to discover the underlying 3-dimensional structure of sensed objects. The approach is model-based using hierarchical, viewpoint independent, 3-dimensional models of the objects which include curved surfaces, concavities and holes. The vision processing provides sparse 3-dimensional data about regions of interest from a stereo algorithm. These regions are then actively explored with a tactile sensor which is mounted on the end of a six degree-of-freedom manipulator. A hierarchical procedure has been developed to integrate the visual and tactile data into accurate 3-dimensional surface and feature primitives which provide geometric and topological information that is used for matching against a database of known objects.

Figure 10.3 is an overview of the system. The system is divided into six main modules: vision sensing, tactile sensing, sensor integration, matching, verification, and the model database. The control flow in the recognition cycle is as follows:

1. The vision system images the scene and analyses all identifiable regions of interest. Stereo matching of edges provides sparse 3-dimensional information about each identified region in the images.

2. Using the 3-dimensional visual cues, the tactile system computes an approach vector and actively explores each region identified from vision.

3. The 3-dimensional data from both tactile and visual sensing are integrated into surface and feature descriptions of the sensed object.

4. The surface and feature descriptions are matched against the model database, invoking all models consistent with the sensory information.

5. The invoked model is verified by further sensing to see if it is correct.

The experimental tactile sensor used in this research was developed by M. Briot at L.A.A.S. in Toulouse, France (Figure 10.4). The sensor is roughly the size and shape of a human index finger and consists of a rigid plastic core covered with 133 conducting surfaces wrapped in a conductive elastomeric foam. The geometry of the sensor is an octagonal cylinder of length 228 mm and radius 20 mm. On each of

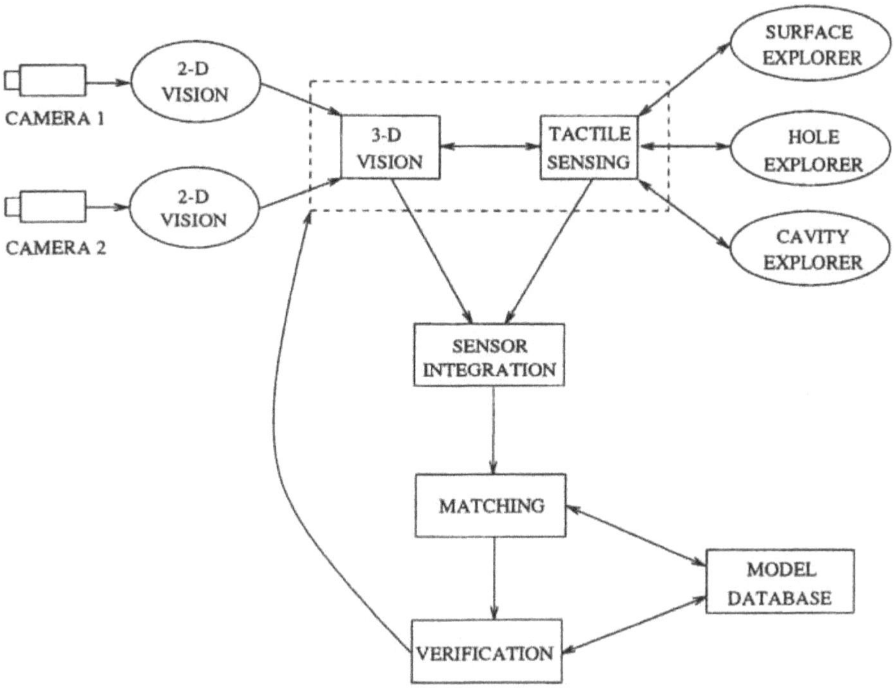

Figure 10.3: *Active vision/tactile system of Allen.*

Figure 10.4: *Single finger tactile sensor.*

the eight sides of the cylinder there are 16 equally spaced conducting surfaces. The tip of the sensor contains one conducting surface, and there are four other sensors located on alternate tapered sides leading to the tip.

The organisation of tactile sensing is on three distinct hardware and software levels (Figure 10.5). The highest level consists of programs on a host computer that provide high-level control information about the regions in space that are to be explored with the sensor. Algorithms have been developed to explore initially the regions isolated from the vision processing with tactile probes to determine if they are surfaces, holes or cavities. Once a region is identified by a tactile probe, it can be further explored by active tactile surface following algorithms that report contact points on surfaces and boundary contours of holes and cavities to the controlling host process. These contacts can be integrated with the 3-dimensional contours from vision to build robust surface and feature descriptions. The intermediate level consists of programs that move the robotic arm based upon feedback from the tactile sensor. The intermediate level receives region exploration parameters from the host which then allows it to invoke a surface exploration, hole exploration or cavity exploration procedure. These procedures use the feedback from the tactile sensor contacts to control arm motion along the exploration path determined by the highest level of processing. The intermediate level communicates with the low-level sensor system via commands that set thresholds for contacts, requests contact interrupts and requests grey level outputs from arbitrary subsets of the sensor's elements.

The classification of a region isolated by the vision system into surface, hole or cavity is performed by an intermediate level tactile exploration program. This program controls the motion of the robotic arm and wrist-mounted tactile sensor as it explores the region. Its input is an approach vector towards the region which establishes the sensor's orientation, and is calculated from the sparse stereo contour data of the region. The tactile sensor's long axis is aligned with this approach vector and the sensor is then moved until it contacts a surface or it moves beyond the plane of the contour points, implying the presence of a hole or a cavity. If the sensor is able to travel its full length beyond this plane without contact, then a hole has been found. If it travels beyond a specified cavity threshold before contact, then it is a cavity. If the region is a surface, a surface exploration program will trace the surface. If it is a hole or cavity, a boundary curve will be traced. The output of these exploration programs is integrated with the 3-dimensional vision data to form surface and feature primitives (described below) that are used in the matching phase.

The integration of vision and touch data for a sensed surface is done by building a Coons' patch description of the surface (Faux & Pratt, 1979). The sparse 3-dimensional contours from vision form the initial patch grid and the description is refined by tactile sensing in the interior of the region. The stereo contours form a set of boundary knot points that are augmented by active surface tracing across the interior of the patch grid. These active traces provide 3-dimensional contact points as well as local curvature data that define the patch.

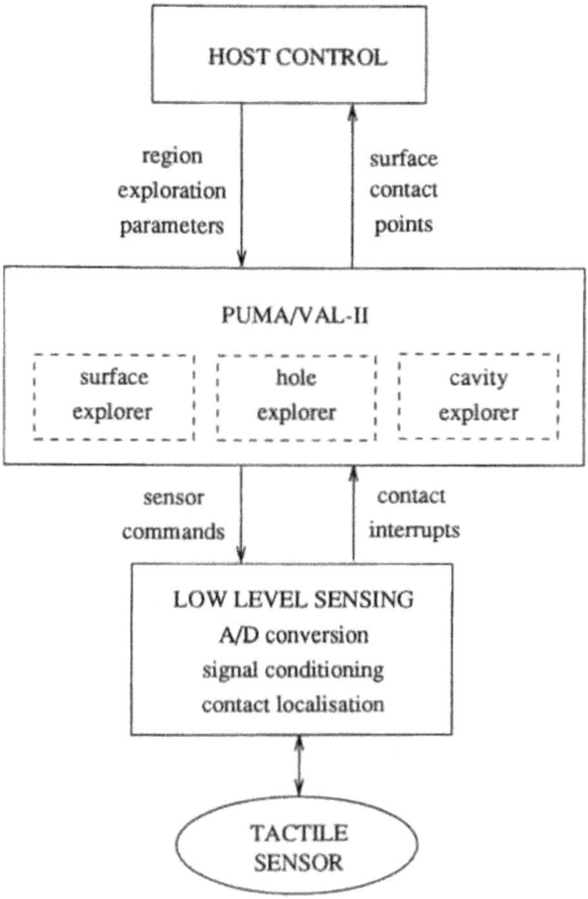

Figure 10.5: *Organisation of tactile sensing.*

The traces begin at the point of surface contact found in the initial exploration of the region guided by the sparse vision data. The sensor then traces in the direction of the boundary contours, using the surface contacts from the tactile sensor to control the robot arm motion.

The movement vector **M** along the surface is determined by:

$$\mathbf{M} = \sum_{i=1}^{3} w_i \mathbf{G}_i \tag{10.8}$$

where w_i are the weights for each of the vectors $\mathbf{G}_i$. $\mathbf{G}_1$ is the unit vector in the direction of the boundary curve midpoint, $\mathbf{G_2}$ is the unit vector formed from the previous two contact points, and $\mathbf{G_3}$ is the unit vector that preserves equal parameterisation of the Coons' surface. $\mathbf{G_1}$ is needed to make progress towards the boundary edge at each movement step. However, with concave and convex surfaces, cycles can occur as the trace progresses. $\mathbf{G_2}$ is used to maintain a path's direction. Once the sensor starts moving in a certain direction it should not stray too far too fast from that path. This vector is an *inertia* vector helping the sensor to stay on a steady course. $\mathbf{G_3}$ is needed to keep the parameterisation of the surface patches uniform, and this vector moves the trace in a direction that preserves equal parameterisation on the knot point grid. This vector is the unit resultant of the vectors from the present contact point on the surface to the endpoints of the boundary curve that the trace is approaching.

The points reported during these tactile traces are combined into cubic least square polynomial curves that are differentiated and scaled to calculate the curvature information required to build composite Coons' patches with C^2 continuity. Thus, each surface region is a lofted composite Coons' surface, formed from 3-dimensional visual contour information and active tactile traces across its interior.

Figure 10.6 shows the surface that results from active tactile sensing of the front surface region of a pitcher. The surfaces are built from sparse amounts of data and the analytic nature of these surfaces allows stable and accurate symbolic descriptions based upon the surface's Gaussian curvature to be computed. These symbolic surface descriptions (i.e., planar, cylindrical or curved) are used by the matcher in recognising the object.

If the region exploring algorithm determines that the region is a hole or cavity, a different tactile tracing routine is used to determine the boundary curve of the feature. The algorithm begins by moving the sensor just beyond the computed least-square plane of a region's contour points, with the sensor aligned with the plane's normal. It then proceeds to move parallel to the plane until it contacts a surface. Once the surface is contacted, the sensor moves along the direction of the surface's tangent plane, recording the contact points and moving until it reaches the starting point of the trace, which constitutes a closed hole or cavity contour.

This can be a noisy procedure as many of the tactile sensor's contacts become activated in a small tight area such as the hole in the handle of a coffee mug. The spatial resolution of the sensor contacts also contributes to this phenomenon. The data is not continuous, but are a set of ordered contact points that need to be smoothed and this is done by approximating the series of linked contour points with a periodic spline curve which matches derivatives at the endpoints.

Once these surfaces, holes and cavities are identified and quantified by active tactile sensing, they provide accurate measures of surfaces' area and curvature, and boundary contour measures of hole and cavities. These measures have been successfully used to label different objects in the object database.

The experimental hardware used is shown in Figure 10.7. The objects to be recognised are rigidly fixed to a worktable and imaged by a pair of CCD cameras.

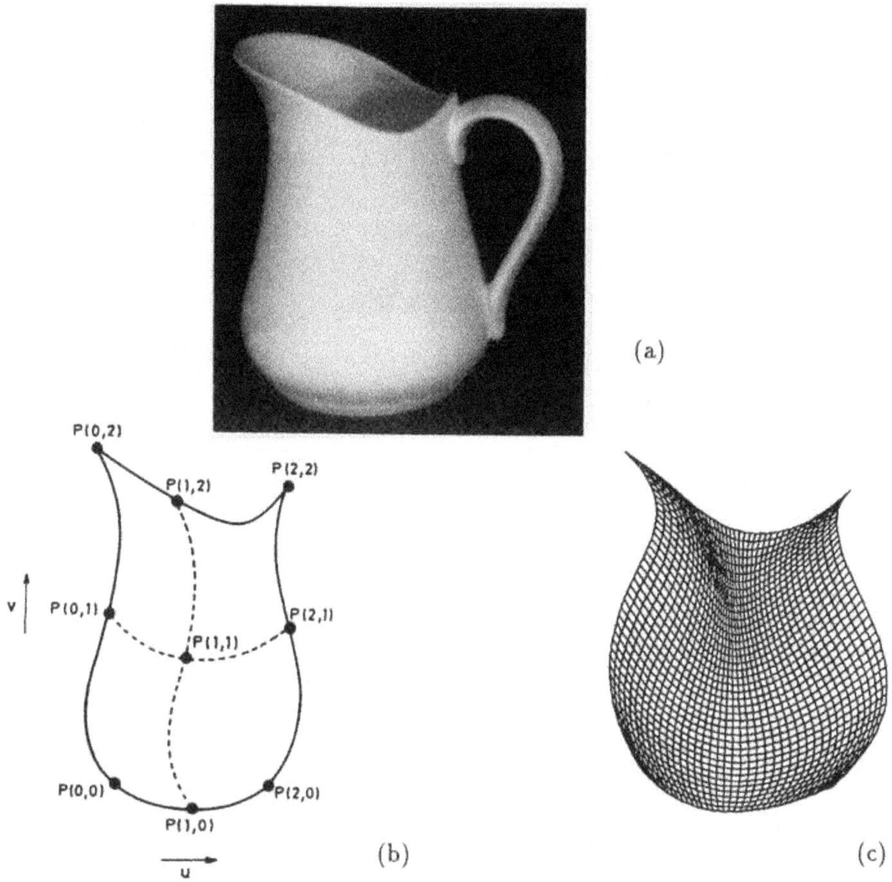

Figure 10.6: *a) Image of pitcher; b) knot points and Coons' patch grid showing tactile traces; c) surface recovered from vision and tactile sensing.*

The tactile sensor is mounted on a 6 degree-of-freedom PUMA 560 manipulator that receives feedback from the tactile sensor and is further controlled by a host processor. The experimental object domain is common kitchen items: mugs, plates, bowls, pitchers, and utensils. The objects are planar as well as volumetric, contain holes and have concave and convex surfaces. These are fairly complex objects which test the modelling and recognition abilities of most existing systems. The objects are homogeneous in colour, with no discernible textures. The lack of surface detail on these objects poses serious problems for many visual recognition systems, since there is a lack of potential features that can be used for matching and depth analysis, motivating the use of active tactile sensing.

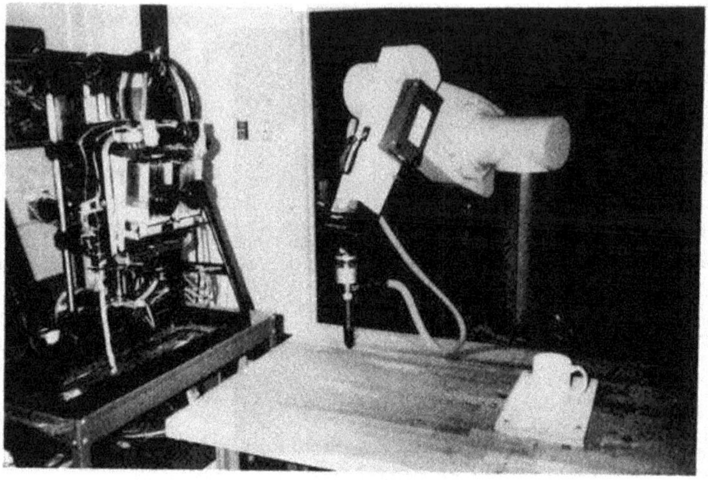

Figure 10.7: *Experimental vision/tactile system.*

10.6.2 Apprehension Using Vision and Touch

Stansfield (1988) has created a robotic perceptual system which utilises passive vision and active touch to perform object apprehension, i.e., the determination of an object's features and the relations among them. This system is based upon models of human perception, particularly in its structure which includes informationally encapsulated, domain specific, knowledge-based modules that perform perceptual tasks. The haptic modules are derived from Lederman and Klatzky's work (Lederman & Klatzky, 1987) in classifying hand movements during object recognition. The haptic nature of the sensing is important, since kinesthetic joint information from a robotic arm is used for active searching with a tactile sensor system attached to the arm that serves as a cutaneous sensor.

Stansfield's system uses visual cues to guide the active haptic exploration of the object. The vision system uses standard edge and region methods along with binocular stereo to determine sparse depth information that can be used to guide the haptic system. In this review, we will focus on the active haptic component of the system.

The hardware used in this system is a Lord LTS-200 tactile sensor with a 10×16 rectangular tactile array pad measuring 1.79×2.87 cm. The tactile array is attached to a force/torque vector sensor that returns six orthogonal components of forces and moments through the contact area of the sensor. The sensor may be moved in six degrees of freedom from a wrist on a PUMA 560 robot arm, and thus be used to actively explore an object of interest.

Stansfield has identified three kinesthetic primitives that can be used for object apprehension:

Compliance is a measure of the hardness of a surface, and is used to measure the amount of deformation a surface undergoes when a force is applied against it. Experiments were run that were able qualitatively to rank relatively hard surfaces (e.g., table tops) versus soft surfaces (e.g., sponges, bubble wrap).

Elasticity is a measure of how a surface recovers its initial shape after being deformed.

Normal-contact position is a measure of surface position and orientation at a single point on the surface. This primitive is extracted by actively positioning the sensor on a surface to maximise the contact area. Since the sensor is a planar tactile array, this establishes a tangent plane to the surface, whose normal is the estimated surface normal.

In addition, four cutaneous primitives are used:

Extended contact is a contact over a large area of the sensing pad that is not elongated. This response is typical if the surface is smooth and larger in size than the sensor.

Point contact is a small, non-elongated contact on the sensor pad. Typically, this results from touching a small object or a sharp part of a larger object.

Edge contact is a region of contact which is highly elongated. (See Muthukrishnan *et al.* (1987) for a discussion of edge detection using a similar sensor.)

Roughness can be estimated from the grey-scale statistics of the contact area, similar to texture classification in computer vision. As in any measure of this sort, results are a function of the sensor array area and the spatial frequency of the surface roughness characteristics.

Stansfield has developed Exploratory Procedures (EPs) to extract these primitives and higher-level EPs to extract features from the objects that can be used in object apprehension. An overview of this system is in Figure 10.8. The feature EPs are more complex than the primitive formation and involve the active exploration of an object using the lower-level primitives.

The design of the system includes a frame-based, slot filler representation of objects. An object is represented as a part-subpart hierarchy of frames. The highest level is the object itself. The lowest level is that of primitive features. Each frame has an associated expert or procedure that knows how to attempt to verify and instantiate the frame.

The frames have fairly simple properties. The major frames for object, body, and part have the property of the three size dimensions of the enclosing volume, given as permitted interval ranges. The object frame also has a list of component types (body or part). A part frame has various symbolic descriptors derived from its gross size and shape: large, small, elongated, extended, etc. A surface patch has

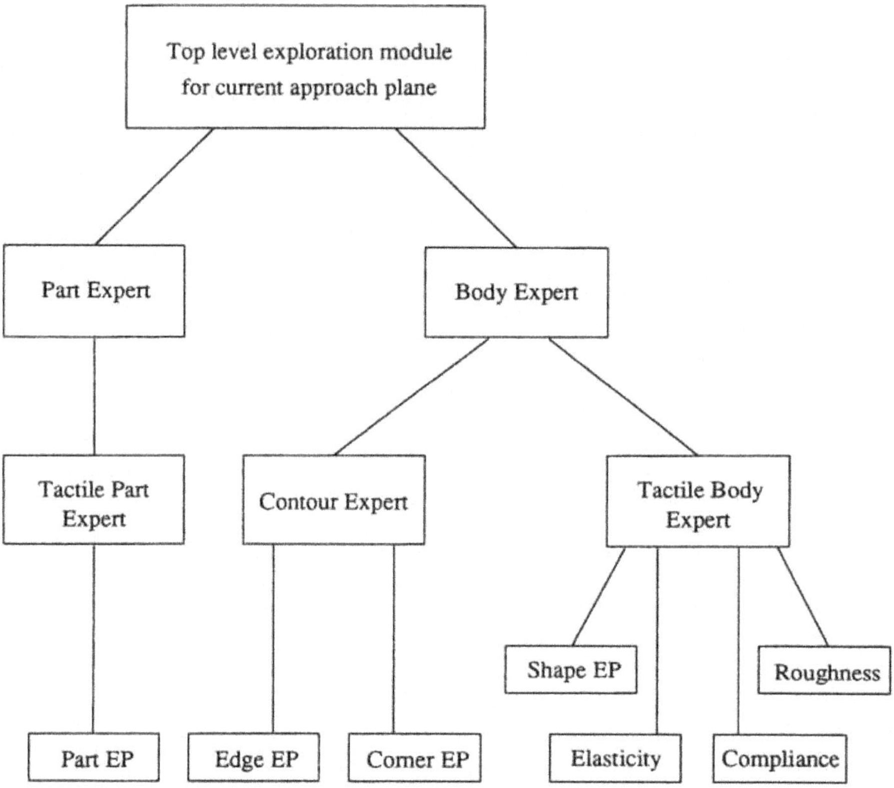

Figure 10.8: *Flow of control for haptic exploration.*

the binary properties of smooth/not smooth, compliant/not compliant, elastic/not elastic, and planar/curved.

The intermediate-level EPs are:

Surface shape: this EP determines the shape of a surface patch by moving locally on the surface and collecting four normal contact values. Using these measures, the patch is classified as planar or curved, using the standard deviation of the normals' directions. Simply shaped surfaces are assumed by this algorithm.

Part: this EP explores a part (defined as a component of a larger entity such as a body) in a certain direction, actively moving along the part.

Contour: defined as a closed series of edge contacts. Using the tactile edge contact EP (see below), this EP is able to distinguish and classify contours, which are quite important in distinguishing an object. Contours come in two types: a rim contour such as on a cup, and a border contour which is the perimeter of a planar surface. The tactile edge contact serves as a trigger to invoke this intermediate-level edge tracker. Since the edge contact creates a direction, the EP can continue to track the edge until it returns to the initial contact point. If the direction of the edge changes radically, then a corner EP is invoked instead.

Edge: this EP is data driven and is invoked by an initial low-level edge contact. It follows an edge until it terminates in a contour or it changes direction to reveal a corner, at which point it continues its exploration.

Corner: this EP determines a corner and extracts the information relevant to the corner by tracing both edges that create the corner.

The system takes a "Look once, and then touch" approach. Since there is no *a priori* information about the orientation of the object, it is approached from several fixed sensing directions. These tactile explorations are used to classify the visual features and to explore the object further and determine its shape and relations between parts. The different explorations are tied together by a spatial polyhedron that these sensing directions form. In this way, the information derived from explorations is integrated into a coherent matching frame.

The system was able to recover shape and other attributes of 16 objects in a variety of poses. This recovery is in the form of filled slots in the object frame, containing both quantitative (e.g., area, volume) and qualitative data (e.g., type of feature, soft or hard). An advantage of this system is that most of the matching involves either comparison of size dimensions or matching of symbolic properties, rather than extended numerical calculation of error functions or rigid-body constraints among different features. It appears that exploration and apprehension of a strong set of robust, distinguishing features will deliver much recognition power, even without use of numerical rigid-body transformations and constraints.

A subsequent work by Stansfield (1990) has extended these results to a multifingered hand. In this work, Interlink tactile pads (Speeter, 1989) were attached to one fingertip of the hand. In addition to implementing a set of haptic EPs as before, the system was able to plan multifingered grasps based upon recognition of an object's structure and properties.

10.7 Summary

This chapter has attempted to outline some of the approaches and problems involved in active tactile sensing. Given the importance of developing intelligent robotic systems that can recognise and manipulate objects, it is surprising that more research has not been done in this area. As stated earlier, a major impediment is the lack of robust and available tactile sensing systems. Most of the sensors described in this chapter are experimental prototypes, with few if any commercially available. Thus, experimentation in this domain is extremely limited.

A further problem is the lack of dextrous multi-fingered hands with many degrees of freedom. It remains a serious open question within the robotics community as to how many fingers (if any!) a robotic grasping device requires. Development of tactile sensors for fingertips assumes an underlying finger/hand structure for mounting them, and these devices are also expensive and not easily available. Chapter 11 describes efforts in haptic object recognition using a multi-fingered Utah-MIT dextrous hand.

An underlying theme of much of the research reported here is that sparse data is adequate for recognition. This is supported by human experience and most of the methods described here use orders of magnitude less data than corresponding vision algorithms.

If we speculate as to where progress may be made in the near future, we can identify three major areas:

Integrated methods using a variety of tactile cues and other sensors seem to be very useful. A potentially promising approach is to activate many haptic experts (such as Stansfield's EPs) in parallel, allowing a system to recover shape from many methods simultaneously and to integrate the results.

Strategies for active sensing of complex 3-dimensional structure are not yet mature. Approaches to such problems as haptic segmentation, parallel finger exploration, and relation of object stability to shape (e.g., "fumbling for recognition") need to be further developed.

Data structures and models of 3-dimensional objects are still very much bound to visual processing paradigms. Unfortunately, multiple fingertip tactile sensor contacts do not map well into rectangular planar matrix arrays as in vision processing. Developing 3-dimensional models explicitly for tactile recognition may be an important research direction.

Finally, we note the recent interest within the scientific community in microstructures and microsensors. The development of even smaller tactile sensors (but robust, cheap and simple in design) may have a large impact on the use of active tactile sensing for object recognition tasks.

References

Allen, P.K. 1987. *Robotic object recognition using vision and touch.* Boston: Kluwer Academic.

Allen, P.K. 1988. Integrating vision and touch for object recognition tasks. *Int. j. robotics research*, **7**(6), 15–32.

Allen, P.K. 1990. Mapping haptic exploratory procedures to multiple shape representations. *Pages 1679–1685 of: IEEE int. conf. on robotics and automation (Cincinnati, OH).* Los Alamitos, CA: IEEE Computer Soc. Press.

Allen, P.K., & Michelman, P. 1990. Acquisition and interpretation of 3-D sensor data from touch. *IEEE trans. on robotics and automation,* **RA-6**(4), 397–404.

Balestrino, A., & Bicchi, A. 1989 (Nov). Adaptive surface following and reconstruction using intrinsic tactile sensing. *Pages 13–17 of: Int. workshop on sensorial integration for industrial robots (Zaragoza, Spain).*

Berkemeier, M.D., & Fearing, R.S. 1990. Determining the axis of a surface of revolution using tactile sensing. *Pages 947–949 of: Proc. IEEE int. conf. on robotics and automation (Cincinnati, OH).* Los Alamitos, CA: IEEE Computer Soc. Press.

Brock, D., & Chiu, S. 1985. Environment perception of an articulated robot hand using contact sensors. *Pages 89–96 of: ASME Pubn. PED-Vol. 15, Robotics and flexible automation.* ASME.

Cameron, A. 1989. Optimal tactile sensor placement. *Pages 308–313 of: Proc. IEEE int. conf. on robotics and automation (Scottsdale, AZ).* Washington, DC: IEEE Computer Soc. Press.

Ellis, R.E. 1987a. Acquiring tactile data for the recognition of planar objects. *Pages 1799–1805 of: Proc. IEEE int. conf. on robotics and automation (Raleigh, NC).* Washington, DC: IEEE Computer Soc. Press.

Ellis, R.E. 1987b. *A tactile sensing strategy for model-based object recognition.* Tech. rept. COINS 87-96. Dept. Computer & Information Science, Univ. Mass. at Amherst.

Faux, I.D., & Pratt, M.J. 1979. *Computational geometry for design and manufacture.* New York: John Wiley.

Fearing, R.S. 1990. Tactile sensing for shape interpretation. *Pages 209–238 of:* Venkataraman, S.T., & Iberall, T. (eds), *Dextrous robot hands.* New York: Springer-Verlag.

Fearing, R.S., & Binford, T.O. 1988. Using a cylindrical tactile sensor for determining curvature. *Pages 765–771 of: Proc. IEEE int. conf. on robotics and automation (Philadelphia, PA).* Washington, DC: IEEE Computer Soc. Press.

Gaston, P.C., & Lozano-Pérez, T. 1984. Tactile recognition and localization using objects models: The case of polyhedra on a plane. *IEEE trans. on pattern analysis and machine intelligence,* **PAMI-6**(3), 257–266.

Grimson, W.E.L. 1986. Sensing strategies for disambiguating among multiple objects in known poses. *IEEE j. robotics and automation*, **2**(4), 196–213.

Grimson, W.E.L., & Lozano-Pérez, T. 1984. Model-based recognition and localization from sparse range or tactile data. *Int. j. robotics research*, **3**(3), 3–35.

Grimson, W.E.L., & Lozano-Pérez, T. 1987. Localizing overlapping parts by searching the interpretation tree. *IEEE trans. on pattern analysis and machine intelligence*, **PAMI-9**(4), 469–482.

Hillis, W.D. 1982. A high resolution imaging touch sensor. *Int. j. robotics research*, **1**(2), 33–44.

Hutchinson, S.A., & Kak, A.C. 1989. Planning sensing strategies in a robot work cell with multi-sensor capabilities. *IEEE trans. on robotics and automation*, **RA-5**(6), 765–783.

Kinoshita, G., Aida, S., & Mori, M. 1975. A pattern classification by dynamic tactile sense information processing. *Pattern recognition*, **7**, 243–251.

Lederman, S.J., & Klatzky, R. 1987. Hand movements: A window into haptic object recognition. *Cognitive psychology*, **19**(3), 342–368.

Mason, M.T., & Salisbury, J.K. 1985. *Robot hands and the mechanics of manipulation*. Cambridge, MA: MIT Press.

Muthukrishnan, C., Smith, D., & Myers, D. 1987. Edge detection in tactile images. *Pages 1500–1505 of: Proc. IEEE int. conf. on robotics and automation (Raleigh, NC)*. Washington, DC: IEEE Computer Soc. Press.

Okada, T., & Tsuchiya, S. 1977. Object recognition by grasping. *Pattern recognition*, **9**(3), 111–119.

Overton, K.J. 1984. *The acquisition, processing and use of tactile sensor data in robot control*. Tech. rept. COINS 84-08. Dept. Computer & Information Science, Univ. Mass. at Amherst.

Ozaki, H., Waku, S., Mohri, A., & Takata, M. 1982. Pattern recognition of a grasped object by unit vector distribution. *IEEE trans. on systems, man and cybernetics*, **SMC-12**(3), 315–324.

Pentland, A.P. 1986 (Dec). *Recognition by parts*. Tech. rept. 406. SRI International.

Roberts, K.S. 1988. A new representation for a line. *Pages 635–640 of: Proc. IEEE computer vision and pattern recognition conf. (Ann Arbor, MI)*. Washington, DC: IEEE Computer Soc. Press.

Roberts, K.S. 1990. Robot active touch exploration: Constraints and strategies. *Pages 980–985 of: Proc. IEEE int. conf. on robotics and automation (Cincinnati, OH).* Los Alamitos, CA: IEEE Computer Soc. Press.

Roberts, K.S. 1991 (May). *Object recognition with a robot hand.* Ph.D. thesis, Dept. Computer Science, Columbia University, New York.

Schneiter, J.L., & Sheridan, T.B. 1990. An automated tactile sensing strategy for planar object recognition and localization. *IEEE trans. on pattern analysis and machine intelligence,* **PAMI-12**(8), 775–786.

Siegel, D.M. 1991. Finding the pose of an object in a hand. *Pages 406–411 of: IEEE int. conf. on robotics and automation (Sacramento, CA).* Los Alamitos, CA: IEEE Computer Soc. Press.

Solina, F. 1987 (Dec). *Shape recovery and segmentation with deformable part models.* Ph.D. thesis, Dept. Computer Science, Univ. Pennsylvania, Philadelphia, PA.

Speeter, T.H. 1989. Flexible piezoresistive touch sensing array. *Proc. SPIE - Int. Soc. Opt. Eng. (Optics, illumination and image sensing for machine vision III, Cambridge, MA),* **1005**, 31–43.

Stansfield, S.A. 1988. A robotic perceptual system utilizing vision and touch. *Int. j. robotics research,* **7**(6), 138–161.

Stansfield, S.A. 1990 (March). *Haptic perception with an articulated, sensate robot hand.* Tech. rept. SAND90-0085 UC-406. Sandia Laboratories, Albuquerque, NM.

Tsujimura, T., & Yabuta, T. 1989. Object detection by tactile sensing method employing force/torque information. *IEEE trans. on robotics and automation,* **RA-5**(4), 444–450.

Chapter 11

Experiments In Active Haptic Perception With The Utah-MIT Dextrous Hand

Peter K. Allen, Paul Michelman and Kenneth S. Roberts

11.1 Introduction

The recent development of a number of multi-fingered, anthropomorphic robotic hands (Mason & Salisbury, 1985; Jacobsen *et al.*, 1986; Tomovic *et al.*, 1987) has provided the research community with a challenge to extend the capabilities of these hands in a number of autonomous environments. Future robotic systems will need to use dextrous robotic hands for tasks such as grasping, manipulation, assembly, inspection and object recognition. The work described here is an effort to incorporate a multi-fingered robotic hand as a central component of an autonomous recognition and grasping system. We describe experiments in active haptic perception[1] with the Utah-MIT dextrous hand performed at the Robotics Laboratory at Columbia University which are part of a larger system we are building for 3-dimensional shape recovery and object recognition using touch and vision methods. It focuses on three Exploratory Procedures (EPs) we have built to acquire and interpret sparse 3-dimensional touch data. These procedures serve as a front end to an integrated shape recovery and object recognition system that can combine these exploratory procedures into strategies that can derive constraints about an object's most probable shape.

Object recognition has traditionally been associated with vision sensor systems.

[1]An important point to be made in incorporating mechanical hands in robot systems is that the human perceptual process of interest is *haptic* perception. By this, we mean the interplay of both the cutaneous system (skin, tactile receptors) and the kinesthetic system (joints, muscle and bone) of the arm (Harmon, 1985).

However, these systems suffer from a number of inherent problems, not the least of which is occlusion. A vision system will be limited to a view that obscures all back-facing areas of the object. Similarly, in robot manipulation tasks, important areas of the work environment are often occluded by the end-effector itself. This difficulty is especially acute during the act of acquiring a grasp on an object, when the contact areas will be occluded. Hence, it is important that object recognition by touch becomes a part of robotic systems.

Little work has been done in using active touch sensing with a multi-fingered robotic hand to acquire scene descriptions, even though this is a well developed human capability (Lederman & Klatzky, 1987). Touch sensing differs from other more passive sensing modalities such as vision in a number of ways. A multi-fingered robotic hand with touch sensors can probe, move, and change its environment. This imposes a level of control on the sensing that makes it typically more difficult than traditional passive sensors in which active control is not an issue. Secondly, touch sensing generates far less data than vision methods; this is especially intriguing in light of psychological evidence (described below) that shows humans can recover shape and various other object attributes very reliably using touch alone.

11.2 A Haptic Perception Framework

In implementing a haptic perception system, it is important to choose an appropriate framework that includes all the necessary modules of an actively controlled robotic hand-arm system. Figure 11.1 is an outline of a framework being implemented at Columbia University for the task of object recognition (discussed in detail in Roberts's thesis (Roberts, 1991)). In this framework, there is a data flow between low-level, data-driven exploration modules that provide sensor input to higher-level reasoning and matching algorithms that try to deduce which object is being sensed. The higher-level modules are also capable of providing input parameters to low-level explorers, based upon hypotheses that are consistent with already sensed data. We briefly describe some of these modules below:

Executive: the function of the Executive is to choose the next move for the robot arm and the multiple robot fingers. From the I Tree (Interpretation Tree) it gets the current status of the identity of the physical object, i.e., what interpretations are currently under active consideration. It chooses a set of hypothesised object components to be verified, the exploration of which will verify (or falsify) one or more of the current set of active interpretations. It receives advice from the Coordinator about what combinations of components may be explored simultaneously.

Coordinator: a hand-arm system contains many degrees of freedom (in our case, the Utah-MIT hand and PUMA arm have 22 degrees of freedom) that need to be controlled. For a given desired set of fingertip motions, there may be an

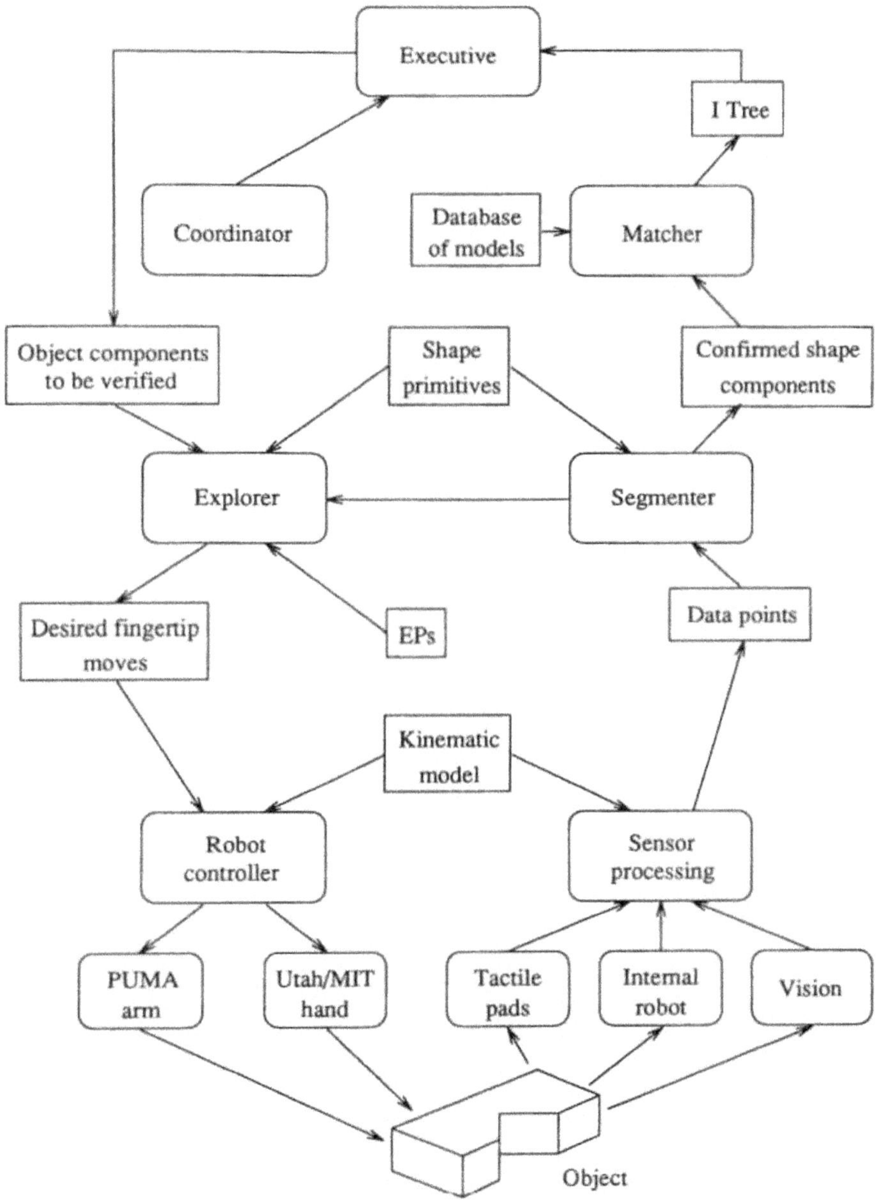

Figure 11.1: *A framework for robotic haptic perception.*

infinite number of hand-arm configurations that achieve them. Or it may be impossible to achieve the desired motions at all. The Coordinator's task is to plan hand-arm configurations that are feasible for the particular active haptic task specified. A non-linear optimisation approach to solving this problem for the Utah-MIT hand is described by Roberts (1990).

Explorer: the function of the Explorer is to receive from the Executive a set of hypothesised object components to be verified, and execute a set of arm and finger moves to carry out the requested explorations. For each component, the Explorer chooses an appropriate Exploratory Procedure (EP), and uses that to generate desired fingertip moves for the Robot Controller to execute.

Exploratory Procedures: the EPs are the identified strategies and procedures to generate desired motions for the fingertips of the hand. The motions are modified based on feedback directly from the sensors. The EPs implemented on the Utah-MIT hand system are the main focus of this chapter and are described in Section 11.4.

Robot controller: the Robot Controller uses the robot and hand kinematic models to convert the desired exploratory motions in world coordinates into desired arm and hand joint angles, and then into actual control signals to the PUMA arm and Utah-MIT hand actuators.

Sensor processing: a hand-arm system may have many different kinds of sensors, e.g., tactile array sensors and internal kinesthetic joint position and tendon tension sensors. Sensor processing is used to report the presence and location of contact between the surface of the fingertip and the surface of the object, and the positions and forces on the fingers and joints of the hand. Kinematic models are then used to convert the sensor readings for each fingertip contact into a data point.

Segmenter: the Segmenter takes sets of data points from the sensors and uses them to compute new shape components (planes, lines, etc.) by fitting the parameters of shape primitives.

Matcher: the Matcher maintains the set of possible interpretations in a data structure called the I Tree. Each interpretation is a matching of the confirmed data components from the Segmenter against components from one of the model objects in the given database of models. When the Segmenter reports a new confirmed component, the Matcher tries to append it to each possible interpretation in the set. Each augmented interpretation is tested against various constraints of adjacency and geometry. If the constraints cannot be satisfied then the interpretation is deleted from the set. When only one interpretation remains, the recognition task is complete, and the Executive reports the identity and location of the explored object.

The framework described above shows the relationship of each module to the overall object recognition task, and allows research to be partitioned among separate, cooperating modules. As new shape primitives, EPs, segmenters, matching techniques, etc. are developed, they can be implemented in the existing framework.

In the remaining sections of this chapter, we describe some of the EPs from this haptic perception framework, and their implementation on the Columbia Hand-Arm System. We also briefly discuss related modules of the framework including the matcher and the segmenter. Section 11.3 is an overview of the hardware/sensing environment we have built and upon which the framework described above is being implemented. Section 11.4 describes the EPs we have implemented for acquiring and interpreting 3-dimensional touch information. Section 11.5 is a summary.

11.3 The Columbia Hand-Arm System

The system we have built consists of a Utah-MIT hand (Jacobsen *et al.*, 1986) attached to a PUMA 560 manipulator. The hand contains four fingers, each with four degrees of freedom. It resembles the human hand in size and shape, but lacks a number of features that humans find very useful. In particular, it has no palmar degree of freedom (closing of the palm) and the thumb is placed directly opposite the other three fingers, with all fingers identical in size (see Figure 11.2). The hand has joint position sensors that yield joint angle data and tendon force sensors that measure forces on each of the two tendons (extensor and flexor) that control a joint. The PUMA adds six degrees of freedom to the system (three translation parameters to move the hand in space and three rotational parameters to orient the hand), yielding a 22 degree-of-freedom system. Clearly, such a system is a nightmare to control at the servo-level in real-time. Our approach is to use the embedded controllers in each of these systems, controlling and communicating with them through an intelligent, high-level controller that links together the movements of arm, hand, and fingers with the feedback sensing of joint positions, tendon forces, and tactile responses on the fingers.

The hardware structure of the system is shown in Figure 11.3. The high-level control resides in a SUN-3 processor. The SUN serves as the central controller, and has access to a full UNIX-based system for program development and debugging as well as a set of window-based utilities to allow graphical output and display of the system's various states. The hand is controlled by an analogue controller that is commanded through D/A boards from a dedicated M68020 system. The SUN is capable of downloading and executing code on the M68020 and can communicate with it through a shared memory interface (Narasimhan *et al.*, 1988). The tactile sensing system is controlled by another dedicated M68020 that monitors the forces on each of the sensor pads. The connection from the SUN to the PUMA is via the VAL-II host control option over a serial interface. We are currently changing the interface to RCCL (Hayward & Paul, 1983) to make the hand-arm interaction

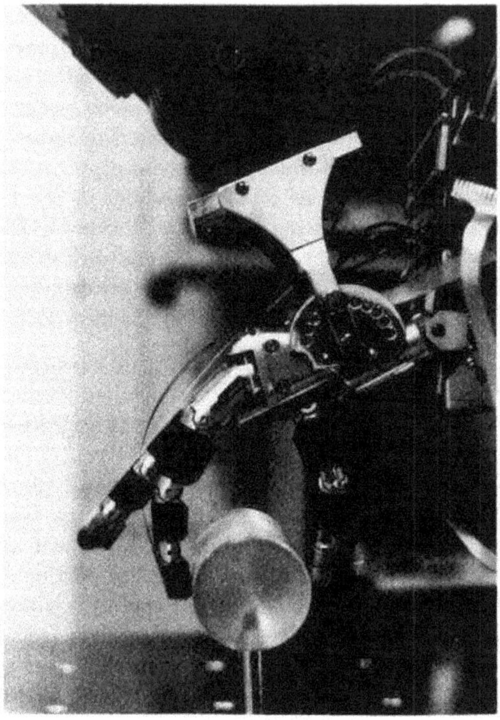

Figure 11.2: *Utah-MIT hand.*

more tightly coupled. The system has been used to perform a number of object manipulation and grasping tasks including pouring liquids from pitchers, removing computer boards from a backplane and removing lightbulbs from sockets. A key component of this system is the use of a programming environment that allows concurrent hand and arm control using task-level scripts (Allen *et al.*, 1990).

11.3.1 Tactile Sensors

While the level of sensing provided by the joint position and tendon force sensors on the Utah-MIT hand is better than earlier implemented hands, it still falls far short of the requirements for a dextrous manipulation system. In particular, what is desired is accurate positional contact information between the hand and a target object, and a measure of the forces exerted by the fingers at these contact points. The sensory feedback provided by the hand does not allow for localisation of contacts. Hence, a requirement for this system is a robust and accurate tactile sensing capability, utilising sensors mounted on the links of the fingers. Tactile sensing differs from traditional vision sensing in its active nature. Thus, a robotic system that employs

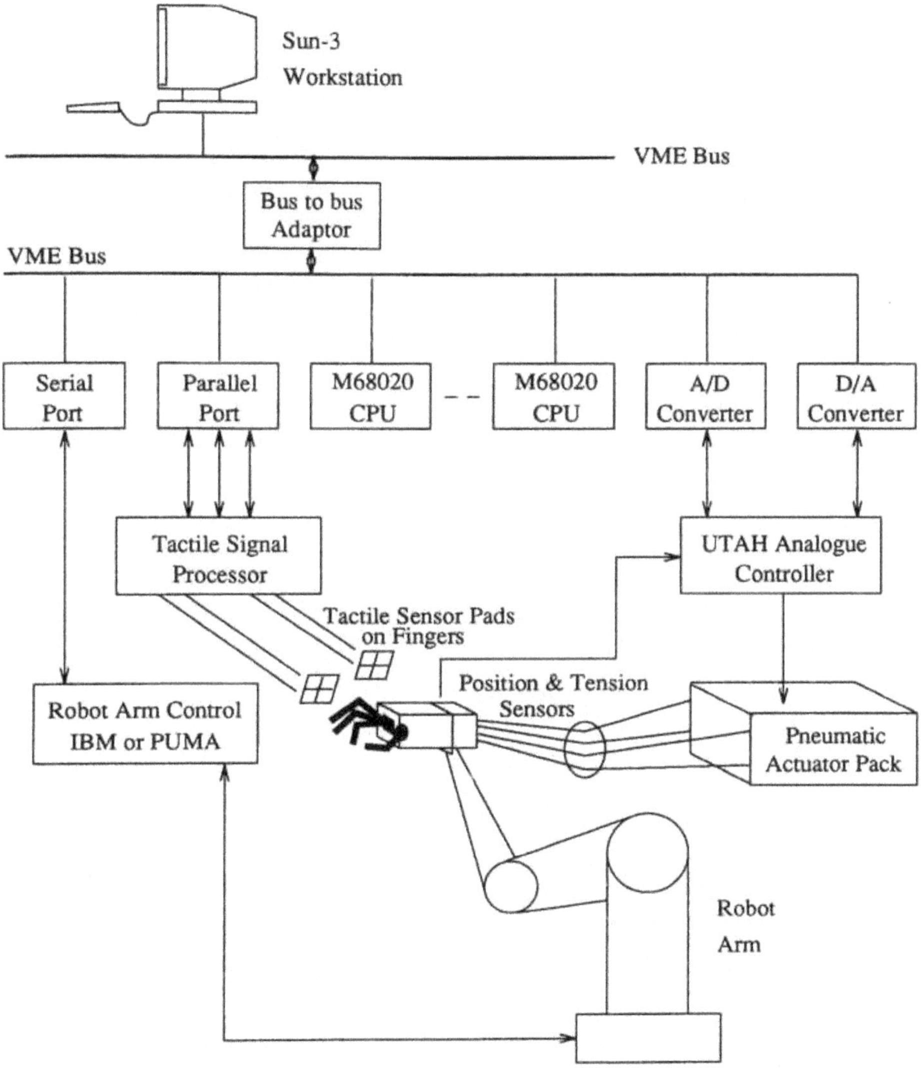

Figure 11.3: *Columbia hand: system hardware.*

tactile sensors on the fingers of a dextrous hand must deal with three related issues: 1) acquisition and interpretation of tactile sensor data from many sites on multiple fingers; 2) low-level control of the dextrous hand using tactile sensor feedback; 3) development of high-level sensing strategies using tactile feedback.

To satisfy the first requirement, we have mounted tactile sensors on each of the hand's fingers. The technology being used is a piezoresistive polymeric material manufactured by Interlink, Inc. (Tise, 1988; Speeter, 1989). In the tactile pad design we are using, the polymer is sandwiched between two pliable sheets of Kapton material that contain electrical etching. The application of forces on the pads provides an increased electrical flow channel between the two sheets as the material within is compressed. The piezoresistive polymer is patterned to form rows on one substrate and columns on the other. The rows and columns form a grid in which each intersection acts as a force-sensitive variable resistance whose value decreases approximately exponentially with normal force. The pads consist of 16 rows by 16 columns, providing a sense resolution of 256 points on a 1.27×2.54 cm pad.

The 256 sites of each sensor pad are addressed independently by analogue circuitry that cancels current flow in all paths of the grid except the one containing the resistive element being measured, using a method developed by van Brussel and Belien (1986). A hardware interface board has been developed to perform this operation at high speed. The interface board performs the analogue-to-digital conversion task by means of an 8-bit flash A/D converter and allows up to sixteen sensor pads to be addressed.

A number of useful digital filters have been implemented including averaging, median and edge detection filters which are very useful in processing noisy tactile data (Muthukrishnan *et al.*, 1987). Intermediate-level primitives have been built to detect lines by using the output of the edge detection procedure in a Hough transform (Ballard & Brown, 1982), and to quantify contact areas using central moment analysis. The contact area and centroid of the contact can be determined using moments, with the second moments determining the eccentricity of the contact region and the principal axes of the contact.

11.4 Exploratory Procedures for Object Recognition Tasks

A number of interesting properties of the human haptic system have been investigated by Lederman and Klatzky and their colleagues (Klatzky *et al.*, 1985; Lederman & Klatzky, 1987; Lederman & Browse, 1988). This work has shown that an important component of the haptic system is its ability to recognise attributes of 3-dimensional objects quickly and accurately. Among these attributes are global shape, hardness, temperature, weight, size, articulation and function. An outcome of this research is the identification of hand movement strategies that are used by humans in discovering different attributes of 3-dimensional objects. They have labelled these EPs, or Exploratory Procedures, and have reported success rates of 96 - 99% in identifying different object properties using two-handed, haptic exploration. We have found it natural to extend these human capabilities to our robotic domain by building robotic analogues of the human EPs.

In designing these EPs, the question of what shape representation to use arises. The disparate nature of these sensing strategies has led us to propose using multiple shape representations. Different shape representations seem to be intuitively better based upon both the task and object domain. Our effort in implementing robotic EPs has led us to find "natural" mappings between some well-known shape representations and EPs. We now describe the three EPs we have implemented on our robotic hand system, and which motivate our choice of shape representation, and describe experiments with real objects.[2]

11.4.1 Exploratory Procedure 1: Grasping by Containment

Grasping by containment is an attempt to understand an object's gross contour and volume by effectively moulding the hand to the object. We have chosen to model objects as superquadrics (Barr, 1981; Pentland, 1986; Bajcsy & Solina, 1987) whose surface 3-dimensional vector $\mathbf{X}$ in spherical coordinates is:

$$\mathbf{X}(\eta,\omega) = \begin{bmatrix} a_1 C_\eta^{\epsilon_1} C_\omega^{\epsilon_2} \\ a_2 C_\eta^{\epsilon_1} S_\omega^{\epsilon_2} \\ a_3 S_\eta^{\epsilon_1} \end{bmatrix}$$

$$-\tfrac{\pi}{2} \leq \eta \leq \tfrac{\pi}{2}, -\pi \leq \omega \leq \pi. \qquad (11.1)$$

C_η and S_ω are $\cos(\eta)$ and $\sin(\omega)$; ϵ_1 and ϵ_2 are the superquadric shape parameters; a_1, a_2, and a_3 are scaling factors along the x, y and z directions. Superquadrics form a rich set of shape primitives that allow a wide degree of freedom in modelling objects. The parameter space is continuous and allows a smooth change from a cuboid to a sphere to a cylinder, with more complex shapes derivable with the

[2]Before describing these EPs, we note that we must be careful in trying to draw too close a comparison between a human hand and devices such as a Utah-MIT hand. Johansson and Vallbo (1983) have reported that there are about 17,000 mechanoreceptors in the skin of the human hand; our robotic hand is more limited with 16 joint sensors, 32 tendon force sensors, and four 16 × 16 fingertip tactile sensors. In addition, a human hand has two main differences in structure from our robotic hand. The first is a highly flexible, opposable thumb that is mounted to the side of the other digits. The thumb of the Utah-MIT hand is identical to the other fingers and is mounted directly opposite the other fingers. The second difference is that a palmar degree of freedom exists in human hands that is missing in the Utah-MIT hand. Humans find this palmar degree of freedom quite useful, especially for encompassing-type grasps where the hand is moulded to an object and it is used as a grasping mechanism in its own right, almost independent of the existence of multi-jointed fingers.

addition of bending and tapering parameters. These "lumps of clay" are deformable by the usual linear stretching and scaling operations and can be combined using Boolean set operations to create more complex objects.

What makes superquadrics particularly relevant for haptic recognition is the following:

1. The models are volumetric in nature, which maps directly into the psychophysical perception processes suggested by grasping by containment.

2. The models can be constrained by the volumetric constraint implied by the joint positions on each finger.

3. The models can be recovered with sparse amounts of point contact data since only a limited number of parameters need to be recovered. There are five parameters related to shape $(\epsilon_1, \epsilon_2, a_1, a_2, a_3)$ and six related to position and orientation in space. Global deformations (tapering, bending) add a few more.

4. In addition to the use of contact points of fingers on a surface, the surface normals from contacts can be used to describe a dual superquadric which has the same analytical properties as the model itself (Barr, 1981).

5. The analytic nature of the model created from the data allows matching and searching strategies in the model space to proceed using both quantitative and qualitative differential geometry measures.

11.4.1.1 Recovery Procedure

For this initial work on recognition, we have used a simplified procedure to gather data points. Our intent is to use the tactile sensors mounted on the finger links to generate contact position data. However, during our initial trials, our tactile sensors were not mounted on the hand. Instead, we opted for a method that used the hand's internal joint angle readings and tendon forces to generate Cartesian positions of contact based upon fingertip contact. Note that in all the experiments described in this chapter, the objects to be explored were rigidly attached to a laboratory table. Exploring objects with active touch sensors can sometimes displace the objects. This is a very difficult problem that is solved at the human level by using two hands, where one hand typically serves as a "fixture". In the absence of a second hand, we have rigidly mounted the objects to allow one hand exploration.

The PUMA arm moves the hand to a position in which it will close around the object. The fingers are spread wide during approach. Then the fingers are closed by position commands until the observed force (estimated by the difference between the flexor and extensor tendon tensions) exceeds a given threshold, which indicates that the finger is in contact with the object. The joint angle positions are read, and kinematic models of the hand and the PUMA arm are used to convert them to (x, y, z) positions in world coordinates. Then the fingers are opened wide again,

Figure 11.4: *Object database.*

and a second containing grasp is executed, with the fingers taking different approach paths. The fingers are spread once again, and the PUMA arm moves the hand to the next position.

The sequence of PUMA positions is given in advance. Once the contact points are determined using the forward kinematics of the hand derived from the joint angle sensors, the sparse sets of point data are injected into the recovery algorithm developed by Solina (1987). This algorithm uses a Levenberg-Marquardt non-linear least squares approximation to fit the superquadric "inside-out function". This is an implicit form of equation 11.1 which records if a sample data point lies inside, outside or on the surface of the superquadric model. By summing the squared distance of each sample data point from the current model, an error-of-fit measure is generated that is minimised by the algorithm.

Equation 11.1 is for a canonical superquadric located at the origin. Since our sensor data can exist anywhere in the world coordinate space, the algorithm must recover the six rotation and translation parameters in addition to the five superquadric shape parameters. In addition, we allow global deformations to include tapering of superquadric forms. The taper is defined to be a linear tapering with two parameters that control the tapering in both the x and y dimensions. The algorithm must recover a minimum of 11 parameters and 13 if the object is tapered.

We tested this procedure against a database of six objects (those shown in Figure 11.4 plus a smaller cylinder). The database included objects that could be modelled as undeformed superquadrics (block, large cylinder, small cylinder) and deformed (tapered) superquadrics (lightbulb, funnel, triangular wedge). The

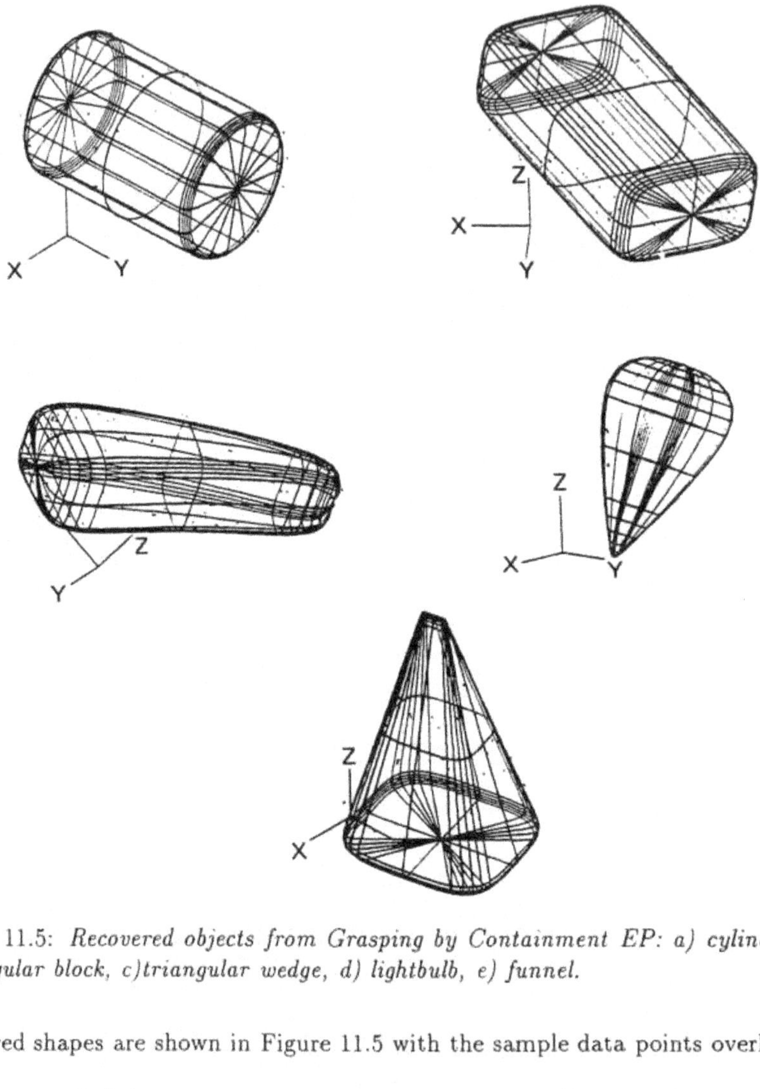

Figure 11.5: *Recovered objects from Grasping by Containment EP: a) cylinder, b) rectangular block, c)triangular wedge, d) lightbulb, e) funnel.*

recovered shapes are shown in Figure 11.5 with the sample data points overlaid on them.

The recovered shapes are an accurate representation of the actual shapes, especially considering the sparse nature of the data and the errors in the derived contact points. These errors are a function of the accuracy and calibration of the robotic arm, the hand joint position sensors, and the kinematic model of the hand itself. The data points are overlaid on the recovered shapes to show the closeness of fit and the sparseness of the data. Each object's shape was recovered with extremely sparse amounts of data, typically 30 – 100 points, depending on the object. It is important to note that this is about two orders of magnitude less than typical range data images which try to recover shape with denser data, that, unlike touch sensing, is limited to a viewpoint that only exposes half the object's surfaces to the sensor. Not all objects can be modelled adequately as superquadrics; however, we feel that in deriving an initial shape estimate from sparse data, they work quite well. The algorithm typically converges after about 10 - 15 iterations, and it takes about 10 seconds on a Sun-4 to recover an undeformed superquadric shape (20 iterations of a 60 contact-point data set). A formal analysis of the convergence properties and accuracy of the recovery method can be found in Solina's thesis (Solina, 1987) and in work by Gross and Boult (1988).

11.4.1.2 Matching

The superquadric recovery process supplies five relevant parameters that are useful in discriminating among an object's shape and size. The superquadric shape parameters ϵ_1 and ϵ_2 are good indicators of gross shape and the scaling parameters a_1, a_2, a_3 are useful in size discrimination. We have written a simple matching algorithm that matches the recovered superquadric parameters against the database of models, and is able to discriminate among the six objects successfully. The algorithm does a simple comparison of the six recovered superquadric parameters against each set of model parameters, and takes a weighted sum of these parameter distance measures as the error of fit. We experimented with ten different sets of sampled data, encompassing the six objects, some in more than one pose. Using this simple procedure on this limited database, we were successfully able to recognise each object uniquely during each trial.

Simple discrimination tests like these are not fully adequate to solve the matching problem. Our intent is to make two important points related to object recognition using sensor data. The first point is that sparse, noisy data can yield good estimates of shape. Second, that this method of haptic shape recovery is viable for a first level of object discrimination. It can effectively prune a much larger set of models, and the remaining feasible models can be actively searched based upon hypotheses generated from the database. The recovery algorithm is useful in active sensory feedback loops where initial coarse estimates can be used to derive the shape, position, and orientation of objects to one level of precision and then subsequent active haptic probes can be used to recover the details of the shape to a finer level.

11.4.2 Exploratory Procedure 2: Planar Surface Explorer

Once a superquadric has been fitted to the initial grasp data, we have a strong hypothesis about an object's shape. Of particular importance are the shape parameters ϵ_1 and ϵ_2. The shape of an object can be inferred from these parameters and used to direct further exploration. For example, if the shape parameters appear to be rectangular ($\epsilon_1, \epsilon_2 \approx 0.1$) then the planar explorer can trace out the plane and perform a least square fit of the trace data to test the surface's planarity.

If the shape parameters appear more cylindrical ($\epsilon_1 \approx 1.0, \epsilon_2 \approx 0.1$) then the planar faces of the cylinder can be explored with this primitive, and the cylinder's contour can be explored and verified with the contour follower EP (described below). A major benefit of using the superquadric analytic shape description is that it supplies orientation and axis data that are necessary for further active probes of the environment with the hand. Instead of a blind search, we can use the recovered orientation parameters to guide the further exploration of the object. Discovering a planar surface can be a very useful constraint in recognition, particularly if two opposing planar faces are grasped. By discovering multiple planar faces on an object, the recovery methods of Grimson and Lozano-Pérez (1983), and Ellis *et al.* (1987) can be invoked, which have proven to be strong constraints on recognition of an object.

The Explorer uses the hand's index finger. While the index finger is held in an extended position, the PUMA arm is moved until the tactile sensors on the index finger contact a surface (if no contact is detected, the procedure terminates). After the initial contact, the Cartesian position of the contact point is noted. The hand and arm then begin an iterative search for the boundaries of the surface by performing the following sequence: (a) lift the finger off the surface until tactile contact is lost; (b) move the arm in a direction parallel to the surface; (c) if the finger is in contact after the movement, note the new contact location, otherwise lower the index finger until it makes contact with the surface again; (d) repeat steps (a) – (c) until the finger fails to make contact in step (c). In step (d), if the finger does not contact the surface, then either the finger has moved beyond the edge of the surface, or the surface is too far away from the finger to be detected. To check for the latter case, the arm must be moved toward the surface. After completing the first collection of data points and finding the edge of the surface, the index finger is moved back to the position of initial contact, and a second mapping of the surface is undertaken in the opposite direction. This procedure continues until a second surface edge is detected. The search now continues as before but in a direction perpendicular to the first two traces. This procedure then is able to map out a set of contact points on the surface, describing its extent. Each time the fingertip contacts the surface, the Cartesian coordinates of the contact are retained. Figure 11.6 shows a pattern of traces on 2 adjacent planar faces of a rectangular block using this EP. Least-square planes were fitted to each of the traces and the computed angle between the recovered planes was 96° (the actual angle is unknown but assumed to be 90°).

11.4.2.1 Segmentation

Implicit in a planar surface explorer EP is the ability to segment an ongoing stream of tactile point responses from many fingers into groupings of planar surfaces. Once planar surfaces are found, they can be intersected to generate hypothesised edges and vertices, which can further constrain an object's shape. Segmentation

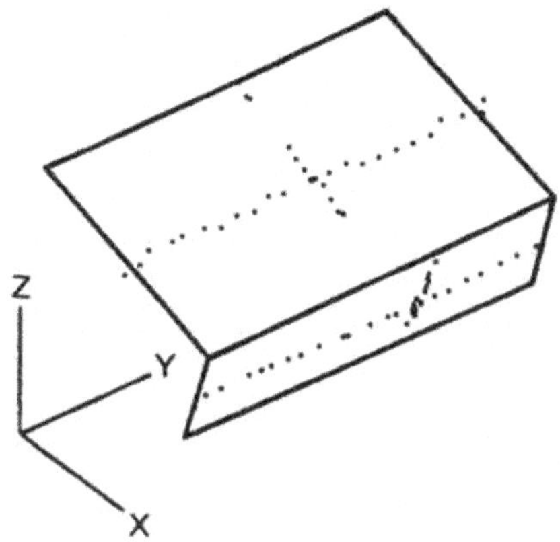

Figure 11.6: *Tactile traces with recovered planar surfaces.*

has been studied extensively in the machine vision literature, but has not been approached from a purely haptic perspective. Roberts (1991) has implemented a haptic segmentation algorithm that tries to take advantage of some of the special characteristics of haptic data that make it different from 2-dimensional and 3-dimensional machine vision data. Some of these differences are:

Incremental: the data from active touch exploration is spread over time. As the fingers move along the surface of the object, they acquire new data points and send them to the Segmenter. When the Segmenter finishes handling one bunch of data points, it reads in the next bunch of data points that has been accumulating.

Sparse: the different fingers will typically be touching different parts of the object, so there will be large gaps in the data. A gap can also result when a single finger is lifted from the surface and moved through free space to another contact point. Even when exploration of the object is finished, its surface may not be densely sampled, because some regions of the surface may be inconvenient to reach.

Local: since each finger will often be tracing continually along the surface of the object, the data points from each finger will tend to be locally clumped.

Surface normal: with many types of touch sensing, it is possible to get an estimate of the surface normal at the point of contact.

Fully 3-dimensional: Vision deals with either a 2-dimensional grey scale image or a range image which has a functional form $z = f(x, y)$, where there is one range value z (at most) for each element (x, y) in the image array. With haptic perception, the fingers can reach around to touch several sides of the object, and the data cannot be organised by a rectangular matrix array.

The algorithm organises points by their spatial proximity and deviation from a hypothesised planar fit. Due to its incremental nature, as points come from active exploration they are incorporated into existing planar fits, or create new seeds if they deviate from the fit. Also, planar segments are merged if possible to create larger segments. An interesting aspect of this algorithm is that points may belong to a number of different planar segments, some of which may be deleted over time.

Using this algorithm and experimentally generated data from the Columbia Hand system, Roberts was able to segment, match, and determine the orientation of seven polyhedral objects in a variety of different poses. The data points were generated by tracing the hand's fingers over the surfaces of the objects in a way similar to humans: incremental and localised. The objects included boxes and wedges, with and without holes and cavities. The segmenter's accuracy in creating planar faces was important to the success of the procedure. The interpretation tree constraints used by Roberts were based upon face, edge and vertex relationships (both geometric and topologic) that were derivable from the Segmenter's output. The Segmenter's ability to create stable groupings also assisted in the accuracy of the recovery of the pose parameters of the objects.

An added benefit of this method is the multi-sensor nature of the algorithm. Data from sources other than touch may be used as part of the Segmenter's input, and higher-order shape primitives such as ellipsoids may be used in addition to planes.

11.4.3 Exploratory Procedure 3: Surface Contour Following

The third EP we have implemented is surface contour following with a two-fingered grasp. This EP will allow us to determine an object's contour, which has been shown to be a strong shape cue from previous vision research (Stevens, 1978; Ponce *et al.*, 1987; Nalwa, 1987). The Contour Follower EP in humans is described by Klatzky and Lederman as "a dynamic procedure in which the hand maintains contact with a contour of the object. Typically, the movement is smooth and non-repetitive within a segment of object contour, stops or shifts direction when a contour segment ends, and does not occur on a homogeneous surface" (Klatzky & Lederman, 1987). It seems natural that this EP will report information that can be used to recover a shape that can be represented as a class of generalised cylinders. The contours we are able to extract from touch are inherently 3-dimensional. This simplifies recovery of shape since the 2-dimensional image projection used in most contour work entails a loss of information. Since we can recover the 3-dimensional contours, we are able to hypothesise a number of different shapes including generalised cylinders and solids of revolution, using the 3-dimensional contour alone.

Generalised cylinders have been proposed by many researchers, beginning with Binford (1971), as a shape modelling primitive. Other researchers have expanded on this idea of a swept volume by creating classes of generalised cylinders or cones, depending upon the nature of the axis curve, sweeping rule and cross-sectional curve. These primitives have special appeal in the recognition of elongated objects and objects that provide strong visual contours.

Fearing (1990) has attempted to recover the shape of a class of these generalised cones (RLSHGC — Right Linear Straight Homogeneous Generalised Cones) using extremely sparse amounts of data. He has characterised the necessary and sufficient conditions for being able to recover the axis and orientation of these cones, given limited, multi-fingered tactile sensor data that includes point contacts, surface normal direction, and surface curvature information.

We have also chosen to use a class of these primitives for shape recovery. The class we are using is surfaces of revolution, which are RSHGC with a circular cross section function (no linear scaling of the cross section is required). These surfaces may be completely described by the rotation of a plane curve about the axis of symmetry. If we take this axis to be the z axis, the surface will intersect the $o-xz$ plane in the plane curve:

$$\mathbf{r}_c(u) = p(u)\mathbf{i} + z(u)\mathbf{k} \tag{11.2}$$

and the surface of revolution has the equation:

$$\mathbf{r} = p(u)\cos\theta\mathbf{i} + p(u)\sin\theta\mathbf{j} + z(u)\mathbf{k}. \tag{11.3}$$

From the definition, we can see that the cross-section curve is circular, but the contour generator curve can be quite arbitrary.

Fearing was able to recover the shape of generalised cylinders from a minimal set of tactile contacts that were static in nature, not dynamic. Our attempt to recover the shape of these objects from touch is based upon receiving less accurate tactile information than the system used by Fearing. In particular, our tactile sensors are not capable of reporting accurate surface normals or surface curvatures, but they are capable of producing localised 3-dimensional positional contacts of fingers to the object. Our experience is that information such as surface curvature is very difficult to sense accurately, particularly with sensors that have low dynamic range or a non-compliant surface, and are used in an active, exploratory manner. Our approach is to use the Contour Follower to recover the 3-dimensional contour curve described above. This procedure maps naturally into a two-fingered Contour Follower EP, in which an object's contours on each side are sensed using the thumb and index finger.

The problem of using a tactile device to trace a surface on an object is a complicated one. Previous work by Allen (1988) used a one-fingered tactile sensor mounted on a PUMA robot arm to trace along a curved surface. This was done by calculating a weighted vector of constraint directions that tried to follow the surface curvature while preserving smoothness of the trace. The method also used a constraint concerned with creating regions bounded by traces that were equivalent in size. Hor (1987) traced contours of planar objects using a planar four-fingered "chopstick" like manipulator. Strain gauge sensors on the fingers of this device would calculate surface normals and move tangentially along a surface, recording the contour. Stansfield (1987) experimented with active tactile sensing to determine an object's shape using a single tactile sensor mounted on a robot arm. Following Klatzky and Lederman, her system also successfully determined several other object properties that included compliance, elasticity and texture.

Our method is now described. First, the PUMA is moved to a location near one end of the explored object, and the thumb and index finger are opened enough to allow them to encompass the object without making contact with it. The exploration axis of the object and the location of the object in space are assumed to be known by the EP. Then the thumb is slowly moved towards the object until the sensors detect contact between the thumb and the object. Next, the index finger follows the same movement. After detecting contact, the positions of the two contact locations are noted, and the fingers are backed away from the object so that they are no longer in contact. The arm and hand are moved a small amount along the axis of the explored object, and the process is repeated. This Exploratory Procedure ends when one of the fingers moves toward the object and fails to make contact.

The detection of contact and conversion to Cartesian coordinates is a process that requires several steps. The fingers are moved towards the object in a number of discrete intervals. After each movement, two checks are performed. First, did the tactile sensor detect contact? Second, did the finger move the entire distance that was commanded (a link of the finger may contact the surface without a tactile response being generated)? If the tactile sensor detects contact, then the location of the centre of the contact region is found. To find the centre of the contact, the first moments of the array are taken. Then a transformation is performed from the fingertip coordinate frame to the hand coordinate frame, and finally, from the hand coordinate frame to world coordinates. The second check may indicate that finger has not moved the distance commanded (and there is no tactile contact). This event would signal that something is impeding a finger from moving. In this case, no centroid of the contact region is found and the data point is thrown out. Currently, after detecting contact that does not involve the tactile sensor, the Exploratory Procedure continues looking for valid contact points along the original search axis.

We have performed a series of experiments that try to recover the shape of a number of different solids of revolution including a wine bottle, a beer bottle, a coke bottle and an Orangina soft drink bottle (a flask-like object). The procedure begins with exploring the object along an exploration axis that is assumed to be perpendicular to the support table (but can be inferred from vision sensing described below). The points generated from these contour traces are then linked into a set of linear contour segments. Circular cross-section curves are then fitted perpendicular to the exploration axis and include trace points from each of the contours. The recovered shapes are shown in Figure 11.7. The shapes are clearly

Figure 11.7: *Recovered solids of revolution from Surface Contour Explorer (left to right: wine bottle, coke bottle, beer bottle, Orangina bottle).*

distinguishable from this sparse data. Additional and important discriminating characteristics are actual 3-dimensional size and volume which are calculable from these representations.

11.4.3.1 Computing the Exploration Axis

Knowing in which direction to trace the object is important to higher-level recovery procedures which need to use this information in the recognition process. Once the hand makes contact with the object, it explores the contour along a known axis which we calculate *a priori*. We have implemented a vision-based technique to determine this axis (Allen, 1990). Our method of visual recovery of the exploration axis uses stereo line matching. Point-based stereo techniques tend to be unreliable in that multiple correspondences between images can cause mismatches and error. More stable matching can occur using larger primitives such as lines. Even using line-based matching, problems can still occur. Matching the endpoints of lines can be prone to errors in the output of the line finder which may break a single line into multiple segments due to differing edge strengths along the line. The problem here is that 3-dimensional depth is being computed, which requires an absolute correspondence of points (whether from point-based or line-based methods).

Our method alleviates this dependence on absolute matching of unstable primitives to generate 3-dimensional depth. All we require of the algorithm is an *orientation* vector in three dimensions. We do not need to have its absolute depth, but need to generate a match between a family of parallel lines sharing the same orientation. This orientation can then be used by the active hand as the exploration axis. The 3-dimensional depth has already been determined from the contact of the hand with the object. Given this 3-dimensional depth from tactile contact, we can follow the 3-dimensional axis determined by the line-based stereo matcher to continue our exploration.

It is important to note that this method is less sensitive to matching errors and baseline measurement, common causes of stereo error. In addition, it is also less prone to the effects of physical point mismatches as the baseline increases, since we are still matching a larger entity, the line itself. Intuitively, the method creates a 3-dimensional plane in space from the camera centre and any two points on the line. This plane and a similar plane from the other camera are all that are needed to create a 3-dimensional intersection line which we can use as the exploration axis.

As there are many lines in a scene, we have to choose a criterion for deciding which lines constitute the axis of the object. For exploration purposes, we simply want to discover a maximum length line which will serve as an axis. In most cases, this is part of the visual occluding contour of the object, which is exactly the axis we desire for active tactile exploration.

11.5 Summary

We have described a set of Exploratory Procedures using touch sensing that can serve as a front end to a multi-sensor object recognition system. The EPs can be used in a coarse-to-fine sensing strategy that tries to build shape descriptions at a number of levels. An important feature of this system is the multiple representations used in recovering and reasoning about shape. The first EP, Grasping by Containment, uses a global volumetric recovery method that is stable and efficient with extremely sparse amounts of data. It can be used as a precursor to more detailed fine shape recovery using either the Planar Surface Explorer or a 3-dimensional Surface Contour EP that can be used to recover solids of revolution.

In the future, we hope to link all the Exploratory Procedures into a fully autonomous system that will be able to use gross object structure as a generator of sensing hypotheses for the finer-level EPs. Each representation is essentially a constraint system that can be used to interpret the scene, and by using multiple shape primitives and constraints, more robust recognition strategies can be developed.

While these experiments show the capabilities of an active haptic perception system, we are continuing our efforts to fuse the haptic system with higher-level vision modules. We believe that future progress in robotic object recognition lies in a multi-sensor approach: we believe no single sensing modality is currently powerful enough to perceive and recognise its environment reliably. Just as humans exploit a multitude of sensor systems, robotic systems need to use multiple sensors for perception.

Acknowledgements

This work was supported in part by DARPA contract N00039-84-C-0165, NSF grants DMC-86-05065, DCI-86-08845, CCR-86-12709, IRI-86-57151, IRI-88-3319 North American Philips Laboratories, Rockwell Inc. and Siemens Corporation.

References

Allen, P.K. 1988. Integrating vision and touch for object recognition tasks. *Int. j. robotics research*, **7**(6), 15–32.

Allen, P.K. 1990. Mapping haptic exploratory procedures to multiple shape representations. *Pages 1679–1685 of: IEEE int. conf. on robotics and automation (Cincinnati, OH)*. Los Alamitos, CA: IEEE Computer Soc. Press.

Allen, P.K., Michelman, P., & Roberts, K.S. 1990. A system for programming and controlling a multi-sensor robotic hand. *IEEE trans. on systems, man, and cybernetics*, **SMC-20**(6), 1450–1456.

Bajcsy, R., & Solina, F. 1987. Three dimensional object representation revisited. *In: Proc. int. conf. on computer vision (London)*. Washington, DC: Computer Science Press.

Ballard, D., & Brown, C. 1982. *Computer vision*. Englewood Cliffs, NJ: Prentice-Hall.

Barr, A. 1981. Superquadrics and angle preserving transformations. *IEEE computer graphics and applications*, **1**(1), 11–23.

Binford, T. 1971. Visual perception by computer. *In: Proc. IEEE conf. on systems and control (Miami, FL)*.

Boult, T., & Gross, A. 1988. On the recovery of superellipsoids. *Pages 1052–1063 of: Proc. DARPA image understanding conf. (Cambridge, MA)*. San Mateo, CA: Morgan-Kauffman.

Ellis, R.E. 1987. Acquiring tactile data for the recognition of planar objects. *Pages 1799–1805 of: Proc. IEEE int. conf. on robotics and automation (Raleigh, NC)*. Washington, DC: IEEE Computer Soc. Press.

Fearing, R.S. 1990. Tactile sensing for shape interpretation. *Pages 209–238 of:* Venkataraman, S.T., & Iberall, T. (eds), *Dextrous robot hands*. New York: Springer-Verlag.

Grimson, W.E.L., & Lozano-Pérez, T. 1983 (Aug). *Model based recognition and localization from sparse range or tactile data*. Tech. rept. AI Memo 738. AI Lab, MIT, Cambridge, MA.

Harmon, L. 1985. Tactile sensing for robots. *Pages 389–424 of:* Hackwood, S., & Beni, G. (eds), *Recent advances in robotics.* New York: John Wiley.

Hayward, V., & Paul, R. 1983. Robot manipulator control under UNIX. *Pages 20:32–20:44 of: Proc. 13th int. symposium on industrial robots (Chicago, IL).* Dearborn, MI: Robotics Inst. of SME.

Hor, M-K. 1987 (Oct). *Control and task planning for a four finger dextrous manipulator.* Ph.D. thesis, Courant Institute, New York Univ.

Jacobsen, S.C., Iversen, E.K., Knutti, D.F., Johnson, R.T., & Biggers, K.B. 1986. Design of the Utah/MIT dextrous hand. *Pages 1520–1532 of: IEEE int. conf. on robotics and automation (San Francisco, CA).* Washington, DC: IEEE Computer Soc. Press.

Johansson, R.S., & Vallbo, A.B. 1983. Tactile sensory coding in the glabrous skin of the human hand. *Trends in neurosciences,* **6**(27), 27–32.

Klatzky, R., & Lederman, S. 1987. The intelligent hand. *Pages 121–151 of:* Bower, G. (ed), *The psychology of learning and motivation,* vol. 21. New York: Academic Press.

Klatzky, R., Lederman, S., & Metzger, V. 1985. Identifying objects by touch: An "expert system". *Perception and psychophysics,* **37**, 299–302.

Lederman, S.J., & Browse, R.A. 1988. The physiology and psychophysics of touch. *In:* Dario, P. (ed), *Sensors and sensory systems for advanced robots.* NATO ASI. New York: Springer-Verlag.

Lederman, S.J., & Klatzky, R. 1987. Hand movements: A window into haptic object recognition. *Cognitive psychology,* **19**(3), 342–368.

Mason, M.T., & Salisbury, J.K. 1985. *Robot hands and the mechanics of manipulation.* Cambridge, MA: MIT Press.

Muthukrishnan, C., Smith, D., & Myers, D. 1987. Edge detection in tactile images. *Pages 1500–1505 of: Proc. IEEE int. conf. on robotics and automation (Raleigh, NC).* Washington, DC: IEEE Computer Soc. Press.

Nalwa, V. 1987. Line drawing interpretation: Bilateral symmetry. *Pages 956–967 of: Proc. DARPA image understanding workshop (Los Angeles, CA).* Los Altos, CA: Morgan-Kauffman Publishers.

Narasimhan, S., Siegel, D.M., & Hollerbach, J.M. 1988. Condor: A revised architecture for controlling the Utah-MIT hand (Philadelphia, PA). *Pages 446–449 of: IEEE int. conf. on robotics and automation.* Washington, DC: IEEE Computer Soc. Press.

Pentland, A.P. 1986 (Dec). *Recognition by parts*. Tech. rept. 406. SRI International.

Ponce, J., Chelberg, D., & Mann, W. 1987. Invariant properties of the projections of straight homogeneous cylinders. *In: Proc. first int. conf. on computer vision (London)*. Washington, DC: Computer Science Press.

Roberts, K.S. 1990. Coordinating a robot arm and multi-finger hand using the quaternion representation. *Pages 1252–1257 of: IEEE int. conf. on robotics and automation (Cincinnati, OH)*. Los Alamitos, CA: IEEE Computer Soc. Press.

Roberts, K.S. 1991 (May). *Object recognition with a robot hand*. Ph.D. thesis, Dept. Computer Science, Columbia University, New York.

Solina, F. 1987 (Dec). *Shape recovery and segmentation with deformable part models*. Ph.D. thesis, Dept. Computer Science, Univ. Pennsylvania, Philadelphia, PA.

Speeter, T.H. 1989. Flexible piezoresistive touch sensing array. *Proc. SPIE - Int. Soc. Opt. Eng. (Optics, illumination and image sensing for machine vision III, Cambridge, MA)*, **1005**, 31–43.

Stansfield, S.A. 1987 (Oct). *Visually-guided haptic object recognition*. Ph.D. thesis, Dept. Computer Science, Univ. Pennsylvania, Philadelphia, PA.

Stevens, K. 1978. Computation of locally parallel structure. *Biological cybernetics*, **29**(1), 19–28.

Tise, B. 1988. A compact high resolution piezoresistive digital tactile sensor. *Pages 760–764 of: Proc. IEEE int. conf. on robotics and automation (Philadelphia, PA)*. Washington, DC: IEEE Computer Soc. Press.

Tomovic, R., Bekey, G., & Karplus, W. 1987. A strategy for grasp synthesis with multi-fingered robot hands. *Pages 83–89 of: Proc. IEEE int. conf. on robotics and automation (Raleigh, NC)*. Washington, DC: IEEE Computer Soc. Press.

van Brussel, H., & Belien, H. 1986. A high resolution tactile sensor for part recognition. *Pages 49–60 of: Proc. 6th int. conf. on robot vision and sensory controls (Paris)*. Kempston, UK: IFS (Pubns).

Chapter 12

Future Trends in Tactile Sensing

Howard R. Nicholls

12.1 Progress To Date

Progress in tactile sensing has continued in two main areas: sensor hardware, and active tactile sensing. Much research has concentrated on developing reliable, robust sensors which can be used in a variety of situations, most significantly for dextrous manipulation. A wide range of transduction methods have been used, from techniques based on measuring change in resistance to photoelastic methods. No single technology has proved to be suitable for all applications, or to satisfy all of the design criteria specified for usable sensors. Transduction techniques which enable sensors to be non-planar are particularly useful, allowing, for example, a sensor to be moulded to the surface of a cylindrical finger of a dextrous hand. Piezoelectric devices based on PVF_2 have proved popular and versatile (e.g., Dario *et al.* (1984)), and piezoresistive sensors using FSR, which can be moulded to non-planar shapes, have been used in practical experiments (e.g., Tise (1988)). Emphasis in tactile hardware research has moved from concentrating on the sensor technology towards issues in tactile system design. This has included concern with the integration of tactile sensors both into control systems and into multi-modal sensing systems.

Early work in tactile data processing concentrated on applying machine vision algorithms to tactile images. Progress, particularly since the late 1980s, has been in the area of active sensing, where measurements are taken from multiple contacts with an object, typically to build up a 3-dimensional representation of the touched object. In its widest form, the processing has incorporated information gathered from fingertip-mounted sensors, and from sensors measuring manipulator joint angles and forces. This haptic sensing has similarities with the human system of touch, and a particular focus for research has been in developing so-called Exploratory Procedures. These are specifically aimed at the extraction of certain object properties, such as local surface shape, hardness, and friction characteristics (e.g., Allen and Michelman (1990); Stansfield (1991)).

12.2 Some Key Trends in Tactile Sensing

Various trends have been identified concerning the development of tactile sensing. With respect to sensors for dextrous manipulation, it is apparent that there is merit in devices which measure changes in contact parameters, rather than just sampling parameter values. A notable example is the stress rate sensor (Howe & Cutkosky, 1989), which measures the rate of change of stresses in the sensor. It is also useful to develop sensors with programmable response ranges or levels of resolution. If these can be altered during a manipulation task, the sensor characteristics can be controlled to suit the operation in progress, providing a trade-off between sensitivity and performance (e.g., Speeter (1990)).

Studies of the human tactile system have demonstrated that there is value in considering both the human tactile sensory organs and the way in which humans use tactile information. There is probably little value in adopting an anthropomorphic approach to tactile sensing, especially in highly constrained environments such as those in industrial robotics. However, there are some conceptual relevances and lessons to be learned from such studies. The human tactile system has a wide range of receptors; some of these receptors are fast adapting, others slowly adapting, some only activate when thresholds are crossed whilst others output signals whatever the state of the stimulus. This rich sensory environment provides a diversity of information to the human cognitive system. Correspondingly, in robotics for unconstrained, unpredictable environments, having a sensor-rich system provides the capability for the detection of a wide range of events and stimuli.

One of the lessons from studying human touch is that tactile sensing is an active modality, and indeed, humans usually apply haptic processing incorporating data not just from the fingertips but also from the whole hand, wrist and arm. Joint sensors, for example, can measure speed, acceleration, position, direction, or detect noxious stimuli. When such information is combined with data from cutaneous sensors, many aspects of the physical world can be derived. This leads to the idea of Exploratory Procedures, and to explicit methods for extracting certain properties of contacted objects. It has been suggested that tactile sensing is best used for determining material properties such as hardness, plasticity, and thermal conductivity rather than geometric properties. However, there is great value in combining vision sensing and touch, since the two modalities complement one another.

Geometric properties are often best determined via vision sensing, but there are problems in this, as there may be occluding objects, inconsistencies in object illumination, etc. Vision may act as the primary geometric sensory modality, and active touch, i.e., haptic sensing, may function as a secondary modality used to gather further data and to resolve ambiguities. One trend is thus the integration of tactile sensors with other sensory modalities, and the method by which this should be done is a major research issue. In attempting to ascertain the effectiveness of particular sensors and processing techniques, the scientific experimental method can be applied, as discussed by Lederman *et al.* in this volume.

In industrial automation, there is a growing requirement for flexibility in manufacturing systems. Production systems need to be more adaptable, since with smaller batch sizes and faster cycle times hard automation is not always an option. Automated equipment needs more and better sensors, and more sophisticated control systems to take account of rapidly changing requirements. Economic considerations are also important, and simple, cheap sensors connected to high level control and supervision systems can provide some of this desired flexibility at justifiable cost. This applies to tactile sensing as well as other modalities. If tactile sensing is to be used in commercial applications, then the sensors must be cheap and versatile. Simple binary probes coupled to a sophisticated supervision system can be economically justifiable, yet still resolve workcell uncertainties. Where machine vision would once have been used, touch sensing now provides an economic alternative.

12.3 Major Outstanding Issues

Allen in Chapter 10 of this volume identified three major areas where progress needs to be made: the development of integrated methods, strategies for active sensing, and data structures and models for representation of 3-dimensional objects. These areas cover many issues, for example, the choice of tactile cues, selection between sensors, integration of multi-sensor data, and the close coupling of such information to a control system. All of these are major topics in their own right, and require considerable research effort.

More work needs to be done on developing Exploratory Procedures. They show promise for exploratory robotics, but more procedures must be developed, and guidelines are needed indicating when and how they should be used. As has been shown, EPs have been identified by studying the human system, and some EPs have been implemented in robot systems, but there is much research still needing to be conducted to determine their scope and suitability.

Much emphasis has been placed on using tactile sensors for shape recognition, but industrial applications often do not need this facility. Much of the research reported here has been motivated by topics in dextrous manipulation and exploratory robotics, and experimentation has been in laboratory conditions without application-specific goals. To complement this, more studies of industrial problems are needed to identify particular areas where tactile sensing can satisfy sensing requirements, so that research will be driven both from an applications viewpoint and from an academic viewpoint.

Some other issues in dextrous manipulation include the determination of how many fingers and hands are appropriate for particular tasks, and the identification of the most versatile manipulator configuration for such tasks. There are very few dextrous manipulators available for research, and there is a need for more robust, usable dextrous hands.

More work is needed on the development of tactile sensors which can be mounted on dextrous manipulators; although various technologies and approaches have been identified, commercial versions are as yet generally unavailable. Good models of tactile sensors are needed, to enable reliable, accurate recovery of data. Another issue that needs further consideration is that of determining when tactile sensors should measure displacement, when they should measure strain, and when other parameters should be sensed.

12.4 Conclusion

This volume has presented a wide spectrum of work in tactile sensing, from planar elasticity to lessons derived from the study of biological touch. There has been a shift away from developing sensors and processing methods in an *ad hoc* way, towards a more rationalised, scientific view of the subject. There are many issues which need to be resolved, and it is hoped that the work presented in this volume has stimulated the reader interested in pursuing this active field of robot sensing.

References

Allen, P.K., & Michelman, P. 1990. Acquisition and interpretation of 3-D sensor data from touch. *IEEE trans. on robotics and automation*, **RA-6**(4), 397–404.

Dario, P., De Rossi, D., Domenici, C., & Francesconi, R. 1984. Ferroelectric polymer tactile sensors with anthropomorphic features. *Pages 332–340 of: Proc. IEEE int. conf. on robotics (Atlanta, GA).* Washington, DC: IEEE Computer Soc. Press.

Howe, R.D., & Cutkosky, M.R. 1989. Sensing skin acceleration for slip and texture perception. *In: Proc. IEEE int. conf. on robotics and automation (Scottsdale, AZ).* Washington, DC: IEEE Computer Soc. Press.

Speeter, T.H. 1990. Control of the Utah/MIT dextrous hand: Hardware and software hierarchy. *J. robotic systems*, **7**(5), 759–790.

Stansfield, S.A. 1991. A haptic system for a multifingered hand. *Pages 658–664 of: Proc. IEEE int. conf. on robotics and automation (Sacramento, CA).* Los Alamitos, CA: IEEE Computer Soc. Press.

Tise, B. 1988. A compact high resolution piezoresistive digital tactile sensor. *Pages 760–764 of: Proc. IEEE int. conf. on robotics and automation (Philadelphia, PA).* Washington, DC: IEEE Computer Soc. Press.

Appendix A

Basic Linear Elasticity

Ronald S. Fearing

A.1 Definitions of Stress and Strain

In this appendix, an introduction to some of the basic relations of linear elasticity are presented. For further information, elasticity texts such as Timoshenko and Goodier (1951), Gladwell (1980), and Johnson (1985) should be consulted.

The state of an elastic body, such as a tactile sensor, can be described by the force distribution per area, *stress*, and the fractional deformation in the body, *strain*. Consider the distributed force $\vec{F}$ applied to the $\hat{z}$ face of a cube with area A as shown in Figure A.1. The *normal stress* on the face, σ_z (in Nm^{-2}), equals the magnitude of the normal force per area:

$$\frac{\vec{F}\cdot\hat{z}}{A} = \sigma_z. \tag{A.1}$$

Similarly, the *shear stresses* on the face, τ_{zx} and τ_{zy} (also in Nm^{-2}), are equal to the magnitude of the tangential force components per area:

$$\frac{\vec{F}\cdot\hat{x}}{A} = \tau_{zx} \tag{A.2}$$

and

$$\frac{\vec{F}\cdot\hat{y}}{A} = \tau_{zy}. \tag{A.3}$$

The state of stress at a location in the body is determined by the six normal stresses and twelve shear stresses on the six faces of an infinitesimal cube. As shown in Figure A.2, for force equilibrium, the normal stresses on opposite faces of the cube must be equal, giving just three normal stresses σ_x, σ_y, and σ_z. For moment and force equilibrium, the shear stresses about any face must be equal, i.e., $\tau_{xy} = \tau_{yx}$, $\tau_{xz} = \tau_{zx}$, and $\tau_{yz} = \tau_{zy}$ (see Figure A.3), giving only three unique shear stresses. Thus the state of stress is completely described by six parameters, a symmetric second order tensor.

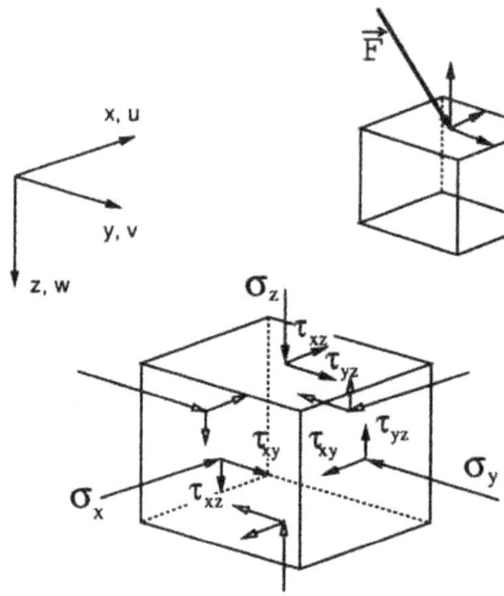

Figure A.1: *Definition of Stress. Normal stress is the force per area acting normal to a face. Shear stress is the force per area acting tangentially to a face.*

Suppose the location of a point in an elastic continuum is given by (x, y, z) before deformation. With applied stress, the continuum is deformed and the point is displaced by $\vec{d} = (u, v, w)$ to a new location $(x + u,\ y + v,\ z + w)$. The normal strain at a point in the continuum is the fractional deformation of the medium (a dimensionless quantity):

$$\epsilon_x = \frac{\partial u}{\partial x}, \quad \epsilon_y = \frac{\partial v}{\partial y}, \quad \epsilon_z = \frac{\partial w}{\partial z}. \tag{A.4}$$

The shear strain corresponds to a change in the angles between faces in a cube (as in Figure A.3), but to first order, the dimensions of the sides are unchanged. For small deformations, the shear strains are given by:

$$\gamma_{zx} = \frac{1}{2}\left(\frac{\partial u}{\partial z} + \frac{\partial w}{\partial x}\right), \quad \gamma_{xy} = \frac{1}{2}\left(\frac{\partial u}{\partial y} + \frac{\partial v}{\partial x}\right), \quad \gamma_{yz} = \frac{1}{2}\left(\frac{\partial v}{\partial z} + \frac{\partial w}{\partial y}\right). \tag{A.5}$$

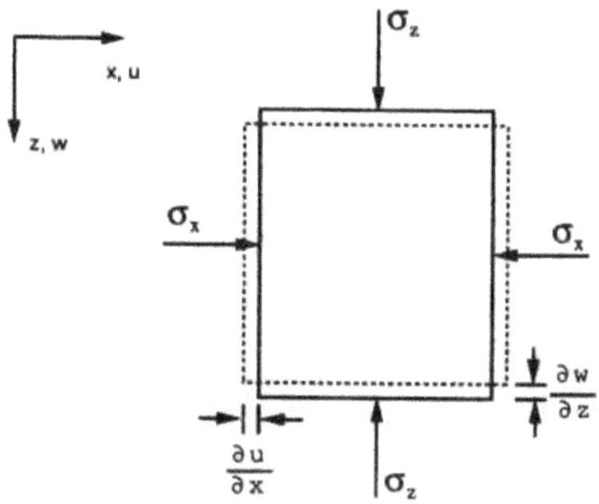

Figure A.2: *Elastic cube (in projection) subject to pure normal stress.*

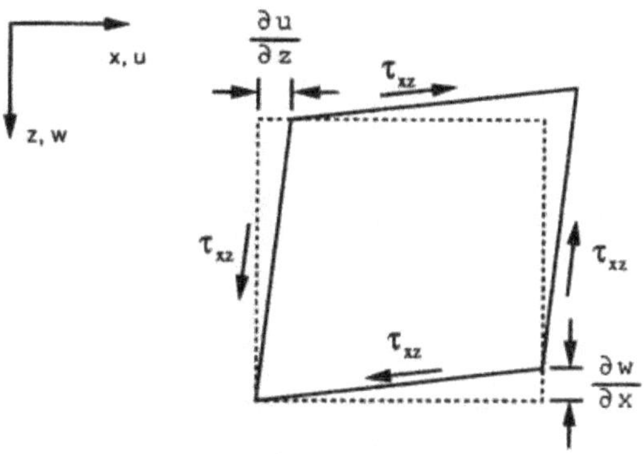

Figure A.3: *Elastic cube (in projection) subject to pure shear stress.*

A.2 Stress and Strain Relations

To keep the elasticity analysis simple, many assumptions about the ideal nature of the elastic body need to be made. In tactile sensor practice, most of these idealisations do not hold, yet the linear elastic theory predicts very well the measured sensor responses, e.g., see Phillips and Johnson (1981) and Fearing (1990). It will be assumed that the body is perfectly elastic (i.e., no time-dependent effects), strains are infinitesimal, and the medium is linear (strain is a linear function of stress). In addition, the elastic medium should have no orientation dependence (i.e., be isotropic), and the material should be homogeneous (i.e., have no inclusions or position dependent changes in elastic parameters). For modelling inhomogeneous or non-isotropic bodies, closed form solutions do not generally exist, and numerical methods must be used.

For materials that are not completely compressible, a contraction along one dimension will be coupled with expansions along the other two. Consider a cube of elastic material under uniaxial tension. Poisson's ratio is defined as the ratio between the strains in the unstressed and stressed directions, and characterises the compressibility of the material. For an incompressible cube which undergoes a strain of 1% along one axis, there must be corresponding strains in the opposite sense of about 0.5% along the other two axes to maintain constant volume. For a medium that does not change in volume with compression, such as rubber, Poisson's ratio can be taken as 0.5. For a completely compressible material, Poisson's ratio is 0.

To first order, for a linear 3-dimensional isotropic medium, the normal strains depend only on normal stresses, and are independent of the shear stresses. The normal strains are (Timoshenko & Goodier, 1951):

$$\epsilon_x = \frac{1}{E}\left(\sigma_x - \nu\left(\sigma_y + \sigma_z\right)\right) \qquad (A.6)$$

$$\epsilon_y = \frac{1}{E}\left(\sigma_y - \nu\left(\sigma_x + \sigma_z\right)\right) \qquad (A.7)$$

$$\epsilon_z = \frac{1}{E}\left(\sigma_z - \nu\left(\sigma_x + \sigma_y\right)\right) \qquad (A.8)$$

where E is Young's modulus of elasticity, given in units of Nm^{-2}, which corresponds to the stiffness of the material, and ν is Poisson's ratio.

To first order, the shear strains at a point in the medium are a function of shear stresses only, given by:

$$\gamma_{xy} = \frac{\tau_{xy}}{2\mu}, \quad \gamma_{xz} = \frac{\tau_{xz}}{2\mu}, \quad \gamma_{yz} = \frac{\tau_{yz}}{2\mu} \qquad (A.9)$$

where μ is the shear modulus (in Nm^{-2}). The shear modulus is defined as the ratio of shear stress to angle change. The two elastic constants E and ν are sufficient to characterise an isotropic elastic material. The shear modulus can be found in terms of Young's modulus and Poisson's ratio by:

$$\mu = \frac{E}{2(1+\nu)}. \qquad (A.10)$$

The relations for normal strains (eq. A.6 – A.8) can then be inverted to yield the stresses as a function of the strains:

$$\sigma_x = \lambda\epsilon + 2\mu\epsilon_x \qquad (A.11)$$

where λ is the Lamé constant:

$$\lambda = \frac{E\nu}{(1+\nu)(1-2\nu)} = \frac{2\mu\nu}{(1-2\nu)} \tag{A.12}$$

and ϵ is the dilatation which is defined as:

$$\epsilon = \epsilon_x + \epsilon_y + \epsilon_z. \tag{A.13}$$

Note that the dilatation ϵ, which corresponds to the change in volume of an infinitesimal element, must be zero for an incompressible material.

The shear stresses are simply:

$$\tau_{xy} = 2\mu\gamma_{xy}, \quad \tau_{xz} = 2\mu\gamma_{xz}, \quad \tau_{yz} = 2\mu\gamma_{yz}. \tag{A.14}$$

There are two simplifying assumptions that can be applied to elasticity problems with the appropriate symmetry. For an infinite-in-extent line force applied to the surface of an elastic half space, the strain in the direction of the line must be zero throughout the body by symmetry (see Figure A.4). This is the plane strain assumption. Alternatively, consider a slice of unit width out of the elastic half space, with a line load acting on the surface. On the faces of this slice, the stresses normal to the face must be zero to satisfy the boundary conditions, that is, the stresses at the surface must be in equilibrium with no applied loading on the side faces. For a thin slice, the stresses normal to the face will also be zero inside. This is the plane stress assumption.

Consider the half space $x > 0$ with a line load applied on the surface along the y-axis. The plane-strain assumption of $\epsilon_y = 0$ is appropriate. Then the stresses (for plane-strain condition) are obtained from the strains via eq. A.11:

$$\sigma_x = \lambda(\epsilon_x + \epsilon_z) + 2\mu\epsilon_x = \frac{2\mu}{1-2\nu}[(1-\nu)\epsilon_x + \nu\epsilon_z] \tag{A.15}$$

$$\sigma_z = \lambda(\epsilon_x + \epsilon_z) + 2\mu\epsilon_z = \frac{2\mu}{1-2\nu}[(1-\nu)\epsilon_z + \nu\epsilon_x] \tag{A.16}$$

$$\tau_{xz} = 2\mu\gamma_{xz}. \tag{A.17}$$

For the plane stress approximation, the substitution $\nu = \nu^*/(1+\nu^*)$ into the plane-strain formulation gives the corresponding stresses within a constant factor (Gladwell, 1980).

A.3 Coordinate Transformations for Stresses

In many cases, it is convenient to measure the stresses or strains in one coordinate system and then transform the stresses or strains to their values in a different coordinate system. This transformation can be done from force balance considerations. For simplicity, consider a rotation of coordinates about the y-axis, with the plane

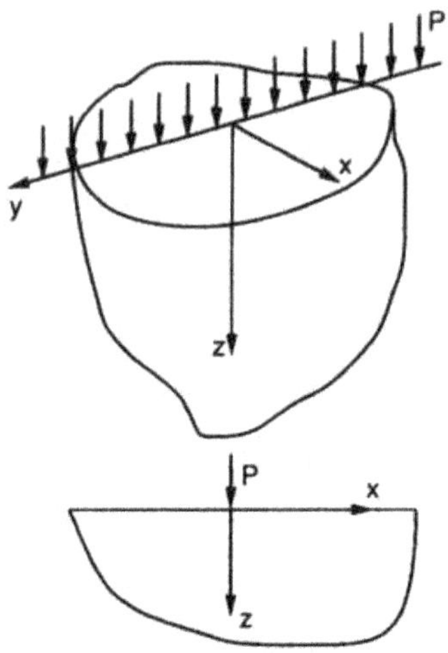

Figure A.4: *Line load on elastic half space.*

stress assumption, allowing a planar analysis. In addition, consider the unit cube of material (in projection) with stresses given in the $x-z$ plane (Cartesian coordinates) as shown in Figure A.5. We wish to find the corresponding stresses in the $r-\theta$ Cartesian coordinate system.

For force equilibrium, the total force on the third face (which has unit depth and length l, hence area l) must balance the forces on the other two faces of the triangular cross-section ($\vec{F_r} + \vec{F_1} + \vec{F_3} = 0$). For each of these two faces, the force on the face must be equal to the product of the stress on that face and its area:

$$\vec{F_1} = (\sigma_x \hat{x} + \tau_{xz}\hat{z})l(\hat{n}\cdot\hat{x}) = \vec{T_1}l(\hat{n}\cdot\hat{x}) \quad (A.18)$$

$$\vec{F_3} = (\sigma_z \hat{z} + \tau_{xz}\hat{x})l(\hat{n}\cdot\hat{z}) = \vec{T_3}l(\hat{n}\cdot\hat{z}) \quad (A.19)$$

where $\vec{T_1}$ and $\vec{T_3}$ are force vectors per unit area on each side of the cube. The stress in the r direction (σ_r) is given by the force in the normal direction per unit area of the side:

$$\sigma_r = \frac{-\vec{F_r}\cdot\hat{n}}{l} = \hat{n}\cdot[\vec{T_1}(\hat{n}\cdot\hat{x}) + \vec{T_3}(\hat{n}\cdot\hat{z})]. \quad (A.20)$$

Similarly,

$$\sigma_\theta = \hat{t}\cdot[\vec{T_1}(\hat{n}\cdot\hat{z}) - \vec{T_3}(\hat{n}\cdot\hat{x})]. \quad (A.21)$$

The shear stress $\tau_{r\theta}$ is given by the component of force in the tangential direction per area of the side:

$$\tau_{r\theta} = \frac{-\vec{F_r}\cdot\hat{t}}{l} = \hat{t}\cdot[\vec{T_1}(\hat{n}\cdot\hat{x}) + \vec{T_3}(\hat{n}\cdot\hat{z})]. \quad (A.22)$$

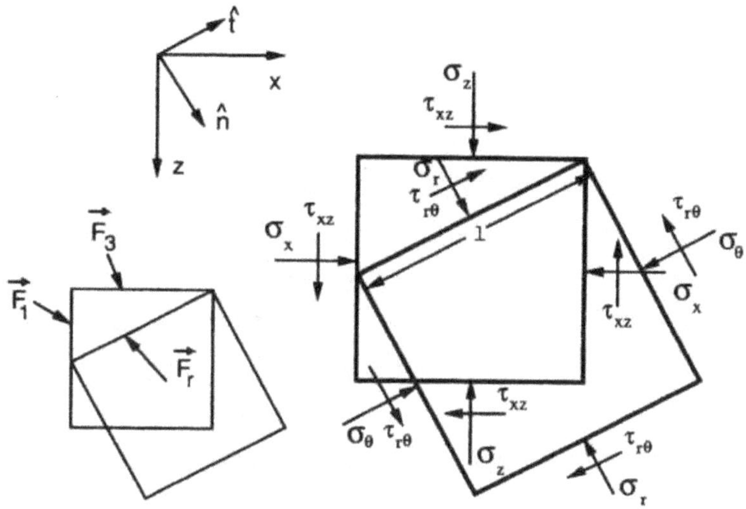

Figure A.5: *Stress coordinate transformation.*

For rotation by an angle θ:

$$\hat{n} = \cos\theta\hat{z} + \sin\theta\hat{x} \quad \text{and} \quad \hat{t} = -\sin\theta\hat{z} + \cos\theta\hat{x}. \tag{A.23}$$

Thus, the rotated stresses are given by:

$$\sigma_r = \hat{n}\cdot[\vec{T}_1\sin\theta + \vec{T}_3\cos\theta]$$

$$= \sigma_x\sin^2\theta + \sigma_z\cos^2\theta + 2\tau_{xz}\sin\theta\cos\theta \tag{A.24}$$

$$\sigma_\theta = \sigma_z\sin^2\theta + \sigma_x\cos^2\theta - 2\tau_{xz}\sin\theta\cos\theta \tag{A.25}$$

$$\tau_{r\theta} = (\sigma_x - \sigma_z)\sin\theta\cos\theta + \tau_{xz}\cos 2\theta \tag{A.26}$$

which is simply a tensor transformation.

A.4 Elastic Equilibrium: Navier's Equation

Given a sensor with a set of forces on the surface, how does one determine the displacements, strains, and stresses within the sensor?

Consider an arbitrary volume V bounded by a surface S. For *balance of linear momentum* (Marsden & Hughes, 1983) we must have:

$$\int_S \vec{T}dS + \int_V \vec{F}dV = \int_V \rho\frac{\partial^2\vec{d}}{dt^2}dV \tag{A.27}$$

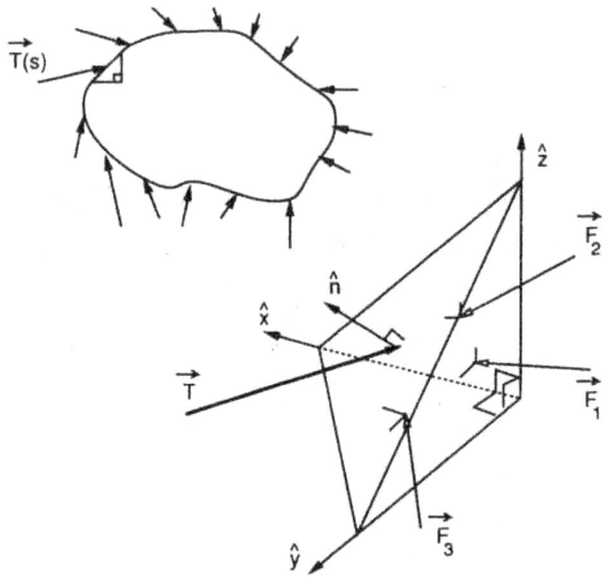

Figure A.6: *Equivalent stresses for surface traction.*

where $\vec{T}$ is the traction (force vector per area) or stress vector on the surface, $\vec{F}$ is the internal body force per unit volume (for example, gravity), ρ is the mass per unit volume, and $\vec{d} = (u, v, w)$ is the displacement of a particle. In elastostatics, there is no time dependent behaviour of the material. (The time-dependent behaviour is important in viscoelasticity.)

For tactile sensors, the internal body force can usually be neglected. Consider a typical rubber finger medium with density $\rho = 1.5 \times 10^3$ kgm^{-3}. The pressure at depth d within the medium is given by $P = \rho g d$. For a 1×10^{-2} m thick skin, (with $g \approx 10$ ms^{-2}), the corresponding pressure is only 150 Nm^{-2}.

For comparison, a sensitive tactile sensor (Fearing, 1990) has a detectable threshold of 600 Nm^{-2}. Thus, it is reasonable to neglect the body force in tactile sensor analysis. However, the internal body force may be significant for larger tactile sensors with very thick but compliant elastic "skins".

For elastostatics and no body force, we know that the sum of the forces acting on the body, i.e., the integral of the traction vector over the whole body must be zero for equilibrium, thus:

$$\int_S \vec{T} dS = \int_S (T_x \hat{x} + T_y \hat{y} + T_z \hat{z}) dS = \vec{0} \qquad \text{(A.28)}$$

and

$$\int_S T_x dS = 0 \, , \int_S T_y dS = 0 \, , \int_S T_z dS = 0 \tag{A.29}$$

where T_x, T_y, and T_z are individual components of traction in the x, y, and z directions respectively.

We next show that the traction field within the body must be divergence free. By calculating internal forces in terms of internal displacement fields, Navier's elastic equilibrium equation will be derived.

Using similar reasoning to the coordinate transformation of the previous section, we note that the force applied to a surface patch of the object must be equal in magnitude and oppositely directed to the tractions applied on a volume enclosing that surface patch. Consider the right tetrahedron with $\vec{T}$ applied to the face with unit area and normal $\hat{n}$ (Figure A.6). Let the net forces acting on the $x-y$, $x-z$ and $y-z$ planes be $\vec{F_3}$, $\vec{F_2}$, and $\vec{F_1}$, respectively. The net force on a face must be equal to the product of the stresses applied and the area of that face. The area of the face is given by the dot product of $\hat{n}$ and the normal of that face, then:

$$\vec{F_1} = (\sigma_x \hat{x} + \tau_{xy} \hat{y} + \tau_{xz} \hat{z})(\hat{n} \cdot \hat{x}) = \vec{T_1}(\hat{n} \cdot \hat{x}) \tag{A.30}$$

$$\vec{F_2} = (\tau_{xy} \hat{x} + \sigma_y \hat{y} + \tau_{yz} \hat{z})(\hat{n} \cdot \hat{y}) = \vec{T_2}(\hat{n} \cdot \hat{y}) \tag{A.31}$$

$$\vec{F_3} = (\tau_{xz} \hat{x} + \tau_{yz} \hat{y} + \sigma_z \hat{z})(\hat{n} \cdot \hat{z}) = \vec{T_3}(\hat{n} \cdot \hat{z}) \tag{A.32}$$

where $\vec{T_3}$, $\vec{T_2}$, and $\vec{T_1}$ are tractions on the $x-y$, $x-z$ and $y-z$ planes respectively of the right tetrahedron. From force equilibrium, traction $\vec{T}$ on $\hat{n}$ face with unit area is given by:

$$\vec{T} = \begin{bmatrix} T_x \\ T_y \\ T_z \end{bmatrix} = -\vec{F_1} - \vec{F_2} - \vec{F_3}. \tag{A.33}$$

Adding components of eq. A.30, A.31, and A.32, we have:

$$\begin{bmatrix} T_x \\ T_y \\ T_z \end{bmatrix} = \begin{bmatrix} \sigma_x \hat{n} \cdot \hat{x} + \tau_{xy} \hat{n} \cdot \hat{y} + \tau_{xz} \hat{n} \cdot \hat{z} \\ \tau_{xy} \hat{n} \cdot \hat{x} + \sigma_y \hat{n} \cdot \hat{y} + \tau_{yz} \hat{n} \cdot \hat{z} \\ \tau_{xz} \hat{n} \cdot \hat{x} + \tau_{yz} \hat{n} \cdot \hat{y} + \sigma_z \hat{n} \cdot \hat{z} \end{bmatrix} = \begin{bmatrix} \vec{T_1} \cdot \hat{n} \\ \vec{T_2} \cdot \hat{n} \\ \vec{T_3} \cdot \hat{n} \end{bmatrix}. \tag{A.34}$$

Defining $d\vec{S} = \hat{n} dS$, we have:

$$\int_S T_x dS = 0 = \int_S \vec{T_1} \cdot d\vec{S} \tag{A.35}$$

$$\int_S T_y dS = 0 = \int_S \vec{T_2} \cdot d\vec{S} \tag{A.36}$$

$$\int_S T_z dS = 0 = \int_S \vec{T_3} \cdot d\vec{S}. \tag{A.37}$$

The surface integral for each component can now be changed to a volume integral by the divergence theorem:

$$\int_S \vec{T_1} \cdot d\vec{S} = \int_V \vec{\nabla} \cdot \vec{T_1} dV = 0 \tag{A.38}$$

where $\vec{\nabla} = \frac{\partial}{\partial x}\hat{x} + \frac{\partial}{\partial y}\hat{y} + \frac{\partial}{\partial z}\hat{z}$. Equation A.38 must hold for any arbitrary volume, thus $\vec{T_i}$ must be divergence free, giving $\vec{\nabla} \cdot \vec{T_i} = 0$.

Now we want to solve for elastic equilibrium in terms of displacements, so using eq. A.11 and eq. A.14 to rewrite stresses in terms of strains (Gladwell, 1980):

$$\begin{aligned}\vec{T_1} &= \sigma_x\hat{x} + \tau_{xy}\hat{y} + \tau_{xz}\hat{z} \\ &= (\lambda\epsilon + 2\mu\epsilon_x)\hat{x} + 2\mu\gamma_{xy}\hat{y} + 2\mu\gamma_{xz}\hat{z} \\ &= \lambda\epsilon\hat{x} + 2\mu(\epsilon_x\hat{x} + \gamma_{xy}\hat{y} + \gamma_{xz}\hat{z}).\end{aligned} \tag{A.39}$$

Recall $\vec{d} = u\hat{x} + v\hat{y} + w\hat{z}$ and note that $\epsilon_x\hat{x} + \gamma_{xy}\hat{y} + \gamma_{xz}\hat{z} = \frac{1}{2}\vec{\nabla}(u) + \frac{1}{2}\frac{\partial\vec{d}}{\partial x}$, thus:

$$\vec{T_1} = \lambda\epsilon\hat{x} + \mu(\vec{\nabla}u + \frac{\partial\vec{d}}{\partial x}) \tag{A.40}$$

and

$$\vec{\nabla} \cdot \vec{T_1} = \lambda\frac{\partial\epsilon}{\partial x} + \mu\nabla^2 u + \mu\frac{\partial\epsilon}{\partial x} = \frac{\mu}{1-2\nu}\frac{\partial\epsilon}{\partial x} + \mu\nabla^2 u = 0 \tag{A.41}$$

with $\nabla\vec{d} = \epsilon$. Similar arguments follow for the y and z components of the traction vector, thus we have finally Navier's Equation for elastostatics with no body force:

$$\frac{\vec{\nabla}\epsilon}{1-2\nu} + \nabla^2\vec{d} = \frac{\vec{\nabla}(\vec{\nabla}\cdot\vec{d})}{1-2\nu} + \nabla^2\vec{d} = 0. \tag{A.42}$$

A general solution of the displacements can be determined using the Papkovich-Neuber elastic potential (Gladwell, 1980). The displacement vector $\vec{d}$ is given as a function of a vector potential $\vec{\psi}$ and a scalar potential ϕ:

$$2\mu\vec{d} = 4(1-\nu)\vec{\psi} - \vec{\nabla}\left(\vec{r}\cdot\vec{\psi} + \phi\right) \tag{A.43}$$

where $\nabla^2\vec{\psi} = 0$ and $\nabla^2\phi = 0$, with

$$\vec{r} = x\hat{x} + y\hat{y} + z\hat{z} \quad \text{and} \quad \vec{\psi} = \psi_1\hat{x} + \psi_2\hat{y} + \psi_3\hat{z}. \tag{A.44}$$

Using this potential function, the displacements, stresses and strains can be found for some simple planar geometries, for example, half space and strip problems, as shown in the next section.

A.5 Solution for Line Load on Half-Space

The boundary conditions for frictionless indentation on the elastic half-space ($z > 0$) with the plane-strain assumption are:

$$\tau_{zx}(z = 0,\ x) = 0, \quad \sigma_x(z = 0,\ x) = 0, \quad \sigma_z(z = 0,\ x) = p(x), \tag{A.45}$$

as shown in Figure A.4. Neglecting any body force, we will solve for displacements in the plane using the potential function solution of eq. A.43 to the elastic equilibrium equations. For the vector potential we choose:

$$\vec{\psi} = \psi_z \hat{z} \tag{A.46}$$

with $\psi_x = 0$. These assumptions lead to the u and w displacements:

$$\begin{aligned} 2\mu w &= 4(1-\nu)\psi_z - \frac{\partial}{\partial z}(z\psi_z + x\psi_x + \phi) \\ &= (3-4\nu)\psi_z - z\frac{\partial \psi_z}{\partial z} - \frac{\partial \phi}{\partial z} \end{aligned} \tag{A.47}$$

$$\begin{aligned} 2\mu u &= 4(1-\nu)\psi_x - \frac{\partial}{\partial x}(z\psi_z + x\psi_x + \phi) \\ &= -z\frac{\partial \psi_z}{\partial x} - \frac{\partial \phi}{\partial x}. \end{aligned} \tag{A.48}$$

We will use Fourier transforms to find harmonic potential functions. A brief review of Fourier transform properties is given. The Fourier transform is defined as:

$$f(x) \rightarrow F(s) = \int_{-\infty}^{\infty} f(x) e^{-j2\pi s x} dx \tag{A.49}$$

where s is the spatial frequency, typically in cycles mm^{-1}. Two properties that will be of use are multiplication and differentiation in the position domain:

$$x f(x) \rightarrow \frac{-1}{j2\pi}\frac{dF(s)}{ds} \tag{A.50}$$

$$\frac{df(x)}{dx} \rightarrow j2\pi s F(s). \tag{A.51}$$

We need to find a function ϕ harmonic in the plane ($\nabla^2\phi = 0 = \frac{\partial^2\phi}{\partial z^2} + \frac{\partial^2\phi}{\partial x^2}$). Consider the 1-dimensional Fourier transform of the potential $\phi(z,x)$ given by $\Phi(z,s)$. Then we have:

$$\frac{\partial^2\phi}{\partial z^2} + \frac{\partial^2\phi}{\partial x^2} \rightarrow \frac{\partial^2\Phi(z,s)}{\partial z^2} - (2\pi s)^2\Phi(z,s) = 0 \tag{A.52}$$

where the derivative rule was applied twice. It follows directly that a solution of the above will be:

$$\Phi(z,s) = \Omega(s)e^{-2\pi|s|z} \tag{A.53}$$

where $\Omega(s)$ is some well-behaved function that is not a function of z. Using the abbreviation $\omega = 2\pi s$, we choose the potential functions to be:

$$\psi_z(z,\ x) \rightarrow A(\omega)e^{-|\omega|z} \quad \text{and} \quad \phi(z,\ x) \rightarrow C(\omega)e^{-|\omega|z} \tag{A.54}$$

where the functions $A(\omega)$, $C(\omega)$ will be found from satisfying the boundary conditions. Thus we need to find σ_z and τ_{zx} on the boundary. Using eq. A.17 for the stress in the z direction we have (in plane strain):

$$\begin{aligned}
\sigma_z &= \frac{2\mu}{1-2\nu}\left((1-\nu)\frac{\partial w}{\partial z} + \nu\frac{\partial u}{\partial x}\right) \qquad (A.55)\\
&= \frac{1-\nu}{1-2\nu}\frac{\partial}{\partial z}[(3-4\nu)\psi_z - z\frac{\partial \psi_z}{\partial z} - \frac{\partial \phi}{\partial z}] + \frac{\nu}{1-2\nu}\frac{\partial}{\partial x}[-z\frac{\partial \psi_z}{\partial x} - \frac{\partial \phi}{\partial x}]\\
&= \frac{(1-\nu)}{1-2\nu}\left(2(1-2\nu)\frac{\partial \psi_z}{\partial z} - z\frac{\partial^2 \psi_z}{\partial z^2} - \frac{\partial^2 \phi}{\partial z^2}\right) + \frac{\nu}{1-2\nu}\left(-z\frac{\partial^2 \psi_z}{\partial x^2} - \frac{\partial^2 \phi}{\partial x^2}\right).
\end{aligned}$$

Now since ψ and ϕ are harmonic ($-\frac{\partial^2 \phi}{\partial z^2} = \frac{\partial^2 \phi}{\partial x^2}$), the above simplifies to:

$$\sigma_z(z,\ x) = \left(2(1-\nu)\frac{\partial \psi_z}{\partial z} - z\frac{\partial^2 \psi_z}{\partial z^2} - \frac{\partial^2 \phi}{\partial z^2}\right). \tag{A.56}$$

Similarly for the shear stress τ_{zx}:

$$\begin{aligned}
\tau_{zx}(z,\ x) &= 2\mu\gamma_{zx} = 2\mu\left(\frac{1}{2}(\frac{\partial w}{\partial x} + \frac{\partial u}{\partial z})\right)\\
&= (1-2\nu)\frac{\partial \psi_z}{\partial x} - z\frac{\partial^2 \psi_z}{\partial z \partial x} - \frac{\partial^2 \phi}{\partial z \partial x}. \qquad (A.57)
\end{aligned}$$

Let the 1-dimensional Fourier transforms of σ_z be given by $\Sigma_z(z,s)$, and τ_{zx} be given by $T_{zx}(z,s)$. Using the potential functions from eq. A.54, we obtain:

$$\Sigma_z(z,\ s) = \left(-2(1-\nu)|\omega|A - z|\omega|^2 A - |\omega|^2 C\right)e^{-|\omega|z} \tag{A.58}$$

and

$$T_{zx}(z,s) = ((1-2\nu)A + z|\omega|A + |\omega|C)(j\omega)e^{-|\omega|z}. \tag{A.59}$$

Using the boundary conditions $T_{zx}(0,s) = 0$ and $\Sigma_z(0,s) = P(s)$, where $P(s)$ is the Fourier transform of $p(x)$, we get:

$$\Sigma_z(0,\ s) = P(s) = -2(1-\nu)|\omega|A - |\omega|^2 C \tag{A.60}$$

and

$$T_{zx}(0,\ s) \ = \ 0 \ = \ (1 - 2\nu)A + |\omega|C. \tag{A.61}$$

Thus:

$$A(s) \ = \ \frac{-P(s)}{|\omega|} \quad \text{and} \quad C(s) \ = \ \frac{(1 - 2\nu)P(s)}{|\omega|^2}. \tag{A.62}$$

A.5.1 Normal Strain for Line Load

We can now determine displacements, strains and stresses anywhere in the half plane. The normal strain ϵ_z is commonly measured in tactile sensors, so:

$$\epsilon_z(z,\ x) \ = \ \frac{1}{2\mu}\left(2(1 - 2\nu)\frac{\partial\psi_z}{\partial z} - z\frac{\partial^2\psi_z}{\partial z^2} - \frac{\partial^2\phi}{\partial z^2}\right) \rightarrow E_z(z,s) \tag{A.63}$$

$$\begin{aligned} E_z(z,\ s) \ &= \ \frac{1}{2\mu}\left(-2(1 - 2\nu)|\omega|A - z|\omega|^2 A - |\omega|^2 C\right)e^{-|\omega|z} \\ &= \ \frac{1}{2\mu}(1 - 2\nu + z|\omega|)P(s)e^{-|\omega|z}. \end{aligned} \tag{A.64}$$

To obtain the result in plane stress, we substitute $\nu = \frac{\nu^*}{\nu^*+1}$, as mentioned in Section A.2, obtaining:

$$\begin{aligned} E_z(z,\ s) \ &= \ \frac{(1 + \nu)}{E}\left(\frac{1 - \nu}{1 + \nu} + z|\omega|\right)P(s)e^{-|\omega|z} \\ &= \ \frac{1}{E}(1 - \nu + (1 + \nu)z|\omega|)\,P(s)e^{-|\omega|z}. \end{aligned} \tag{A.65}$$

We define $H(z = d_o,\ s) = \frac{1}{E}(1 - \nu + (1 + \nu)2\pi d_o|s|)e^{-2\pi|s|d_o}$ as the *spatial frequency response* of the elastic medium for sensor depth d_o.

Taking the inverse Fourier transform of $H(s,\ d_o)$, one obtains (for a unit load $p(x) = \delta(x) \rightarrow P(s) = 1$):

$$\epsilon_z(d_o, x) \ = \ h(d_o, x) \ = \ \frac{2d_o}{\pi E r^4}(d_o^2 - \nu x^2) \tag{A.66}$$

with $r^2 = d_o^2 + x^2$. We will refer to $h(x)$ as the *impulse response* of a particular sensor element located beneath the surface at (d_o, x). Note that the impulse response is a real and even function as is the frequency response.

A.5.2 Surface Displacement for Line Load

A tactile sensor could also measure the normal displacement of the surface, $w(x, z = 0)$, which is easily obtained from:

$$2\mu w \ = \ (3 - 4\nu)\psi_z - z\frac{\partial\psi_z}{\partial z} - \frac{\partial\phi}{\partial z} \tag{A.67}$$

$$2\mu W(z,\ s) \ = \ e^{-|\omega|z}P(s)\left(z + \frac{2(1 - \nu)}{|\omega|}\right). \tag{A.68}$$

A.6 Solution for Load on Half Space

Planar elastic analysis is a simple starting point and is a reasonable approximation for some contact types, for example, grasping polyhedra with edges parallel to the finger length. For problems without this symmetry, a 3-dimensional elastic solution is needed. For a thick skin, a half-space model may be appropriate. The problems of a concentrated normal force and concentrated tangential force on a half-space were solved by Boussinesq and Cerruti in the 19th century, and are given by Johnson (1985). For a concentrated normal load, the displacements are:

$$u = \frac{P}{4\pi\mu}\left(\frac{xz}{r^3} - (1 - 2\nu)\frac{x}{r(r + z)}\right) \tag{A.69}$$

$$v = \frac{P}{4\pi\mu}\left(\frac{yz}{r^3} - (1 - 2\nu)\frac{y}{r(r + z)}\right) \tag{A.70}$$

$$w = \frac{P}{4\pi\mu}\left(\frac{z^2}{r^3} + \frac{2(1 - \nu)}{r}\right) \tag{A.71}$$

where $r^2 = x^2 + y^2 + z^2$ is the distance from the load on the surface.

The normal strain can be easily calculated from $\epsilon_z = \frac{\partial w}{\partial z}$. Noting that $\frac{\partial r}{\partial z} = \frac{z}{r}$, we have:

$$\epsilon_z = \frac{P}{4\pi\mu}\left[\frac{2z}{r^3} - \frac{3z^3}{r^5} - 2(1 - \nu)\frac{z}{r^3}\right], \quad \text{or}$$

$$\epsilon_z = \frac{Pz}{4\pi\mu r^3}\left(2\nu - \frac{3z^2}{r^2}\right). \tag{A.72}$$

References

Fearing, R.S. 1990. Tactile sensing mechanisms. *Int. j. robotics research*, **9**(3), 3–23.

Gladwell, G.M.L. 1980. *Contact problems in the classical theory of elasticity.* Netherlands: Sijthoff and Nordhoff.

Johnson, K.L. 1985. *Contact mechanics.* Cambridge, UK: Cambridge University Press.

Marsden, J.E., & Hughes, T.R. 1983. *Mathematical foundations of elasticity.* Englewood Cliffs, NJ: Prentice-Hall.

Phillips, J.R., & Johnson, K.O. 1981. Tactile spatial resolution III. A continuum mechanics model of skin predicting mechanoreceptor responses to bars, edges, and gratings. *J. neurophysiology*, **46**(6), 1204–1225.

Timoshenko, S., & Goodier, J.N. 1951. *Theory of elasticity.* New York: McGraw-Hill.

Index

A-fibres, 126
action potential, 127
aliasing, 87-88
anchoring fibrils, 130
architecture, 98
assembly sequence specification, 114
automated assembly, 109

balance of linear momentum, 283
basement membrane, 124, 129, 131, 138, 140
bin-picking, 95
blood-nerve barrier, 124
Boussinesq problem, 79, 290

C-fibres, 126
capacitive transduction, 23-26, 38, 232
carbon fibres, 15
centroid, 57
classification of afferent nerves, 125
Columbia Hand-Arm System, 253
conduction speed, 126
contact sensing, 50
control architecture, 96
control, 221, 229, 231, 236-237
creep, 17
cutaneous sensitivity, 160
cutaneous, 8, 60, 152, 194, 208, 213, 240-241, 249

deflection, 80-81, 85-86
dermal papillae, 129-131
dermis, 129
desmosomes, 130
dimensionality, 5
discrimination, 222-223
displacement, 77, 81-83, 85, 287, 289-290
dynamic nuclear bag fibre, 142

encapsulated terminals, 131
encapsulation, 134
epidermis, 129
EPs, see Exploratory Procedures
ESPRIT, 95, 97, 106
EUREKA/FAMOS, 108
experimental
  bias, 156
  hypotheses, 156
  method, 155
  statistical analysis, 157
  study, 156
Exploratory Procedures, 64-65, 198, 200-203, 205, 241-244, 249, 252-253, 257, 261-262, 264-266, 268, 273-275
exploratory robotics, 194
exteroceptors, 123

FAI, 134
FAII, 134
fascicles, 124
fast adapting receptor, 128
flexible manufacturing, 97
flexible assembly, 108
force
  angle, 86-87
  sensing, 50
force-closure, 51
forcel, 5
form closure, 51
Fourier transform, 58, 82, 86, 287-288

friction, 65-66, 209
FSR, 18-19, 273

general somatic
afferents, 124
innervation, 123
Golgi tendon organ-like endings, 141
Golgi tendon organs, 139, 143
grasping, 51-54, 97-98

haptic
cues, 213
perception, 194, 207
sensing, 8-9, 60, 64, 152, 194, 221, 224, 240, 243-244, 249-250, 253, 256, 261, 263, 268, 274
hardness, 65-66, 202-203
hemidesmosomes, 130, 138
hyperacuity, 164
hypodermis, 129

impulse response, 78, 80, 289
InFACT
machine, 109, 111
project, 108
inspection, 10
integrated potential, 127
intermediate ridges, 131
internode, 125, 128
interpretation tree, 62, 223, 225, 227-228, 250, 264
intrinsic tactile sensing, 54-55, 228-232

joint capsule, 141
joints, 140

kinesthesia, 140, 142
kinesthetic, 8, 60, 152, 194, 240, 249
knowledge-based system, 97

limb proprioception, 140
limiting ridges, 131
localisation, 50, 62, 223, 232
locational uncertainty, 115, 118-119

magnetic transduction, 26-30, 39
magnetoelasticity, 26-29, 39
magnetoresistivity, 28-29, 39
mechanical crosstalk, 8, 90
mechanical transduction, 29, 39
mechanoreceptor, 131, 134, 140-142, 209, 211
Meissner's corpuscles, 134, 137
Merkel cell, 135
Merkel cell-related endings, 133-135
model matching, 62-63
moments
definition of, 57
central, 57
normalised central, 57
motor control, in humans, 208
muscle spindles, 142
myelin sheath, 124

Navier's equation, 283, 286
nerve
conduction, 126
fibre, 124
impulse, 126
neurophysiology, 152
nociceptors, 131, 133
node of Ranvier, 125, 127
nuclear chain fibres, 142

object recognition
in robotics, 55-56, 59, 62, 222, 23-234, 240, 249, 253, 261, 268
in humans, 205-206
optical transduction, 31-35, 40, 99
orientation, 96-98, 100, 104

Pacinian corpuscle, 129, 132, 138, 141
pattern recognition in humans, 174
perineurium, 124
periosteum, 141
peripheral nerve, 124
photoelasticity, 35
piezoelectric transduction, 20-23, 38, 273
piezoresistive transduction, 18, 99, 273

planar elasticity, 75-93, 277-290
Poisson's ratio, 280
pressure, 85-86
    sensitivity, in humans, 161
proprioceptors, 123
psychophysics, 152
pulse-echo ranging, 36
PVDF, 20
PVF_2, 20-23
pyroelectric transduction, 20-23, 38

receptor potentials, 127
refractory period, 127
resistive transduction, 15-19, 38
resting potential, 126
rete pegs, 129-131
Ruffini corpuscles, 139, 143
Ruffini-type corpuscles, 141

SAI, 134
SAII, 134
Schwann cells, 124, 138-140
scientific method, the, 155
segmentation, 263
sensor integration, 111
sensor, definition of, 2
shape description, 61-62
shape recovery, 50, 232-233, 243
shear modulus, 280
shear, 28-29, 34-35
    strain, 280
    stress, 22, 81, 277, 281-282
skin acceleration sensor, 22, 53 173, 212
skin
    glabrous, 130-131
    hairy, 130
    mechanics in humans, 158
    structure of, 129
slip, 22, 25, 30, 51-53, 212
slowly adapting receptor, 128
sodium channels, 127
spatial frequency response, 289
spatial sensitivity, in humans, 162
spatiotemporal transfer function, 168
speed of conduction, 128
static nuclear bag fibre, 142
strain gauge, 229
strain, 77-78, 81, 85-87, 159, 277
    normal, 82 278, 280, 289-290
    plane strain assumption, 281
    plane, 79
stress rate sensor, 22, 53, 173, 274
stress, 34, 77-78, 159, 277, 282
    normal, 81, 277
    plane stress assumption, 77, 281
    plane, 79, 87, 289
superquadric, 233, 257-259, 261-262
surface curvature, 230, 232-233, 236, 238, 265
surface normal, 66, 223-227, 230, 232-233
surface traction, 81, 83

tactel, 5, 13
tactile
    domain, 5-7
    image, 98, 100, 102, 105
    sensing, 4, 49-50, 75, 85, 107, 151-185, 273-275
        active, 8, 60, 221-223, 232, 238, 243
    sensors, 4, 13-40, 87, 273-275
    subroutines, 65
    systems, 14, 19, 273
task generics, 114
taxel, 5
temporal sensitivity in humans, 165
texture, 22, 65, 66, 169, 173, 183, 202-203
thermal sensing, 37, 180-181
thermoreceptors, 131, 133
tight junction, 124
tonofibrils, 130
torque, 28-29, 34-35
touch probe, 107108, 112-113
touch, 151, 195

traction, 284-285

unencapsulated terminals, 131
Utah-MIT hand, 225, 228, 233, 244, 250, 252-254, 257

vibration sensing, 165-167
virtual sensors, 111
vision, 151, 177, 179, 195, 204, 222, 234, 236, 240, 244, 249-250, 263, 267-268
visual domain, 5-7

www.ingramcontent.com/pod-product-compliance
Lightning Source LLC
LaVergne TN
LVHW060543180926
848329LV00007B/45
9789810208707